GREEK SCIENCE
OF THE
HELLENISTIC ERA

We all want to understand the world around us, and the ancient Greeks were the first to try to do so in a way we can properly call scientific. Their thought and writings laid the essential foundations for the revivals of science in medieval Baghdad and renaissance Europe. Now their work is accessible to all, with this invaluable introduction to almost a hundred scientific authors active from 320 BCE to 230 CE.

The book begins with an outline of a new socio-political model for the development and decline of Greek science. Eleven chapters of fully translated source material follow, with the disciplines covered ranging from the science which the Greeks saw as fundamental – mathematics – through astronomy, astrology and geography, mechanics, optics and pneumatics, and then on to the non-mathematical sciences of alchemy, biology, medicine, and "psychology." Each chapter contains an accessible introduction on the origins and development of the topic in question, and all the authors are set in context with brief biographies.

No other one-volume survey is as up to date, has such broad yet detailed coverage, or offers as many primary sources – several of which are not available elsewhere. With clear, accurate translations, and numerous illustrations, this is an essential resource for students of the history of science in general, and ancient science in particular.

Georgia L. Irby-Massie studied mathematics and classics at the University of Georgia, and took her PhD in classics at the University of Colorado. Her publications include works on ancient religion, Roman epigraphy, and ancient science. She currently teaches classics at Louisiana State University.

Paul T. Keyser studied physics and classics at St. Andrew's School, Duke, and Boulder. He did further study at the Center for Hellenic Studies, and has also taught classics. His publications include work on gravitational physics, and on ancient science. He is currently crafting Java for IBM's Watson Research Center.

GREEK SCIENCE
OF THE
HELLENISTIC ERA

A Sourcebook

Georgia L. Irby-Massie and
Paul T. Keyser

London and New York

First published 2002
by Routledge
11 New Fetter Lane, London EC4P 4EE

Simultaneously published in the USA and Canada
by Routledge
29 West 35th Street, New York. NY 10001

Routledge is an imprint of the Taylor & Francis Group

Typeset in Baskerville and Syntax by
BOOK NOW Ltd

British Library Cataloguing in Publication Data
A catalogue record for this book is available
from the British Library

Library of Congress Cataloging-in-Publication Data
Irby-Massie. Georgia L. (Georgia Lynette). 1965–
Greek Science of the Hellenistic Era : a sourcebook / Georgia L. Irby-Massie and Paul
T. Keyser.
p. cm.
Includes bibliographical references and index.
1. Science–Greece–History–To 1500. 2. Science. Ancient. I. Keyser. Paul T. (Paul
Turquand). 1957– II. Title.
Q127.G7 172 2002
509.38–dc21   2001041999

ISBN 0-415-23847-1 (hbk)
ISBN 0-415-23848-X (pbk)

BRIEF CONTENTS

CONTENTS

CONTENTS

CONTENTS

hemisphere is seen, and the part of the sphere that is seen
appears as a circumference; 51 If, when several objects move at
unequal speed, the eye also moves in the same direction, the
objects moving with the same speed as the eye will seem to
stand still, those moving slower will seem to move in reverse, and
those moving faster will seem to move ahead; 52 When some
objects are moved, and one is obviously not moved, the unmoved
object will seem to move backward; 54 When objects move at
equal speed, the more remote seem to move slower; 55 If the eye
remains at rest, while things seen are moved, the more remote
of the things seen will seem to be left behind; 57 When things
lie at the same distance and the edges are not in line with the
middle, it makes the whole figure sometimes concave, sometimes
convex

CONTENTS

CONTENTS

PREFACE

More of Greek "science" survives than does any other category of ancient Greek literature, and yet much of that is obscure even to classicists. This source-book aims at treatment of that problem, not cure. No consensus on the nature of modern "science" exists, much less on the nature of ancient Greek "science" – we have therefore preferred inclusion to omission. Although the *kosmos* is vast and time endless, books and lives are not – these selections are a tiny fraction of the texts. Selection and translation distort and disappoint – but a warped mirror and dim candle are better than no view at all. A book twice or more the length of this one (as Cohen and Drabkin 1958) and in Greek would be a far better representation – and much less read. Most experts will find some favored passage absent.

Division and definition whether of objects or concepts remains a deep conundrum, but our chapter divisions separate segments of Greek science at joints we believe natural. Terminology is a famous difficulty, but various Greek terms did exist which more or less cover the semantic range of "science," as well as most of the chapter titles (except "psychology," which is thus always in quotation marks). In their synthesis (on which see Chapter 1), mathematics are foundational and unite astronomy, geography, and physics (mechanics, optics, and pneumatics), while the non-mathematical sciences of natural phenomena (alchemy, biology, medicine, and "psychology") stood apart until astrologers proposed planetary "sympathy" with body-parts, plants, and stones. Since the soul and the heavens were both seen as made of *aither*, such a framework provided a path to harmony and fulfillment, both individual and social. Other chapter distributions are possible: e.g., fusing astronomy and astrology, and even geography (or else separating cosmology from astronomy); fusing mechanics and pneumatics, and even optics (or else separating out music); fusing biology and medicine, and even "psychology" (or else splitting biology into botany and zoology), etc. These and other independent possibilities allow for hundreds of choices – we cannot please every reader.

Within each "field" (chapter) we eschew fine distinctions of topic as misleading and arbitrary: each reader will find their own relevant passages. We have preferred fewer longer extracts and a broader range of authors: those readily accessible elsewhere are less prominent here. The authors are presented in chronological order (dates are mostly given as the midpoint of a range with the ± representing "almost certain" limits); instead of the BC/AD or even BCE/CE system, we prefer the system used by historians of astronomy, giving dates before the turn of the era as negative numbers (thus 2 BCE is −1, and 123 BCE is −122, etc.). For Galen, Heron, Hipparchos, Ptolemy, and Theophrastos we present the works in chronological order as well. Our date limits reflect the model of Chapter 1, thus strictly excluding Aristotle (and the pseudo-Aristotelian corpus) at one end, and all but one Aristotelian commentator at the other. (Hanno and Putheas may stretch the bounds before, while Anatolios and Zosimos extend a bit later.)

Most material is reprinted, but we include over three dozen authors and works hardly found in English (in every chapter save that on pneumatics). Explanatory material in the chapter prolegomena and notes to authors and texts does not argue but states, and experts will readily see where choices were made. (We also assume readers will know or can easily access data on Greek and Greco-Roman culture.) We have also preferred to include more text at the expense of some notes, believing that there is greater value in having more material available, and given that even book-length commentaries on some of these texts would not suffice. The bibliography includes only works in English, appropriate to our intended readership (though experts will recognize that we have studied scholarly literature in German, French, and Italian).

All reprinted translations originally published before 1976 (and a few more recent), except for one papyrus horoscope, have been revised, the better to accord with the Greek (especially those from Stevenson [1932], Cohen and Drabkin [1958], and Bruin and Vondjidis [1971], which are practically new versions). One work is extant only in Armenian, which was checked by G. M. Browne; nine are extant only in Arabic, eight of which were checked by G. M. Browne and Petra M. Sijpesteijn. We have carefully considered their reports but final responsibility for the translations rests with us. (One Arabic text has never been printed, and we were unable to obtain a photograph: Menelaus, *Densities.*) G. J. Toomer kindly provided us with corrections to the translation of Diokles. Words added *metri gratia* in verse translations are *italicized* (a font used elsewhere for other purposes). Transliterated Greek words are italic in roman (upright) text, but where italic is used to distinguish material that is present in only one source for an extract, such words are underlined. Underlining is also occasionally used to indicate metrical stress. In the new translations (marked by an * preceding the citation of the modern edition at the end of each text) we have tried to balance accuracy and clarity. Most names have been transliterated directly from Greek (rather than first passing them through Latin, as often occurs); though we employ "ch" for *chi* rather than "kh," and a few very familiar names (Aristotle, Plato, Plutarch, Ptolemy, Pythagoras, and Strabo) have been left in their traditional inaccurate form (rather than Aristotelês, Platôn, Ploutarchos, Ptolemaios, Puthagoras, and Strabôn). (Given the tendency of manuscripts to garble numerals, we have rendered all numerals as such, and all numbers written out in words as such.)

Some figures are reprinted, but most were redrawn at the authors' directions based on published manuscript drawings (i.e., they too are translations), a few are reconstructed. Those in Chapters 2, 3, and 5 are by Paul A. Whyman (who also provided graphics support for greyscale figures in Chapters 6 and 9); the maps and the figures in Chapter 4 are by Keith Massie (who also provided graphics support for all figures in Chapter 10); those in Chapter 6 are by Paul T. Keyser; those in Chapters 7 and 8 are by Klyph Bohm; and those in Chapter 9 are by Susan Guinn-Chipman. All diagrams employ upper-case Greek letters as labels, which is probably closest to the original labeling.

The original concept for the book is due to Paul T. Keyser (1988–9), who is also responsible for all verse rend(er)ings and the explanatory material (including Chapter 1, largely written at the University of Alabama 1997–8, and twice given as a paper in 1998, at Loyola in New Orleans and at the University of Texas, San Antonio).

We give thanks to the following people who helped in ways great and small (listed in reverse alphabetical order): Sam Weber, Robert L. Trapasso MD, Richard Stoneman, Jordan Siverd, Eckart Schütrumpf, Alex Schubert, Michael P. Perrone, Hayden N. Pelliccia, Toni Pardi, Keith Massie, Stephan Heilen, David Cinabro, Kathryn S. Chew, W. M. Calder III, Laurel M. Bowman, and Bard Bloom. The following university libraries provided needed material and support: Alberta, Colorado, Cornell, Illinois-UC, LSU, Princeton, and Vassar; important support was also provided by the Center for Hellenic Studies and the New York Public Library.

Paul T. Keyser would like to dedicate this book to his maternal grandfather, J. P. Dionne (excluded by his class from the education he deserved).

Note added in proof

Although dated 1999, A. Jones, "Geminus and the Isia," *Harvard Studies in Classical Philology* 99 (1999), pp. 255–267, dating Geminus to –55 ± 30, only appeared while this book was in press.

ACKNOWLEDGMENTS

We wish to thank the following publishers for granting permission to reprint excerpts from previously published translations:

Akademie-Verlag
  deLacy, P. H., *Galen: On the Doctrines of Hippocrates and Plato* CMG V.4.1.2 (© 1984). Excerpts reprinted with the kind permission of the publisher.

American Philosophical Society Excerpts from the following publications reprinted with kind permission of the *American Philosophical Society*:
  Goldstein, B. R., "The Arabic version of Ptolemy's Planetary Hypotheses," *Transactions of the American Philosophical Society*, 57.4 (1967)
  Jones, A., "Ptolemy's First Commentator," *Transactions of the American Philosophical Society*, 80.7 (1990)
  Neugebauer, O., and H. B. vanHoesen, *Greek Horoscopes* (1959)
  Smith, A. M., *Ptolemy's Theory of Visual Perception* (1996)

The American University of Beirut
  Bruin, F., and A. Vondjidis, *The Books of Autolycus*, pp. 10–11, 16–17, 57–59. © 1971 by the American University of Beirut. Reprinted by permission.

Aris & Phillips
  Phillips, A. A., and M. M. Willcock, *Xenophon and Arrian, On Hunting*. © 1999. Extracts reproduced by permission of Aris & Phillips, Ltd.

E.J. Brill Excerpts from the following works are reprinted by permission of the publisher:
  Behr, Charles A., *P. Aelius Aristides: The Complete Works*, 2 vols. © 1986
  Steckerl, F., *The Fragments of Praxagoras of Cos and his school*. © 1958
  Todd, R. B., *Alexander of Aphrodisias on Stoic physics*. © 1976

Cambridge University Press Excerpts from the following publications are reprinted with the permission of Cambridge University Press:
  Barker, A., *Greek Musical Writings*, vol. 2 (© 1989)

Brain, P., *Galen on Bloodletting* (© 1986)

Gow, A. S. F., and A. F. Scholfield, *Nicander: The Poems and Poetical Fragments* (© 1953)

Heath, T. L., *Works of Archimedes* (1897/1953)

Heath, T. L., *The Thirteen Books of Euclid's Elements*, 2nd edn (1926, 1956)

Kidd, D., *Aratus' Phaenomena* (© 1997)

Kidd, I. G., *Posidonius*, vol. 3 (© 1999)

Long, A. A., and D. N. Sedley, *The Hellenistic Philosophers*, vol. 1 (© 1987)

von Staden, H., *Herophilus: The Art of Medicine in Early Alexandria* (© 1989)

Charles C. Thomas Publisher

Excerpts from Green, R. M., *A Translation of Galen's Hygiene* (1951). Courtesy of Charles C. Thomas Publisher, Ltd., Springfield, Illinois.

The Continuum International Publishing Group, Ltd.

Dicks, D. R., *Geographical Fragments of Hipparchus*. © 1960 by Athlone Press.

Duckworth

Toomer, G. J., *Ptolemy's Almagest*. ©1984 by G. J. Toomer.

Edwin Mellen Press

Knoefel, P. K., and M. C. Covi, *A Hellenistic Treatise on Poisonous Animals*. © 1991 by Edwin Mellen Press. Extracts reproduced by kind permission of the publisher.

Hackett

Walzer, R., and M. Frede, *Three Treatises on the Nature of Science* (1985)

Harvard University Press   Excerpts from the following texts are reprinted by permission of the publisher and the Trustees of the Loeb Classical Library, Cambridge, Mass: Harvard University Press.

Cherniss, H., *Plutarch's Moralia*, vol. 13.1 (1976)

Cherniss, H., and W. C. Helmbold, *Plutarch's Moralia*, vol. 12 (1957)

Einarson, B., and G. K. K. Link, *Theophrastus: Causes of Plants*, 3 vols (1926, 1990, 1990)

Furley, D. J., *Aristotle: On the Cosmos* (1955)

Gulick, C. B., *Athenaeus Deipnosophistae*, vol. 2 (1927), vol. 3 (1927), vol. 4 (1930), vol. 7 (1941)

Hort, A. F., *Theophrastus Enquiry into Plants*, 2 vols (1916, 1926)

Jones, H. L., *Strabo*, 8 vols: 1 (1917), 2 (1923), 7 (1930)

Minar, E. L., *Plutarch's Moralia*, vol. 9 (1961)

Paton, W. R., *Polybius*, vol. 1 (1920)

Pearson, L., and F. H. Sandbach, *Plutarch's Moralia*, vol. 11 (1965)

Robbins, F. E., *Ptolemy Tetrabiblos* (1940)

Sandbach, F. H., *Plutarch's Moralia*, vol. 15 (1927/1969)

Scholfield, A. F., *Aelian: On the Characteristics of Animals*, 3 vols (1958, 1958, 1959)

Thomas, I., *Selections Illustrating the History of Greek Mathematics*, 2 vols (1939, 1941)

Johns Hopkins University Press

Temkin, O., "Galen's 'Advice for an Epileptic Boy,'" *Bulletin of the Institute of the History of Medicine* 2 (1934), pp. 179–189.

Temkin, O., trans., *Soranus' Gynaecology*, pp. 8–11, 34–38, 63–64, 145, 165–6, 178–9. © 1991. [Johns Hopkins] Reprinted with permission of the Johns Hopkins University Press.

The Journal of Chemical Education Excerpts from the following articles used with kind permission of the *Journal of Chemical Education*:

Caley, E. R., "The Leyden Papyrus X," *Journal of Chemical Education* 3 (October, 1926), pp. 1149–1166.

Caley, E. R., "The Stockholm Papyrus," *Journal of Chemical Education* 4 (August, 1927), pp. 979–1002.

Journal of the History of Astronomy

Shapiro, A. E., "Archimedes's Measurement of the Sun's Apparent Diameter," *Journal of the History of Astronomy* 6 (1975), pp. 75–83. Excerpts reproduced with kind permission of the author and the journal.

Karger

Siegel, R. E., *Galen on Psychology, Psychopathology, and Function and Diseases of the Nervous System; an Analysis of His Doctrines, Observations and Experiments* (© 1973). Excerpts reprinted with the kind permission of Karger, Basel.

Munksgaard

Drachmann, A. G., *Mechanical Technology of Greek and Roman Antiquity* (© 1963). Excerpts reprinted with kind permission of the publisher.

The Optical Society of America

Burton, H. E., "Euclid's Optics," *Journal of the Optical Society of America* 35 (1945), pp. 357–372. Excerpt reprinted with the kind permission of the Optical Society of America.

Orion

Heath, T. L., *Greek Astronomy* (1932)

Original Books

White, Robert J., *The Interpretation of Dreams: Oneirocritica by Artemidorus* (Noyes Press, 1975, 1990)

ACKNOWLEDGMENTS

Oxford
Heath, T. L., *Aristarchus of Samos* (1913/1966/1997)
Marsden, E. W., *Greek & Roman Artillery: Technical Treatises* (1971/1999)
Singer, P. N., *Galen: Selected Works* (1997)
Vallance, J. T., *The Lost Theory of Asclepiades of Bithynia* (1990)

Phanes
Condos, T., *Star Myths of the Greeks and Romans* (© 1997). Excerpts reprinted with kind permission of the publisher.

Princeton
Casson, L., *The Periplus Maris Erythraei.* © 1989 by Princeton University Press. Reprinted by Permission of Princeton University Press.

Routledge*
Longrigg, J., *Greek Medicine: From the Heroic to the Hellenistic Age: A Source Book* (1998)

Sauer-Verlag (Teubner)
Pingree, D., *Dorothei Sidonii Carmina.* Leipzig: B. G. Teubner 1976. © K. G. Saur München and Leipzig. Excerpts reproduced with permission of the publisher.

Scholars Press
Terian, A., *Philonis Alexandrini de Animalibus.* © 1981 by Scholars Press. Excerpts reproduced with permission of the publisher.

Springer-Verlag
Toomer, G. J., *Diocles on Burning Mirrors* (1976)
Toomer, G. J., *Ptolemy's Almagest.* © 1984 by G. J. Toomer.

Undena
Burstein, S. M., *The Babyloniaca of Berossus*, SANE 1/5 (© 1978). Excerpt reprinted with kind permission of the publisher.

University of Notre Dame
Excerpts from *Theophrastus De Ventis*, edited with introduction, translation, and commentary by Coutant, Victor, and Val L. Eichenlaub. © 1975 by the University of Notre Dame Press.

Van Gorcum & Comp, Publishers and Printers
Coutant, V., *Theophrastus De Igne* (© 1971). Excerpts reprinted with kind permission of the publisher.

Wellcome Historical Medical Museum
Singer, C. S., Galen *Anatomical Procedures*. © 1956. Extracts first published by the Wellcome Historical Medical Museum in 1956 are reproduced by kind permission of the Wellcome Trust.

Wistar Institute
Goss, C. M., "On Anatomy of Nerves by Galen of Pergamon," *American Journal of Anatomy* 118 (1966), pp. 327–335. Extracts are reproduced by kind permission of the Wistar Institute, Philadelphia, PA.

W.W. Norton & Co.
Lloyd, G. E. R., *Greek Science After Aristotle*. ©1973 by W.W. Norton & Co. Extract reproduced by kind permission of W.W. Norton & Co.

We have made good faith efforts to contact publishers regarding the following:
Ayasofia Museum:
Prager, F. D., *Philo of Byzantium: Pneumatica* (Reichert, 1974)
College of Physicians of Philadelphia:
Scarborough, John, and Vivian Nutton, "The Preface of Dioscourides' *Materia Medica*: Introduction, Translation, and Commentary," *Transactions and Studies of the College of Physicians of Philadelphia*, new series, 4 (1982)
History of Technology:
Murphy, S., "Heron of Alexandria's On Automaton-Making," *History of Technology* 17 (1995)
Islamic Research Institute:
Rescher, N., and M. E. Marmura, *The Refutation by Alexander of Aphrodisias of Galen's Treatise on the Theory of Motion* (1965)

We have also excerpted from the following works which fall under the Public Domain:
Adams, F., *The Extant Works of Aretaeus the Cappadocian* (1856)
D'Ooge, M. L., *Nicomachus of Gerasa: Introduction to Arithmetic* (1926)
Schoff, W. H., *The Periplus of Hanno* (1912)
Stevenson, E. L., *Geography of Claudius Ptolemy* (1932)
Stratton, G. M., *Theophrastus and the Greek Physiological Psychology before Aristotle* (1917)

Extracts from Cohen, M. R., and I. E. Drabkin, *A Source Book in Greek Science* (1958) are Professor Drabkin's original translations, revised.

The rights to Caley, E. R., and J. F. C. Richards, *Theophrastus on Stones* (1956) have reverted to the editors, both of whom are deceased.

FIGURE AND MAPS CREDITS

Chapter 5

Figure 5.2   Based on Kidd, I. G., *Posidonius* vol. 3 (Cambirdge: Cambridge University Press 1999), p. 183.

Figure 5.5   Corsica, Europe Map 6, MS DeRicci 97, f. 64v, Renaissance and Medieval Manuscripts Collection, *ca.* 850–*ca.* 1600. Manuscripts and Archives Division, The New York Public Library.

Figure 5.6   Sardinia, Europe Map 7, MS DeRicci 97, f. 66v, Renaissance and Medieval Manuscripts Collection, *ca.* 850–*ca.* 1600. Manuscripts and Archives Division, The New York Public Library.

Chapter 6

Figure 6.2   Marsden, E. W., *Greek & Roman Artillery: Technical Treatises* (Oxford: Oxford University Press 1971/1999), p. 177, Figs 15 and 16.

Figure 6.3   Marsden, E. W., *Greek & Roman Artillery: Technical Treatises* (Oxford: Oxford University Press 1971/1999), p. 93, Fig. 4.a.

Figure 6.4   MS. Laur. Plut. 74.7, f. 200 (modified from the reproduction in E. D. Phillips, *Aspects of Greek Medicine* [Philadelphia: Charles Press 1973/1987], plate 14).

Figure 6.5   MS. Laur. Plut. 74.7, f. 203v (modified from the reproduction in E. D. Phillips, *Aspects of Greek Medicine* [Philadelphia: Charles Press 1973/1987], plate 13).

Figure 6.7   Schneider, R., *Griechische Poliorketiker* v. 3: *Athenaios* (Berlin: Weidmanns 1912), Tafel V, top (facing p. 68).

Figure 6.9   Nix, L. and W. Schmidt, *Heronis Alexandrini Opera Omnia* vol. 2 (Leipzig: Teubner 1900/1976), p. 296.

Figure 6.10   Schoene, H., *Heronis Alexandrini Opera Omnia* vol. 3 (Leipzig: Teubner 1903/1976), p. 205.

Chapter 9

Figure 9.4   MS Marcianus Gr. 2327, f. 188–V (modified from the reproduction in M. Berthelot, *Collection des Anciens Alchimistes Grecs* vol. 1 [Paris: G. Steinheil 1887], p. 132).

Chapter 10

Figures 10.1, 10.2, 10.3, 10.4: Gunther, R. T., *The Greek Herbal of Dioscorides* (London, New York: Hafner Pub. Co., 1934/1968); reproduced from MS Vindob. med. Gr. 1, for which see (no author) *Codex Vindobonensis med. Gr. 1 der Österreichischen Nationalbibliothek* (Graz: Akademische Druck- u. Verlagsanstalt, 1965–1970), 2 v (Codices selecti phototypice impressi, vol. 12).

Figures 10.5, 10.6, and 10.7: Thompson, D. W., *Glossary of Greek Fishes* (London: Oxford University Press 1947) respectively pp. 68, 28, and 208; which were taken from William Yarrell, *A History of British Fishes* 2nd edn (London: J. Van Voorst, 1841) respectively vol. 2, p. 377; vol. 1, p. 310; and vol. 1, p. 170.

TIMELINE OF AUTHORS
EXCERPTED

Our 98 to 100 authors are listed in the order of the midrange of their known period of activity (Ailian the Platonist may be the same as Ailian of Praeneste; the two alchemical papyri may be by the same man). Authors whose lifetimes are known (marked "lived") are assigned a period of activity beginning in their 20th or 25th year. Precision is hardly possible. (The symbol "#" marks the 43 authors in Gillispie 1970–1980.)

Author	Dates	Chapter
Hanno of Carthage	about –500 (published about –300)	geography
Putheas of Massalia #	late fourth century BCE	geography
Theophrastos of Eresos #	lived –370 to –286	pneumatics, alchemy, biology, medicine, "psychology"
Autolukos of Pitanê #	about –300	astronomy
Eukleidês ("Euclid") of Alexandria #	about –300	mathematics, mechanics, optics
Sotakos	about –300	alchemy
Praxagoras of Kos #	–300 ± 25	medicine
Epikouros of Samos #	lived –340 to –269	mechanics, optics, alchemy, "psychology"
Klearchos of Soloi (in Cyprus)	lived –330 to –270	astronomy
Aristarchos of Samos #	–286 to –269	astronomy
Straton of Lampsakos #	–286 to –269	geography, mechanics, pneumatics, "psychology"
Herophilos of Chalkedon #	–280 ± 25	medicine, "psychology"
Erasistratos of Keos #	–280 ± 25	medicine

Author	Dates	Chapter
Berôsos of Babylon	–270 ± 10	astrology
Aratos of Soloi #	lived about –315 to –240	astrology
Melampous	250 ± 50	"psychology"
Archimedes of Syracuse #	lived about –285 to –211	mathematics, mechanics, optics, pneumatics
Chrusippos of Soloi #	lived about –280 to –206	astronomy, mechanics, alchemy, "psychology"
Eratosthenes of Kurênê #	lived about –285 to –193	mathematics, astrology, geography
Andreas (of Karustos?)	assassinated –216	medicine
Philon of Buzantion #	about –200	mechanics, pneumatics
Apollonios of Pergê #	–200 ± 10	mathematics, astronomy
Hegesianax of Alexandria (near Troy) #	–195 ± 20	astronomy
Diokles #	–185 ± 5	optics
Bolos of Mendes #	–180 ± 30	alchemy, biology
Glaukias of Taras	–175 ± 20	medicine
Agatharchides of Knidos	lived about –215 to about –140	geography, optics, biology, medicine
Biton of Pergamon	–160 ± 20	mechanics
Hupsikles of Alexandria #	–160 or –150	astronomy
Seleukos of Seleukia on the Red Sea	about –150	geography
Demetrios of Bithunian Apamea	date very uncertain: –140 ± 120	medicine
Polubios of Megalopolis	–140 ± 20	geography
"Petosiris" #	–140 ± 5	astrology
Hipparchos of Nikaia #	–146 to –126	mathematics, astronomy, astrology, geography, mechanics
Nikandros of Kolophon	–130 ± 20	biology, medicine
Aristokles	–110 ± 10	pneumatics
Theodosios of Bithunia #	around –100	astronomy
Ptolemaïs of Kurênê	probably –100 ± 60	"psychology"

Author	Dates	Chapter
Leonidas of Buzantion #	−100 ± 20	biology
Asklepiades of Bithunia #	−100 ± 20	medicine
pseudo-(**Skumnos** of Chios)	−85 ± 10	geography
Poseidonios of Apamea (in Syria) #	−100 to −50	astronomy, geography, optics, alchemy, biology
Herakleides of Taras	−75 ± 20	medicine
Alexander of Ephesos	around −60	astronomy
Athenaios of Kuzikos	first century BCE	mechanics
Maria the Jewess	first century BCE	alchemy
author of *Kosmos*	around −50	astronomy, geography
Apollonios of Kition	−40 ± 20	mechanics
Damostratos (possibly C. Claudius Titianus "Demostratus")	−40 ± 20	biology
Anaxilaos of Larissa #	−40 ± 10	alchemy
Apollonios Mus	−20 ± 30	medicine
Xenarchos of Kilikian Seleukeia	lived about −75 to 17	astronomy
Strabo of Amaseia #	lived −63 to about 21	geography, alchemy, biology
Thrasullos (of Mendê?)	died 36, active 10 ± 25	"psychology"
Alexander of Mundos #	15 ± 25	biology
Imbrasios of Ephesos	25 ± 75?	astrology
Philon of Alexandria	30 ± 25	biology
Dioskourides of Anazarbos #	about 40 to 50	alchemy, biology, medicine
"**Isis**"	first century CE	alchemy
Kleopatra	first century CE	alchemy
Geminus (of Rhodes?) #	mid-first century CE	astronomy, astrology
Aristokles of Messênê	50 ± 100	astronomy
Dorotheos of Sidon #	about 25 to 75	astrology
author of *Voyage on the Red Sea*	58 ± 12	geography
Balbillos (Ti. Claudius Balbillus)	60 ± 20	astrology

Author	Dates	Chapter
Heron of Alexandria #	55 to 68	mathematics, geography, mechanics, optics, pneumatics
Thessalos of Tralles	65 ± 15	astrology
Aretaios of Kappadokia # (compare below)	70 ± 20 (or 170 ± 20)	medicine, "psychology"
Xenokrates of Aphrodisias	75 ± 20	medicine
Apollinarius (of Aizanoi?)	80 ± 50	astronomy
Menelaus of Alexandria #	96 ± 2	mathematics, alchemy
Plutarch (Ploutarchos) of Chaironeia	lived about 50 to about 120	astronomy, geography, mechanics, optics, alchemy, biology, medicine
Nikomachos of Gerasa #	about 100	mathematics
Rufus of Ephesos #	108 ± 7	medicine
Marinos of Tyre	110 ± 5	geography
T. Pitenius	early second century CE	astrology
Manethon	120 ± 20	astrology
Soranos of Ephesos #	120 ± 20	medicine
Dionusios of Philadelphia (or "Periegetes," the guide)	125 ± 15	biology
Antonius Polemon of Laodikaia on the Lukos	lived 88 ± 3 to 144	"psychology"
Theon of Smurna #	130 ± 10	astronomy, geography
Arrian of Nikomedia	lived about 86 to about 160	geography, biology
Ptolemy (Claudius Ptolemaeus) of Alexandria #	lived about 100 to about 175	mathematics, astronomy, astrology, geography, mechanics, optics, pneumatics, "psychology"
Diophantos of Alexandria #	150 ± 80	mathematics
Ailian the Platonist (same as Ailian of Praeneste?)	second or early third century CE	mechanics
Artemidoros of Ephesos	160 ± 30	"psychology"

Author	Dates	Chapter
Aelius Aristides	160 ± 20	"psychology"
Sosigenes	about 165	astronomy
Aretaios of Kappadokia (compare above) #	170 ± 20? (or 70 ± 20)	medicine, "psychology"
Oppian of Anazarbos	175 ± 5	biology
Galen of Pergamon #	lived 129 to 210 ± 5	mechanics, optics, medicine, "psychology"
Philoumenos of Alexandria	around 180	medicine
Antigonos of Nikaia	180 ± 40	astrology
Vettius Valens of Antioch	180 ± 5	astrology
Ailian (Claudius Ailianus) of Praeneste	lived 168 ± 3 to 233 ± 3	biology
Alexander of Aphrodisias #	203 ± 5	alchemy
Anonymous of Apamea (pseudo-Oppian)	215 ± 5	biology
Zosimos of Panopolis #	about 250	alchemy
author of *Leyden Papyrus "X"*	around 250	alchemy
author of *Stockholm Papyrus* (same as previous?)	around 250	alchemy
Anatolios of Alexandria #	250 ± 30	mathematics

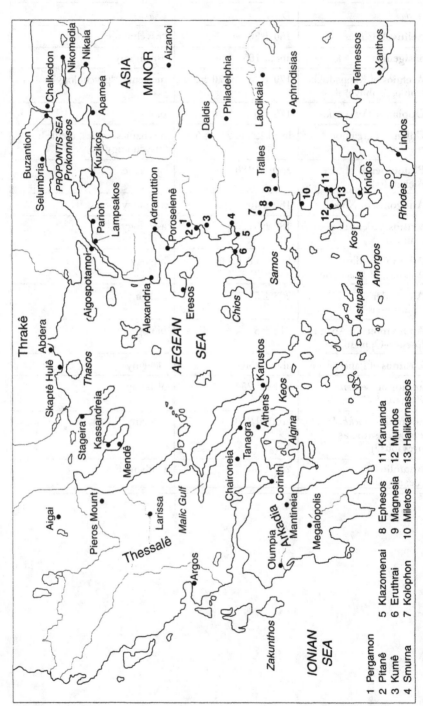

Map 1 Greece and the Aegean region.

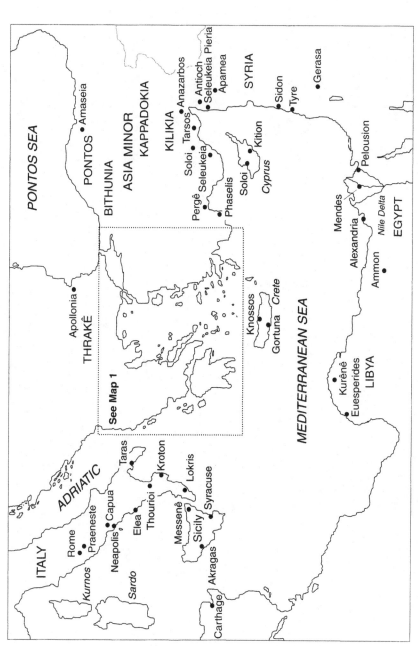

Map 2 The Mediterranean region.

1

INTRODUCTION

Every person naturally seeks to know.

(Aristotle, *Metaphysics* 1[980a22])

Greek poets invoked divine Muses to guide their tale, and often began, as before the walls of Troy, in mid-stream. Neither Klio of history nor Ourania of astronomy has appeared for this work, but the story does begin in its middle, after the most-studied period of Greek science (up through Aristotle). Numerous works exist on the early philosophers who speculated about the natural world, as well as on Hippokrates, Plato, and Aristotle. The focus here is the era between Aristotle and the "late-antique" commentators on Aristotle, during which a wide variety of writings were composed, but no single synthesis was standard.

Certain ideas were, however, standard or at least broadly received; and "science" was on the whole studied in a certain context. First, what is covered in this source-book as 'science' cannot be precisely distinguished from other intellectual activities. Science borders on philosophy (metaphysics, ethics, and epistemology are excluded), it lies near technology (most material found in books such as Humphrey, Oleson and Sherwood [1998] is excluded), it has affinity with magic (theurgy and all incantations are excluded, but astrology and alchemy are included), and touches on theology (divine cosmogonies and the theology of the soul are excluded). Those distinctions, almost impossible to make for Greek authors before around –400, become clearer in Aristotle, and after. That conceptual territory, adjacent to but not within magic, philosophy, technology, and theology, is what this book undertakes to cover. Next, what was the socio-political context of those Greeks writing such works in this era? Again, precision is impossible, primarily due to lack of evidence. Almost all were male (though note Kleopatra, Maria, Ptolemaïs, and probably the author of *Isis to Horus*: cf. Irby-Massie [1993]), not unusual in that time and place. Many of these men worked at Alexandria, sponsored in some way by the Ptolemaic kings of Egypt; others seem to have been supported in some way by other wealthy rulers. But many, perhaps most, lived and worked far from those prestigious centers, and depended on their own wealth, that of private patrons, or on fees they charged for teaching and

1

practicing their skill: astrologers, mechanics (especially artillery-makers), and doctors are known to have done so; indubitably others also did. City-states and kingdoms found medicine, mechanics, and even astronomy useful, while all of the topics here treated had some intellectual cachet (i.e., they were among the things one could validly study and debate). Their readers we know far less about, though evidently there was enough interest among the literate classes to elicit these books and many more that are lost.

The earlier period of Greek science (up through Aristotle) had created a legacy of shared assumptions, beliefs, concepts, definitions, goals, problems, and results, the full study of which has filled many a book. It was on these that workers in the period here covered built, modifying them as they saw fit. A brief sketch of some of these developments is in order, which can be no more complete or precise than would be a survey of similar length over Aristotle or Plato (compare Rihll [1999]).

From Eukleidês on, Greek mathematics (Chapter 2) was systematized as a logico-deductive system, employing lengths as its fundamental entity (to represent both geometrical figures and numbers), and describing its project largely in terms of certain problems (Knorr [1986]). Arithmetic as manifested in Heron and Diophantos either developed later, or at least is absent from our earlier sources. The difficulties surrounding the parallel-lines postulate were neither ignored nor seen as critical; infinitesimals were rejected, and irrational lengths (such as the diagonal of the unit square) were regularly handled as a case separate from rational lengths. Among the standard problems were "to double a volume" (i.e., to find what we would call a cube root), to "square" the circle (i.e., to calculate its area), to describe the properties of curved lines constructed in certain ways (e.g., the "parabola"), and to explain the various arithmetical properties of numbers (e.g., divisibility). There was definite influence from Babylonian arithmetic, but its exact extent and nature is unclear.

Once the geocentric hypothesis and the concept of the spherical earth became accepted (well before the period covered in this book), much of the remaining "project" of the study of the sky (*astronomia* or *astrologia*, Chapters 3 and 4) involved working out the detailed consequences of that model (Evans [1998]). It was assumed that there would be some 'simple' and mathematical picture to explain the observed phenomena, always seen as fundamentally regular. The shift from a concentric-sphere model (as in Aristotle and earlier) to the circular-orbit model is crucial; another key move was the development of trigonometry (the mathematics needed to work with that model). One chief motivation for astronomy was theoretical – the desire to understand the phenomena; but another was calendrical – to regulate the complex Greek luni-solar religious calendars. The desire to predict both lunar and solar eclipses was never fully realized, but lunar eclipses could reliably be forecast from the time of Hipparchos, and increasingly precise predictions of planetary positions were available (culminating with Ptolemy's tables).

A third chief motivation for astronomy was astrology (Chapter 4) – the belief that the motions of the planets affected the earth (most recently in English see Barton [1994] 9–31, 92–113, 179–197). The planets were associated with certain gods, and people remained certain that some kinds of divine influence emanated from them to us, though the kind and extent of that influence was greatly debated. It was more and more widely believed that the *kosmos* was connected by "sympathy" between its different parts: i.e., that a change in one part caused a *corresponding* change in others. Both astronomy and astrology are known to have been influenced by Babylonian learning: Hipparchos used their data, their style of prediction (somehow received as Egyptian) became one of the standard modes of Greek astrology, and the divine associations of the planets and many constellations are Babylonian.

The greatly-widened geographical horizon of the Greeks after Alexander contributed to knowledge of much more of the globe, and in the era of this book one of the chief problems of Greek geography (Chapter 5) was how best to map that data about the globe. A variety of techniques was employed, and eventually geometrical projections of a sphere onto a plane were developed. Another chief problem was the determination of the size of the earth, from astronomical observations. Likewise the problem of determining latitude (always expressed in terms of longest and shortest days) and even longitude continued to occupy geographers. The influence of latitude and other geographical factors on the local character of plants, animals, and people also remained an interest. Perhaps the largest bulk of geography in this time was descriptive, which tended to stress the marvelous and extraordinary; Greece remained the ideal central mean between extremes (Romm [1992]).

In the related areas of mechanics (Chapter 6), optics (Chapter 7), and hydrostatics/pneumatics (Chapter 8), the goal was to explain the phenomena through geometry. Each type of behavior offered "paradoxes" which were to be understood as in fact somehow regular. Thus the movement of large weights by small forces, or the causing of small weights to have large impacts was studied by workers in mechanics. Opticians sought to explain optical illusions of various kinds, and to produce astonishing sights, a work encouraged by the leading role of vision in Greek philosophy. Pneumatics workers debated the nature and role of the void in fluids, and applied their art to water-supply, the floating of ships, and the often surprising motion of fluids. In all three fields, one persistent goal was the production of marvels, either as a demonstration of power or of the range of the model, or as a means to elicit financial support.

The project in alchemy (Chapter 9) was to explain the transformations of materials on the basis of the standard four-element model; colors and all other properties were believed to be secondary qualities dependent on elemental composition (Keyser [1990]). A competing model was that substances are composed of atoms, whose shapes and arrangements created secondary qualities. In both models, color seems to have been conceived as a mutable surface property, and substances were believed to have powers to affect and even transform one another. Within that framework, the alchemists achieved broad success, both explanatory and performative, since they were able to "produce" silver, emeralds, dyes, etc. They continued to believe that the production of gold was also possible, but were less able to show success.

In Biology (Chapter 10), one of the chief early goals had been to explain the origin and diversity of life (Lloyd [1983]). In the period of this book, little reference to that is made, apart from the widespread belief that some kinds of small plants and animals were "spontaneously" generated out of non-living matter. It continued to be assumed that animal behavior was in many respects *like* human behavior. One chief problem or locus of debate was the extent to which animals could be said to have human rationality. Just as for non-living matter, animals and especially plants were seen as having powers.

Greek medical thought (Chapter 11) typically assumed that health was a balance of factors, or a mean between extremes, without agreement on which (Longrigg [1993]). Many medical writers of this period assumed the factors were the four "humours" – this spelling is used for distinction – but others speak of the balance of blood and *pneuma*, or of the motion of corpuscles. To a greater extent than in any other area of Greek science, medical thought was segmented into "schools" – patterns of shared concepts to which a medical writer typically declared his allegiance. The "dogmatists" asserted the primacy of Hippokrates and the possibility of rational and theoretical models of health and disease; they typically adhere to the "four humours" model. The Empiricists denied the possibility of rational accounts, and prescribed treatment on the basis of their own and others' successful experience. Herophilos started a school, which emphasized pharmacy for therapy, the pulse for diagnosis, and dissection for studying the body. Erasistratos and his followers sought cures in balancing the flow of blood and *pneuma* in the body (although he rejected venesection). Standard treatments continued to include dietary and activity regulation, as well as venesection, enemas, and emetics (to evacuate "bad" material), with surgery more rarely resorted to. Drug remedies, believed to change the state of the body in useful ways, were primarily herbal (with some mineral or animal components), and became increasingly complex mixtures, as the doctors sought to combine and balance the powers of the ingredients (Keyser [1997a]). In addition to the new use of the pulse in diagnosis, and the increasing use of drug mixtures, another change was the practice of human dissection (previous knowledge of human anatomy was based on experience on the battlefield and analogy with animals).

The "psychology" chapter (Chapter 12) is something of a mixed bag, but unified as Greek thought on the physiology, or bodily and physical basis, of the soul (Roccatagliata [1986]). Many Greeks believed that some or all aspects of rational activity (sensation, etc.) could be explained on a physical basis. One persistent problem was the physical location of the soul – some advocated the heart and others the brain. Another was the degree to which character could be understood or even predicted on the basis of the body and its phenomena. Just as for health and illness of the body, it was widely assumed that the health or illness of the soul was determined or affected by the balance of factors in the body.

Such were the concepts and the actors in Greek science of this era. But what was the larger socio-political context, and can we explain why science had the character it did? The origins of Greek science have been much debated, but the most appealing model is that due to Lloyd, who argues that it flourished in the peculiar political context of Greece – city-states with a relatively large political elite possessing a tradition of active and productive debate on significant political questions (Lloyd [1979] 226–267; [1987] 50–171; 1991; 1992).

In fact, it is an extension of that model that seems best to explain the trajectory of Greek science in this period. Again, this can be no more than a sketch of an argument. In the earliest Greek science in the period up to about –370 (as briefly described in the prolegomenon to each following chapter), what is notable is the persistence of certain notions – and the great variety of the systems proposed. This account is focussed on cosmology and medicine, as relevant to the purpose and as better represented in the surviving sources. That there is *kosmic* stability beneath *kosmic* change, that health is a kind of stability, and that the human being and body is an integral part of the *kosmos*, seem to have been inevitable assumptions. And each system developed to explain the *kosmos* indeed claims to explain it all – and yet leaves much unexplained and open. These competing and open systems co-existed, and thereby fostered debate – and investigation. In all these early thinkers one finds an over-arching desire for a unifying theory of everything, though details are often wanting. For example, just how do the four elements, or the four humours, in fact explain actual changes in the world or the body? Or again, how do mere rearrangements of atoms explain even the differences between water and wine? That was not the primary goal of these authors. Rather, they were seeking meaning in the natural world. This is perhaps clearest in the writings of the more poetic of the authors, Xenophanes of Kolophon, Parmenides of Elea, and Empedokles of Akragas. Xenophanes writes primarily of the nature of the true divine mind who orders the world properly and rightly – and stably. The earth, he sang, stretched down to infinity, and all things came to be from earth and sea. Parmenides insisted that existence itself, reality as a whole, is divine, uncreated, and immortal. Such Being had for Parmenides certain properties – among them stability and sphericity. For Empedokles, the four "roots" (Earth, Water, Air, and Fire) were divine and imperishable, and the true basis of all existence.

5

For all of these authors, the goal was to perceive the world as an ordered whole – i.e., as a *kosmos*. Their theories tended to validate various traditional notions, and also produced a kind of presumptive synthesis, which undergirded most later systems. Despite this prevenient set of assumptions, there was no prevailing "standard model" but rather an "almost-free market" of competing – and interacting – models. A brief list of some of the key concepts and relevant assumptions of this set of systems will be helpful. First of all is the assumption that it is a meaningful and answerable question to ask what is the essential nature of reality, including the nature of ourselves. Second, that we are also able to ask what is the underlying stability in a world of flux. Third, that we can ask for an account of how one thing follows another, that is, an account of the nature of causation. Beyond those three fundamental questions, was a group of paradigmatic notions: (1) the world was an ordered whole, (2) that order included a natural balance and a natural hierarchy, (3) geometric structure had explanatory power, (4) there were persisting simple elements which constituted stuff, and (5) there existed between certain pairs of things a natural harmony or "sympathy." In particular, with reference to cosmology, it was assumed that all heavenly bodies moved in eternal circles about the central earth; and with reference to the body, it was assumed that a balanced diet and activity ("regimen") would guard the balanced stability of health, and that a body fallen into imbalance could be restored to balance primarily via diet, drugs, and activity. For those thinkers, there was no contradiction between the notion of a "natural hierarchy" and a "natural harmony" – they saw a harmony between upper and lower, and between rulers and ruled – a perspective we have long since abandoned as illusory.

This more or less coherent collection of rarely or never-questioned assumptions, beliefs, concepts, and definitions formed the basis of all later Greek explanatory syntheses. Examining mathematics, geography, biology, and so on, would add a few items but not substantially alter the picture. Moreover, despite wide agreement on this presumptive synthesis, there was no single "standard model" nor even any grand unified theory of health and disease or of the *kosmos*. Instead there was a variety of competing models, none dominant, none claiming explanatory totality, and none formalized into a school or self-perpetuating tradition.

The intellectual situation corresponds exactly with the political situation in Greece, as emphasized by Lloyd – contentious public debate, constantly shifting politics, multiple autonomous states. The essence of the argument is that cultural forms and norms become engrained, endemic, and therefore implicit in the thought-process of members of the society. The dominant political paradigm subtly forms and informs the other cultural paradigms, in a mutually reinforcing system.

But in the era roughly –360 to –300, a different tendency developed (of course no historical transition is precise and abrupt). For a few generations,

Greek thinkers built upon this open-ended collection of ideas a variety of syntheses, each of which had the goal of demonstrating the world as an ordered whole. But, in contrast to the earlier period, the systems developed in this era sought a synthesis of all knowledge, and claimed explanatory totality. A crucial second contrast with the earlier period was that each of these systems did involve the founding of a self-perpetuating school.

It was Plato who was responsible for the first attempt at such a synthesis and school (his was the "Academy"). Plato indeed believed that he had found in his theory of the Forms an explanation for all reality – but in all his dialogues (and especially in the *Timaios*, written at the end of his life), he takes care to manifest the difficulties and loose ends of his account. For Plato, the ultimate explanatory entity was the Form, the true and abstract essence of any existing thing, and each existing thing did either more or less fully embody its essence, i.e., its Form. Each Form was in fact the cause of all its embodiments, since it provided the goal towards which every one of its embodiments was striving. Thus geometry and astronomy together became the divine path which could lead one up from mere instances to the ultimate, true, and eternal essences. Plato's ordered whole was especially hierarchical, as a glance at the *Republic* will show. Plato's account of the physical world seeks to show the hidden operation of reason, and uncovers a mathematical structure. The sensible world is composed of the usual four elements, which are themselves formed from purely geometrical triangles. His account of the human body in health and disease is aggressively teleological – every detail exists for a higher purpose (*Timaios* 69–90). We have, he says, three souls – the lowest is in the liver, the seat of desires, and is responsible for our physical appetites; the middle soul is in the heart which listens to reason and masters the body; and the highest soul is in the brain which reasons. The anatomy of the rest of the body is subordinated to these three organs. His metaphor of the three souls is inherently political, and is expressed in explicitly political language – so it is no surprise that his account of disease makes it an ethical or moral matter (*Timaios* 72–73, 86–87). Plato holds that disease derives from three causes: (1) an imbalance of the elements, or (2) the production of corrupt residues or humors, or (3) the blockage of the movement of bodily air (*Timaios* 81–86). The restoration of the ordered governance of the body is accomplished by the enforced functioning of its organs and members through diet and activity (and preferably not drugs: *Timaios* 87–91).

Aristotle studied with Plato, but later founded his own school (the "Lukeion") and synthesis; he believed that, in his system, philosophy had come close to completion. He promulgated a synthesis of great complexity, which included all the common notions listed above, but was founded on a doctrine of four distinct and hierarchically ordered causes. These are the familiar "material," "efficient," "formal," and "final" causes, in terms of which Aristotle believed that any possible actual thing or event could always be explained. Unlike all predecessors, so far as we are aware, he attempted in his own work – and left it as a research program for his students and successors – to investigate in detail how these four causes operated in every part of the *kosmos*. Aristotle's own project in biology was to make manifest the operation of the "formal" and "final" causes in all animals, though none of his extant works focuses on the human body. (The formal cause is roughly that which gives structure to some thing, and the final cause is the purpose for which the thing exists.)

A third variety of synthesis was offered by Zenon (the founder of what came to be called the Stoic school), for whom there were two cooperating fundamental principles of being: the active *pneuma* (there really is no acceptable English translation), and the passive matter. The *pneuma* is a kind of very fine yet corporeal stuff which completely penetrates all matter everywhere in the *kosmos* without either losing its identity or causing matter to lose its identity. Moreover, this *pneuma* is coextensive with matter, and cannot be separated from matter. This *pneuma* serves to bind together unified wholes, such as living beings and souls, so that they possess an inner coherence or sympathy. Such a sympathy causes effects on one part of a unified whole to be transmitted to all its other parts. Moreover, the *kosmos*, like the body and the soul, is also a living whole, unified by the all-pervasive *pneuma*. Although Zenon founded his school after Aristotle's death, few of his words survive, and his work properly belongs before the era of this book, so he himself does not figure in any of our chapters.

The school of Epikouros (the "Epicureans"), though in many ways almost a negation of this synthetic trend, also originates towards the end of the fourth century BCE. Epikouros sought to explain all of nature on two principles – atoms and the void. Atoms were the indivisible and changeless building blocks of all stuff, which moved through the otherwise utterly empty void. Although Epikouros, like Zenon, properly belongs to the era before our book, we have chosen to include a few passages, for several reasons: very little survives of later writers from his school (so that omitting him would in effect distort the picture), and the distinction of the eras is not of course absolute or precise.

Two other systems probably originated at this time, though in neither case do we have the name of their founder. These are the Hippokratic school of medicine, and the organized "neo-Pythagorean" movement.

There is good reason to believe that Hippokrates actually lived and was a doctor; he had died by around –375. There are certainly no references to any school or system of Hippokratic medicine before that time. The earliest reference to a systematic and Hippokratic way of thinking is found in Plato, *Phaidros* 270c–d, where he explains that Hippokrates taught that one must first understand the nature of the "whole" in order to understand the body, seemingly a reference to a belief in some sympathy between the body and the *kosmos*. Plato uses it as a point of departure for a discussion of a formal method of inquiry. Aristotle's student Menon wrote up a history of medicine, of which a kind of digest survives; there it is stated that Hippokrates taught that disease is caused by the "breaths" or "gases" which arise from digesting improper food, and which displace the air we inhale. We can only conclude that, around –360 or a bit later, someone put together a picture of "Hippokrates" and collected some books under his name. By late in the fourth century or early in the following century, there is evidence of a Hippokratic school or schools (one at Kos, one at Knidos), which continued to exist for some centuries. Diokles of Karustos was a product of that era and movement, and a sketch of his life and work will serve as an outline of what "the Hippokratic School" was intending. He seems to have lived from about –375 to about –295, and was for much of his life resident in Athens. Diokles is attested to have written the first book of anatomy, a probable influence on Aristotle. His only surviving work is a letter to King Antigonos (reigned –305 to –300), in which he divides the body into the four parts head, chest, innards, and bladder. Disorders in each of these parts are described by listing all symptoms; for each region, Diokles prescribes a regimen for cure. He adds a regimen arranged by the six seasons of the year: e.g., at the spring equinox, eat the juiciest foods and most pungent. The approach is reminiscent of the Hippokratic work *Regimen*; also related to this is a third contribution of Diokles, the development of the medically and dietetically oriented systematic classification of plants and animals, a system further elaborated by Aristotle. Diokles seems to have begun the systematic collection of prior opinions, and summarized and integrated the medical achievements of his predecessors. Like his coeval Theophrastos, Diokles studied plants; he, however, wrote up his researches as the earliest Greek herbal.

Turning next to the "Pythagorean School," we first recall that like all great teachers, Pythagoras himself never published anything, and the evidence that he founded any sort of self-perpetuating school is very dubious. (He did run a secret political society in his own lifetime.) Early writers who refer to Pythagoras mention only his sayings, and make no statement whatsoever about a school. The evidence of Aristotle and Aristoxenos, who wrote books on the Pythagoreans, suggests that it was in the mid-fourth century that the movement was reorganized as a formal school (the "neo-Pythagoreans": some scholars restrict the term "neo-Pythagorean" to a later period). It is at this time that the characteristic Pythagorean doctrines of medical import are first attested: advocacy of vegetarianism, and denial of surgery and abortion. It has been argued that the so-called "Hippokratic Oath" is in fact a neo-Pythagorean production of around –350. The neo-Pythagoreans emphasized the fundamental role of number in the *kosmos*, and focused their thought especially on astronomy and music.

What is significant about the rise of these synthetic systems is that all of them originate in one brief era, roughly –360 to –300: none before – and none for centuries thereafter. These two generations were exactly when, under the sway of Philip, Alexander, and their successors, the entire Greek world was, for the first time ever, unified, and even united with much of the Eastern Mediterranean world. During this period, the old democratic traditions of debate were gradually stifled. The transition in Athens from "Old Comedy" with its ribald and explicit political commentary to the purely private and domestic "New Comedy" parallels this exactly. No longer was there a socio-political context of open and free debate among the literate elite – now instead there was a centripetal, harmonizing, and synthesizing tendency in socio-political thought and action. This brief cultural synthesis ended in –300 with the foundation of the four competing "successor" states (as they are called): the Ptolemaic empire, the Seleukid Empire, and the kingdoms of Macedon and of Thrakê.

There are two great differences between, on the one hand, the philosophical systems of this era (temporarily leaving aside the "Hippokratic synthesis"), and on the other hand, those of the previous era. First is that the competing systems now claimed to explain the entire *kosmos* in detail as an ordered whole, on the basis of a few principles. And second, the adherents of these systems sought to perpetuate themselves by founding schools. That is, each system grasped for the same sort of universal predominance in the intellectual field that Philip and Alexander were seeking in the political field. (Moreover, there was a strong tendency in this era for the leading intellectuals to congregate in Athens, whereas previously, Greek scientific thinkers came from and worked in all parts of the Greek world.)

The locus of active investigation of natural phenomena in Hellenistic Greek science (i.e., from the period covered by this book, about –300 to about 230) lay in the contested ground between the competing syntheses – and in the loose ends within each synthesis. Such investigation was, however, limited by the fact that most of their disputes were not resolvable by means within their power. For example, the followers of Zenon (the Stoics) argued that matter is fundamentally passive and quality-less, but that it is also permeated by *pneuma*, which generates all observable qualities or attributes. In contrast, the followers of Aristotle claimed that, first, four fundamental underlying but inseparable principles (arranged as two pairs of opposites – hot/cold and wet/dry) combined to form the four Empedoklean elements, which then mixed to form observable stuffs. For atomists such as Epikouros, material properties were determined by mechanical rearrangements of atoms, themselves quality-less, and for Platonists, observable qualities existed because of the participation of the individual instance in the eternal Form. While all of these agreed that material qualities (color, density, etc.) were secondary, there was no agreement about how they were produced – and no prospect of being able to determine how. A similar story could be told about many of the open questions of the era – any conceivable test or experiment lay beyond the technical expertise of the ancient world. Nevertheless, some details were added to one synthesis or another, or even to the body of knowledge agreed to by all participants in the debates.

First, a brief description of such expansions of knowledge, two in medicine and one in cosmology, which may be attributed to work done in the contested ground between these syntheses. Then a sketch of some work done on the loose ends within one synthesis.

To early Greeks, it had seemed as if there were differing forces or powers within their bodies, to which they gave a variety of names. Plato described three competing "souls," one located in the liver (the soul which desires), one located in the heart (the soul which emotes), and one located in the head (the soul which reasons). Most other early Greek thinkers located most or all of the human soul in the heart, as did Aristotle and Zenon, and it remained part of the collection of ideas. But the discussion did allow for the possibility that the head might be involved – at least Alkmaion and Plato had thought so. And in the third century BCE, the physician Herophilos, who came from the provincial town of Chalkedon, but was working in Alexandria of Egypt, showed via dissection that a very good case could be made that all sensation, will, and thought were to be located in the brain. Herophilos discovered the bodily structures we call nerves, distinguished between sensory nerves and motor nerves, traced all of them to the brain, and described the cerebellum as the command center of the brain. Particularly clear is the evidence that he traced the optic nerve from the eye to the brain and mind. He seems to have attempted to explain the mode of operation of the nerves by hypothesizing that they contained or transmitted *pneuma*. Herophilos thus seems to have adopted Aristotelian notions of the desirability of close observation of bodily structures, Platonic or Alkmaionic notions of a hegemonic brain, and Stoic notions of the decisive role of *pneuma*, to produce what may fairly be called one of the enduring advances of Hellenistic science.

Erasistratos of Keos studied in Athens around –310, and by fifteen years later had an international reputation; he was dead by mid-century. In anatomy, Erasistratos correctly inferred on the basis of comparative dissections that the degree of convolution of the brain is correlated with the relative intelligence of species; he also distinguished the cerebrum from the cerebellum. More-over, he conceived the function of the heart to be that of a pump (like the force-pump invented around that time by Ktesibios in Alexandria): the left ventricle contains *pneuma*, the right blood, and the semilunar valves prevent reflux while the bicuspid and tricuspid valves control efflux. He maintained that veins and arteries originate from the heart, and that veins contain blood, but the arteries only *pneuma*.

According to Erasistratos, cut arteries bleed because the escaping *pneuma* draws blood from the veins via something like capillaries, too small to be seen. This in fact was also the origin of most disease – blood to excess in the veins leaks through to the arteries and is there compressed by the *pneuma* as it obstructs the free flow of the *pneuma*, leading to many symptoms, especially fever. Humoural pathology is notably absent or restricted in application, and in particular phlebotomy was entirely eschewed. He preferred prevention, on which he was the first to write a dedicated treatise.

His point of departure in physiology was the contemporary theory found in Theophrastos and others: pockets of void are scattered throughout all things, and are the explanation of transparency, compression, and mixing. Physiological events occur due to bodily liquids "following the void." Consistent with this outlook are his metaphors for bodily structure – each part is a three-fold weave of vein, artery, and nerve. The body in fact accretes or weaves itself together in this manner as it grows, nourished by the arterial *pneuma*. Nutrition itself takes place analogously, the stomach actually compressing the nutriment out of the food.

Greek thinkers tended to assume that the motions of the lights in the sky (the stars, the moon, and the sun) were in some important sense regular. Plato, like many before him, had insisted that these lights must be divine beings whose motions ought to be somehow composed of the best and most divine motion, the circle. Aristotle, like others before him, concocted an elaborate system of dozens of spheres centered on the Earth. These notions were part of the common corpus and formed the basis of Hellenistic astronomy. But the problem remained – how exactly to explain the observed motions, some of which can seem quite irregular. In the early second century BCE, two provincial mathematicians, Apollonios of Pergê and Hipparchos of Nikaia, worked out a system by which, for the first time ever, the motions of all the planets could in principle be predicted, on the basis of a model which provided a picture of their motions. Moreover, this system allowed the prediction of the motions of the sun and moon (the actual prediction of the motions of the other wandering stars was not accomplished for several centuries). This system did not question the geocentric hypothesis, but modified it by allowing planets to move on "geocentric epicycles" – that is, circles not centered on the Earth, but centered on a second point, which itself was moving about the Earth. (To clarify – from the sun's point of view, our moon moves on a heliocentric epicycle, to wit, its orbit about the Earth.) Methodologically, this was important, for it showed that seemingly complex motions of celestial bodies could in fact be explained in terms of fundamentally simple motions. Prediction itself had been accomplished by the Babylonians over 500 years before, but they offered no picture of how the motions could be regular. Hipparchos also made use of his own and earlier observations to show that the position in the sky of the sun at the time of the equinoxes had slowly been changing (or, as we would say, that the axis of the rotation of the Earth slowly wobbles) – this is the phenomenon called the precession of the equinoxes. The work of Hipparchos thus shows an eclectic use of material from different syntheses – Platonic notions about the circularity of planetary motion, Aristotelian notions about the complexity of the heavenly phenomena, as well as, again, the need for careful observation. There is also reason to believe that Hipparchos employed Babylonian data.

These three, Hipparchos, Erasistratos, and Herophilos, are seen to have been working in the gaps between the competing syntheses. There were also those who worked on the loose ends within a synthesis. The best examples of this in cosmology happen to be Theophrastos and Straton, who worked within the tradition established by Aristotle; a good example from medicine is the physician Praxagoras of Kos.

Theophrastos of Eresos succeeded Aristotle as head of the "Lukeion," and from that position wrote a plethora of works, only a small fraction of which survive. From the little we have, it would seem that Theophrastos started, as Aristotle had advised, by collecting data on what was known and believed by influential writers, about a subject of research. Aristotle's goal in much of that work had been to discover and demonstrate the operation of the "formal" and "final" causes in the subject being studied (above). But Theophrastos did not proceed as Aristotle had – rather he often is found to be demonstrating difficulties with the standard model. Most significantly, he suggests that it may be that the operation of a final cause is not the source of certain phenomena (such as deer antlers or male breasts). He also pointed out that Aristotle's "Prime Mover" (the eternal and unchanging cause of the movement of the sphere of the stars) somehow fails to cause the earth itself to rotate, and that fire, supposedly an element as fundamental as water or air, in fact behaves very differently (being, for example, generated only from a fuel). Theophrastos continues Aristotle's project of making and assembling copious observations, but he is not as inclined to theory – he questions Aristotle's synthesis, but does not reject it, and provides no alternative. His work served to show that there were many productive loose ends even in Aristotle's comprehensive system.

Straton of Lampsakos succeeded Theophrastos as head of Aristotle's "Lukeion" – and went much further in questioning the Aristotelian synthesis. Although none of his writings survives intact, the fragments make it clear that he denied, in part, a key tenet of Aristotle's system. Demokritos had claimed that the universe is made of only two things, atoms and the void (empty space). This concept of void Aristotle was zealous to deny: there could not ever be, he asserted, any such thing – it was a self-contradictory concept. Straton hypothesized that void could exist, at least in principle (so it was not self-contradictory), and he hypothesized that void did exist, in actuality, as tiny pockets within all matter. This theory, called by him the theory of "disseminate" void (and called by modern scholars the "micro-void" theory), he used to explain how things could be transparent, how sound could propagate, how mixing could occur, and how some stuffs could be compressible – for none of which observations could Aristotle's theory offer a convincing account. Second, Aristotle had insisted that the four elements had each their natural place, and each their natural motion towards that place;

then, in order to explain the circular (not vertical) natural motion of the heavens, he hypothesized the existence of a fifth heavenly element (called *aither*). Now Straton absolutely denied the existence of this form of matter, thus broaching the absolute separation of the eternal Aristotelian heaven from the changing Aristotelian earth. Moreover, he went beyond Theophrastos and not only questioned but denied Aristotle's theory of a "Prime Mover" and of final causes. That is, he removed the operation of purpose from the *kosmos* as a whole.

Praxagoras of Kos was active around –300, so that his birth date has been set at around –340 (making him roughly a contemporary of Zenon the Stoic); he was a teacher of Herophilos. Instead of the standard four humours proposed by Polubos, Praxagoras proposed fully eleven humours to explain health and disease: Sweet, Mixed, Clear, Sour, Soapy, Salty, Bitter, "Leek-green," Yolky, Corrosive, Clotting, and Bloody. Just as we commonly refer to four flavors (bitter, sweet, salty, and sour), but acknowledge that such a system scarcely explains the taste, for example, of carrots, so Praxagoras' system seems intended as a trenchant criticism of the four-humour theory.

Of more lasting significance, Praxagoras made a number of connected discoveries and suggestions related to the role of the heart and blood in movement and life. He is the first to direct attention to the pulse as a diagnostic tool, examining arterial pulsations and specifying varieties of pathological movements. He also first distinguished veins and arteries (perhaps initially on the basis that the pulse could be sensed only in the arteries), and is responsible for the theory that veins contain blood but arteries *pneuma* (thus he employs a Stoic concept). The heart is the seat of the soul and thought, and the brain merely an excrescence or appendage of the spinal cord. Finally, the nerves are a refinement of the arteries, which control the movements of the fingers and hands, and in fact these nerves originate at the heart. This is a sensible and coherent theory – the soul resides in the heart, out from which run arteries filled with *pneuma*, pulsing with that contained life, and thinning down to the nerves which control bodily movements. Empedokles had sought to explain nutrition by positing the presence in food of essential elements of some kind (the four "roots"); Praxagoras took another line entirely and made the principle be the inhaled *pneuma*. Digestion, on the other hand, though necessary was essentially putrefaction and rotting.

The same sort of socio-political parallelism can be seen here again; for in the Hellenistic era, though scarcely repeating the era of Greek city-states, there was once more active debate and political opportunity, but in novel forms. For several generations after the death of Alexander the Great, the various "successors" and their kingdoms struggled for control of pieces of his empire, and in that context, there was great opportunity for qualified men from all parts of the Greek world to gain power, as "friends" of the kings, or as mercenaries, or as technical experts. Many Greek cities not wholly controlled by any of the kingdoms formed federated states, and frequent embassies and envoys traveled across the Greek world. The ruling classes were everywhere mobile (even if the kingship itself was securely hereditary). By the end of the third century BCE, this had somewhat faded – and there are somewhat fewer Greek science writings from this period. But soon the early Roman involvement in south Italy, Greece, and the eastern Mediterranean led to political settlements that preserved distinctions between states and were selfishly arbitrated by the Romans. That itself gave new stimulus to public debate and political activity, especially in the form of competing embassies to Rome (and Rome's policy was to allow these client states freely to regulate their own internal affairs).

The end of the period covered by this book is represented by the turn to a new kind of science or at least systematic synthesis, manifested in the indefatigable efforts of the Aristotelian commentators to harmonize Plato and Aristotle, in the neo-Platonism of Plotinus, and even already a little in Galen and Ptolemy. This final synthesis left no loose ends, and claimed finality and wholeness: all prior systems were either subsumed as special cases or partial versions, or else were rejected and abandoned. Plato's Forms, Stoic beliefs about *pneuma* and the soul, Pythagorean notions of the hidden and immanent role of number in the construction of the universe, and Aristotle's system of elements and causes, were fused together.

Although Rome had completed her conquest of the lands around the Mediterranean basin by –30, it was precisely in the second century CE, especially in the reign of Hadrian (117–138), that the form of the Roman Empire became itself an ordered whole of the sort to encourage hyper-synthesis. Until the very end of the first century CE, the empire could better have been described as an articulated whole, composed of political entities of many sorts. There were client kingdoms (such as Kommagênê and Armenia Minor), client leagues (such as the Lukian League), and various city-states, called "free" or "confederate" (such as Athens, Massalia [Marseilles], Rhodes, and Sparta). Such entities, though under the rule of Rome as far as foreign policy went, were both in theory and in practice free to regulate their internal affairs according to their own wishes and beliefs.

All that died in the second century CE. Already early in the century the historian Tacitus could speak of the "immense body of the empire" (*Histories* 1.16); around the same time, Iosephos, speaking as "Agrippa," describes the awesome and all-pervading Roman power (*Jewish War* 2.16[345–401]). And indeed, provincials internalized the Imperial ideology of a direct relationship with the emperor, as the guarantor of imperial benefits. As the orator Aelius Aristides remarked in his mid-century panegyric *To Rome* 60, the Roman emperors have made the world into "a single city, a single house, a single family." Similarly, the historian Appian describes the unity of the empire as subject to the emperor (*pr.* 24–28). Many changes brought about this new organic and corporate wholeness. The first large grant of Roman citizenship to subjects of Rome living outside Italy was made by the emperor Vespasian in 74 to a province in Spain, but it was during the second century CE that Roman citizenship was gradually granted to groups throughout the empire, and in the end, every free man in the empire was made a citizen by an edict of the emperor Caracalla in 212. The demise of the system of client kingdoms began in 106, when the emperor Trajan formed the province of Arabia (a region near and in modern Jordan), by annexing and abolishing several client kingdoms. And all the other client states, including the city-states called "free" or "confederate," suffered a similar amoebic absorption within the second century CE. During the same period (especially under Hadrian) all the exterior borders of the empire were more sharply defined (often through the building of a wall, as if around a city). The same emperor promoted the first systematization of Roman law (some of that effort survives in the *Institutes* of Gaius), and deeply reorganized the imperial civil service into a true and more centralized bureaucracy. Beginning already under Trajan at the start of the century, that bureaucracy was more and more the micro-manager of the civil and financial affairs of the cities and provinces of the empire. Local assemblies ceased to function within the second century, as the imperial *curatores* and (more ominously) the *correctores* exerted more and deeper control. Ultimately, all imperial revenue came to be regarded as the emperor's private fund, a situation made explicit by the emperor Seuerus (193–211) who allocated all taxes to a fund called the *res priuata*, the emperor's "private affair." Roman culture had long displayed a tendency to prefer handbooks and encyclopedias summarizing useful knowledge (such as the *Natural History* of Pliny) – now that preference became irresistible.

A few parallels from other ages may be mentioned. The Arabic interest in and systematic translation of Greek science seems to be due to a similar collocation of socio-political factors in Baghdad around 750–950 (Gutas [1998]). The north Italian renaissance and its scientific developments certainly owed a great deal to the political situation there. The British Enlightenment was due in large measure to a context of debate and political openness. And the leading role of science in the second half of the twentieth century in the United States is clearly due to the socio-political context.

Historical interpretation is perpetually bedeviled with the uncertainty and bias of partial knowledge, and with the risk of jumping from a *post hoc* observation to a *propter hoc* conclusion. The human tendency to assimilate our thinking to our environment is well known (and what part of our environment is more affective than our political and cultural environment?) – it could even be said to be an essential part of the human condition. And at three critical moments or periods, from the fourth century BCE through the early third century CE, that human tendency seems to have directed the overall course of ancient Greek science. Political monopoly promoted intellectual synthesis, while political pluralism promoted intellectual debate and productivity. And in the end, the pinnacle of political uniformity fostered the creation of a hyper-synthesis which promised a view of the body and the universe as an ordered and meaningful whole, with no openings for productive questions. Any system in which, by definition, there are no loose ends can only stifle enquiry, for it can never be confronted either by its own defects nor by the way that the world is. But for some people (then and now), ordered and meaningful wholeness is preferable to loose ends and open questions.

2

MATHEMATICS

We seek patterns and number to order the world around us, counting days and people and much that we see. The earliest societies knapped stone blades more symmetric than functionality required, cut ordered rows of notches as records, and painted geometric grids among their cave-animals. Geometric order surrounded us in circular huts and woven clothes; and in Mesopotamia, long before literature, writing began with numerical tally-marks and abstract geo- metric counters representing trade-goods (Schmandt-Besserat [1992]).

Most peoples practiced land-mensuration; the Egyptians also raised geometric pyramids. Mesopotamian cities were laid out on grids, and their ziggurats drew on elementary geometry. Out of their tally-marks, Mesopotamians developed a numerical system in base-60, on which they built arithmetic and linear algebra, e.g., areas of figures and relations between sides of triangles (including "Pythagoras' theorem"); they always represented square roots as approximations to a fraction.

Our reliable knowledge of Greek mathematics begins with Zenon of Elea (−455 ± 20) who asserted, through paradoxes about plurality, the impossibility of infinite division. Around this time, Hippodamos of Miletos imported grid-based city-plans, laying out Peiraios near Athens and Thourioi in southern Italy. In response to Zenon, Demokritos of Abdera (−410 ± 30) raised a difficulty about cones: are two adjacent faces of the plane-section of a cone equal (if so, a cone is a cylinder) or not (if not, a cone is a pile of discs)? The question arose when he worked out the ratio of the volume of a cone to that of its enclosing cylinder. Hippokrates of Chios (−420 ± 20) – as reported by Eudemos of Rhodes a century later – offered the earliest known proofs (of the equality of area of certain figures) in his work on "squaring the circle."

Sometime during the fifth century BCE other concepts became known, especially certain theorems (later attributed to Thalês) and the incommen- surability of the diagonal of the unit square. The Thalean theorems include that equilateral triangles are equiangular and that the opposite angles formed by two crossing lines are equal. The era's method of approximating curvilinear areas and volumes using a series of rectilinear figures may have been based upon the work of Demokritos. Certain problems became established as standard for mathematicians, including determining the area of ("squaring") a circle, and doubling the volume of a cube (which they knew to be equivalent to finding two means in continued proportion, i.e., to finding a cube root). Many Greeks came to believe that numbers were powerful and unlocked the secrets of the universe; this conceptual system came to be associated with Pythagoras of Samos (–530 ± 30) about whom almost nothing is known despite a plethora of data. Philolaus of Kroton (–420 ± 10) in his book on the Pythagorean *kosmos* also thought it significant to classify numbers as odd, even, and "even–odd" (i.e., divisible by two but not by four).

Theaitetos of Athens (died –368) is said to have studied incommensurables and the five regular polyhedra. Archutas of Taras (–375 ± 25), a follower of the teachings of Pythagoras, offered a solution to the duplication of the cube which employed an intersecting cylinder, torus, and cone, and asserted that mathematics was the key to reality (especially in astronomy, mechanics, and music). His aristocratic friend Plato wrote dialogues (–365 ± 20) inculcating the doctrine that numbers had explanatory and perhaps generative power: in the *Meno* 81–84 mathematics serves as the exemplary teaching (as also in *Republic* 7); the *Theaitetos* (146–148) attests the understanding that mathematics must be founded upon a properly-constituted set of definitions; the *Parmenides* (142–145) raises difficulties about unity and plurality; the *Timaios* constructs the entire *kosmos* from numbers and shapes; and the possibly posthumous *Epinomis* reasserts the crucial role of number in the world and in our understanding thereof. Around the same time, Leon synthesized what was known about mathematics, including solvability conditions, in his *Elements*; another similar effort by Theudios of Magnesia has also perished. A bit later Archutas' pupil Eudoxos of Knidos (–355 ± 10) invented a theory of proportion which handled both rational and irrational magnitudes and became standard in Greek mathematics (Eukleidês, *Elements* 5), and systematized the Demokritean method of approaching a limit while avoiding infinitesimals. Menaichmos of Prokonnesos (described as studying under Eudoxos, –345 ± 20) wrote on the duplication of the cube and on conic sections, treating mathematics as an exercise in logic, not an investigation of essential realities.

Speusippos of Athens (nephew of Plato and head of the Academy –346 to –338) based his mathematics on "axioms," statements he regarded as indisputably true, and wrote *Pythagorean Numbers* about the properties of numbers and shapes, especially "10" and the five regular polyhedra; for him mathematics enabled comprehension of reality but was not its essence. Xenokrates of Chalkedon (a student of Plato and head of the Academy –338 to –313) asserted the reality of all mathematical entities (including infinitesimal lines), and worked on combinatorics (all his works, such as *On Numbers*, are lost). Aristotle's system (–335 ± 10) radically denied that mathematics underlay physical reality (*Metaphysics* 5, *Physics* 2.2 [193b22–194b15], and *On Heaven* 3.7 [305b27–306b2]); the sole mathematical work attributed to him, *Indivisible Lines*, argues against Xenokrates' infinitesimal lines.

2.1 Eukleidês ("Euclid") who worked in Alexandria under Ptolemy I (i.e., *ca.* −300), wrote what became for millennia the standard textbook on mathematics (he wrote others on astronomy and optics and music). Books 1–6 concern plane geometry; books 7–10 arithmetic, number-theory, and irrationals; and 11–13 solid geometry.

Elements

1 Definitions

1. A *point* is that which has no part.
2. A *line* is length without breadth.
3. The extremities of a line are points.
4. A *straight line* is a line which lies evenly with the points on itself. [Plato, Parmenides 137e]
5. A *surface* is that which has length and breadth only.
6. The extremities of a surface are lines.
7. A *plane surface* is a surface which lies evenly with the straight lines on itself.
8. A *plane angle* is the mutual inclination of two intersecting and non-collinear lines in a plane.
9. And when the lines containing the angle are straight, the angle is called *rectilinear*.
10. When a straight line set up on a straight line makes the adjacent angles equal one to another, each of the equal angles is *right*, and the straight line standing on the other is called a *perpendicular* to that on which it stands.
11. An *obtuse angle* is an angle greater than a right angle.
12. An *acute angle* is an angle less than a right angle.
13. A *boundary* is that which is the extremity of anything.
14. A *figure* is that which is contained by any boundary or boundaries.
15. A *circle* is a plane figure contained by one line such that all the straight lines falling on it from one point among those lying within the figure are equal one to another.
. . .
22. Of quadrilateral figures, a *square* ["tetragon"] is that which is both equilateral and right-angled; an *oblong* is right-angled but not equilateral; a *rhombus* is equilateral but not right-angled; and a *rhomboid* has its opposite sides and angles equal one to another but is neither equilateral nor right-angled; and let quadrilaterals other than these be called *trapezia*.
23. *Parallels* are straight lines which, being in the same plane and extending indefinitely in both directions, do not meet one another in either direction.

Postulates [1–3 assert that space is continuous and infinite]

1. Let the following be postulated: to draw a straight line from any point to any point.

2. To produce a finite straight line continuously in a straight line.
3. To describe a circle with any center and diameter.
4. All right angles are equal one to another. [space is homogenous]
5. If a straight line intersecting two straight lines makes the interior angles on the same side less than two right angles, the two straight lines, extended indefinitely, intersect on the side of the angles less than two right angles. [compare Tóth 1969]

Common notions [equality of geometrical figures means either in length or area as appropriate]

1. Things which are equal to the same thing are equal one to another.
2. If equals are added to equals, the wholes are equal.
3. If equals are subtracted from equals, the remainders are equal.
4. Things which coincide with one another are equal one to another.
5. The whole is greater than the part. [denying Zenon's paradox]

(Thomas [1939] 437, 439, 441, 443, 445)

1.47 ["Pythagoras' Theorem"]

In right-angled triangles the square on the side subtending the right angle is equal to the squares on the sides containing the right angle. [known but never proved by Mesopotamians]

Let ABΓ be a right-angled triangle having the angle BAΓ right; I say that the square on BΓ is equal to the squares on BA, AΓ (Figure 2.1).

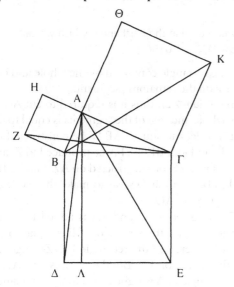

Figure 2.1

For let there be constructed [literally: "drawn up"] on BΓ the square BΔEΓ, and on BA, AΓ the squares HB, ΘΓ [quadrilaterals are often indicated by their diagonal corners; *Elements* 1.46], and through A let AΛ be drawn parallel to either BΔ or ΓE, and let AΔ, ZΓ be joined. Then, since each of the angles BAΓ, BAH is right, it follows that with a straight line BA and at the point A on it, two straight lines AΓ, AH, not lying on the same side, make the adjacent angles equal to two right angles; therefore ΓA is in a straight line with AH [*Elements* 1.14]. For the same reasons BA is also in a straight line with AΘ. And since the angle ΔBΓ is equal to the angle ZBA (for each is right), let the angle ABΓ be added to each; the whole angle ΔBA is therefore equal to the whole angle ZBΓ. And since ΔB is equal to BΓ, and ZB to BA, the two sides ΔB, BA are equal to the two sides BΓ, ZB, respectively; and the angle ΔBA is equal to the angle ZBΓ. The base AΔ is therefore equal to the base ZΓ, and the triangle ABΔ is equal to the triangle ZBΓ [*Elements* 1.4]. Now the parallelogram BΛ is double the triangle ABΔ, for they have the same base BΔ and are in the same parallels BΔ, AΛ [*Elements* 1.41]. And the square HB is double the triangle ZBΓ, for they have the same base ZB and are in the same parallels ZB, HΓ. Therefore the parallelogram BΛ is equal to the square HB. Similarly, if AE, BK are joined, it can also be proved that the parallelogram ΓΛ is equal to the square ΘΓ. Therefore the whole square BΔEΓ is equal to the two squares HB, ΘΓ. And the square BΔEΓ is constructed on BΓ, while the squares HB, ΘΓ are constructed on BA, AΓ. Therefore the square on the side BΓ is equal to the squares on the sides BA, AΓ.

Therefore in right-angled triangles the square on the side subtending the right angle is equal to the squares on the sides containing the right angle. Q.E.D.

(Thomas [1939] 179, 181, 183, 185)

2.11 [constructing the proportion used in the Parthenon, which we call (sqrt(5)+1)/2 = 1.618...]

To cut a given straight line so that the rectangle contained by the whole and one of the segments is equal to the square on the remaining segment.

Let AB be the given straight line (Figure 2.2); thus it is required to cut AB so that the rectangle contained by the whole and one of the segments is equal to the square on the remaining segment. For let the square ABΔΓ be constructed on AB; let AΓ be bisected at the point E, and join BE; let ΓA be produced to Z, and let EZ be made equal to BE; let the square ZΘ be constructed on AZ, and let HΘ be produced to K. I say that AB has been cut at Θ so as to make the rectangle contained by AB, BΘ equal to the square on AΘ.

For, since the straight line AΓ is bisected at E, and ZA is added to it, the rectangle contained by ΓZ, ZA together with the square on AE is equal to the square on EZ. But EZ is equal to EB; therefore the rectangle ΓZ, ZA together with the square on AE is equal to the square on EB. But the squares on BA, AE are equal to the square on EB (for the angle at A is right): therefore the rectangle ΓZ, ZA together with the square on AE is equal to the squares on BA, AE. Let the

22

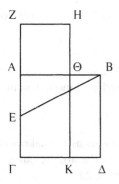

Figure 2.2

square on AE be subtracted from each; therefore the remaining rectangle ΓZ, ZA is equal to the square on AB. Now the rectangle ΓZ, ZA is ZK, for AZ is equal to ZH; and the square on AB is AΔ; therefore ZK is equal to AΔ. Let AK be subtracted from each; therefore ZΘ which remains is equal to ΘΔ. And ΘΔ is the rectangle AB, BΘ, for AB is equal to BΔ; and ZΘ is the square on AΘ; therefore the rectangle contained by AB, BΘ is equal to the square on ΘA.

Therefore the given straight line AB has been cut at Θ so as to make the rectangle contained by AB, BΘ equal to the square on ΘA. Q.E.D.

(Heath [1926/1956] 1.402–403)

3.16 [the infinitesimal "horn-angle" between a circle and a tangent line]

The straight line drawn perpendicular to the diameter of a circle from its end will fall outside, and another straight line cannot be interposed between it and the circumference; and its angle with the circumference is greater, and the remaining angle less, than any acute angle.

(Heath [1926/1956] 2.37)

7 *Definitions* [numbers are usually represented by the length of a line-segment]

1. A *monad* is that in virtue of which each existing thing is called one.
2. A *number* is a multitude composed of monads. [so "one" is only equivocally a number]
3. A lesser number is a *part* of a greater number, when it measures the greater,
4. But *parts*, when it does not measure it.
5. The greater number is a *multiple* of the lesser when it is measured by the lesser.
6. An *even number* is one that is divisible into two equal parts.

7. An *odd number* is one that is not divisible into two equal parts (i.e., differs from an even number by a monad).

. . .

12. A *prime number* is one that is measured by the monad alone.
13. Numbers *relatively prime* are those which are measured by a monad alone as common measure.

. . .

17. And when two numbers have multiplied each other so as to make some number, the resulting number is called *plane*, and its sides are the numbers which have multiplied each other. [compare Nikomachos, 2.8]
18. And when three numbers have multiplied each other so as to make some number, the resulting number is *solid*, and its sides are the numbers which have multiplied each other.
19. A *square number* is equal multiplied by equal, or one that is contained by two equal numbers.
20. And a *cube* is equal multiplied by equal and again by equal, or a number that is contained by three equal numbers.
21. Numbers are *proportional* when the first is the same multiple, or the same part, or the same parts, of the second as the third is of the fourth. [for incommensurable numbers like the diagonal of the unit square, see *Elements* 5.Def.5]
22. *Similar plane* and *solid* numbers are those which have their sides proportional.
23. A *perfect number* is one that is equal to [the sum of] its own parts. [*Elements* 9.Prop.36; the first four are 6, 28, 496, 8128]

(Thomas [1939] 67, 69, 71)

9.20 [prime numbers]

Prime numbers are more than any given multitude of prime numbers [Golomb 1985]

Let A, B, Γ be the given prime numbers; I say that there are more prime numbers than A, B, Γ. [start with any quantity of primes]

For let the least number measured by A, B, Γ be taken, and let it be ΔE; let the monad ΔZ be added to ΔE. Then EZ is either prime or not. First, let it be prime [start with 2, 3, and 5, then EZ is 31; with 2, 3, 5, and 7, EZ is 211; and with 2, 3, 5, 7, and 11, EZ is 2311]; then the prime numbers A, B, Γ, EZ have been found which are more than A, B, Γ.

Next, let EZ not be prime; therefore it is measured by some prime [start with 2 through 13 and EZ is 30031 = 59 · 509]. Let it be measured by the prime H. I say that H is not the same as any of A, B, Γ. For, if possible, let it be so. Now A, B, Γ measure ΔE; therefore H also will measure ΔE. But it also measures EZ. Therefore H, a number, will measure the remainder [*Elements* 7.28], the monad ΔZ: which is absurd. Therefore H is not the same as any one of the numbers A, B, Γ. And by hypothesis it is prime.

24

Therefore the prime numbers A, B, Γ, H have been found which are more than the given multitude of A, B, Γ. Q.E.D.

(Heath [1926/1956] 2.412)

9.35 [sum of geometric series]

If as many numbers as we please are in continued proportion [successive pairs have the same ratio], and if numbers equal to the first are subtracted from the second and the last, then, as the excess of the second is to the first so will the excess of the last be to all those before it. [sum of geometric series; Aristotle *Physics* 3.6 (206b3–34) acknowledges the potential sum; Archimedes, *Squaring the Parabola* 22–23 sums the infinite series in continued proportion ¼ by a method valid for any proportion less than 1/1]

Let there be as many numbers as we please in continued proportion, A, BΓ, Δ, EZ, beginning from A as least, and let there be subtracted from BΓ and EZ the numbers BH, ZΘ, each equal to A (Figure 2.3); I say that, as HΓ is to A, so is EΘ to A, BΓ, Δ [summed].

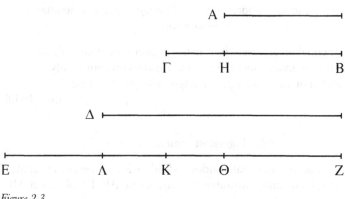

Figure 2.3

For let ZK be made equal to BΓ, and ZΛ equal to Δ. Then, since ZK is equal to BΓ, and of these the part ZΘ is equal to the part BH, therefore the remainder ΘK is equal to the remainder HΓ. And since, as EZ is to Δ, so is Δ to BΓ, and BΓ to A, while Δ is equal to ZΛ, BΓ to ZK, and A to ZΘ, therefore, as EZ is to ZΛ, so is ΛZ to ZK, and ZK to ZΘ. By "separation" [*Elements* 7.11], as EΛ is to ΛZ, so is ΛK to ZK, and KΘ to ZΘ. Therefore also, as one of the antecedents is to one of the consequents, so are all the antecedents to all the consequents [*Elements* 7.12]; therefore, as KΘ is to ZΘ, so are EΛ, ΛK, KΘ [summed] to ΛZ, ZK, ΘZ [summed]. But KΘ is equal to ΓH, ZΘ to A, and ΛZ, ZK, ΘZ to Δ, BΓ, A [summed]; therefore, as ΓH is to A, so is EΘ to Δ, BΓ, A. Therefore, as the excess of the second is to the first, so is the excess of the last to all those before it. Q.E.D.

(Heath [1926/1956] 2.420)

10 *Definitions* [commensurable and incommensurable numbers: Fowler [1992]]

1. Magnitudes measured by the same measure are called *commensurable*, and *incommensurable* those allowing no common measure.
2. Straight lines are *commensurable in square* when the squares on them are measured by the same area, and *incommensurable in square* when the squares on them allow no area as common measure.
3. With these hypotheses, it is proved that there exist infinitely many straight lines both commensurable and incommensurable (some in length only and others in square also) with a given straight line. Let the given straight line be called *rational* [*rhêtê*], and straight lines commensurable with it (in length and square or only in square) *rational*, but those incommensurable with it *irrational* [*alogos*].

(Heath [1926/1956] 3.10)

10.1 [Eudoxos' approach to the limit bypassing infinitesimals ("exhaustion")]

Given two unequal magnitudes, if from the greater more than half be subtracted, and from that remainder more than its half, and so on continually, a magnitude will be left less than the smaller given magnitude. [see 12.2 below]

(Thomas [1939] 453)

10.3 [greatest common divisor]

Given two commensurable magnitudes, to find their greatest common measure.
Let the two given commensurable magnitudes be AB, ΓΔ of which AB is the less; thus it is required to find the greatest common measure of AB, ΓΔ. [the familiar and very efficient algorithm is given]

(Heath [1926/1956] 3.20)

12.2 ["exhaustion" used to determine circular area]

Circles are to one another as the squares on their diameters. [proof omitted]

(Thomas [1939] 459)

12.10 [Demokritos' theorem]

Any cone is a third part of the cylinder having the same base and equal height. [proof omitted]

(Heath [1926/1956] 3.400)

2.2 Archimedes of Syracuse (Surakusê) (*ca.* −285 to −211), was close to the ruling family of Syracuse, visited Alexandria, and wrote on mathematics, mechanics, optics, and pneumatics, addressing his works to Konon of Samos, Eratosthenes of Kurênê, and others.

Area of the Circle [surviving only in a paraphrased version; see
Dijksterhuis [1956/1987] 222–240]

1. The area of any circle is equal to a right-angled triangle in which one of the sides about the right angle is equal to the radius, and the other to the circumference, of the circle.
2. The area of a circle is to the square on its diameter as 11 to 14. [i.e., "pi" is approximately 22/7]
3. The circumference of any circle is greater than three times the diameter, and the excess is less than a seventh part of the diameter but more than ten seventy-firsts. [demonstrated using a 96-agon; Heron of Alexandria, *Metrica* 1.26 says that Archimedes showed in *On Prisms and Cylinders* (lost) that "pi" is greater than "211875:67441" = 3.14163 . . ., and less than "197888:62351" = 3.17377. . ., erroneous ratios which are presumed copying errors.]

(Thomas [1939] 317, 319, 321, 333)

Spirals

Definitions 1 [the figure discovered by Archimedes, based on which he built
the screw and the auger; see Dijksterhuis [1956/1987] 264–285]

If a straight line is drawn in a plane, and one end remains fixed while it uniformly revolves arbitrarily often back to its starting-place, and if as the line revolves it carries a point uniformly along itself beginning from the fixed point, the point will draw a spiral [*hêlix*] in the plane.

(Thomas [1941] 183)

Sand-Reckoner

3.1–4 [expressing large numbers in base-one-hundred-million: see Dijksterhuis
[1956/1987] 360–373; compare the base-ten-thousand system of Apollonios
of Pergê's lost *Okutokion* in which he calculated "pi"; Archimedes in this work
is replying to the Greek proverb that the grains of sand are innumerable,
as in Pindar, *Olympians* 2.98]

1 I think it would be useful to explain the naming of the numbers to be used, so that, as in other matters, those who have not come across the book sent to Zeuxippos [otherwise unknown] may not be confused because there had been no preliminary discussion of it here.

2 Now we already have names for the numbers up to a myriad [10^4], and beyond a myriad we can count in myriads up to a myriad myriads [10^8]. Therefore, let those numbers up to a myriad myriads be called first-order numbers, and let a myriad myriads of first-order numbers be called a monad of second-order numbers [from 10^8 to 10^{16}], and let monads of the second-order numbers be countable, and from the monads let there be formed tens and hundreds and thousands and myriads up to a myriad myriads [Pythagoreans called 10 the monad of the second course, 100 the monad of the third course, etc.]. Again, let a myriad myriads of second-order numbers be called a monad of third-order numbers [from 10^{16} to 10^{24}], and let monads of third-order numbers be countable, and from the monads let there be formed tens and hundreds and thousands and myriads up to a myriad myriads.

3 In the same manner, let a myriad myriads of third-order numbers be called a monad of fourth-order numbers [from 10^{24} to 10^{32}], and let a myriad myriads of fourth-order numbers be called a monad of fifth-order numbers [from 10^{32} to 10^{40}], and let the process continue in this way until the labels reach myriad-myriad times a myriad myriads [the myriad-myriadth order: $10^{800,000,000}$]. It is sufficient to know the numbers up to this point, but we may go beyond it. [he computes that the *kosmos* could contain fewer than myriad-myriad eighth-order monads, or 10^{63} grains]

4 For let the numbers now mentioned be called numbers of the first period [1 to $10^{800,000,000}$], and let the last number of the first period be called a monad of second-period first-order numbers [$10^{800,000,000}$ to $10^{800,000,008}$]. And again, let a myriad myriads of second-period first-order numbers be called a monad of second-period second-order numbers [$10^{800,000,008}$ to $10^{800,000,016}$]. Similarly let the last of these numbers be called a monad of second-period third-order numbers [$10^{800,000,016}$ to $10^{800,000,024}$], and let the process continue in this way until the labels of numbers in the second period reach myriad-myriad times a myriad myriads [the myriad-myriadth order of the second period: $10^{1,600,000,000}$].

Again, let the last number of the second period be called a monad of third-period first-order numbers [$10^{1,600,000,000}$ to $10^{1,600,000,008}$], and let the process continue in this way up to a myriad myriad units of numbers of the myriad-myriadth order of the myriad-myriadth period [$10^{80,000,000,000,000,000}$].

(Thomas [1941] 199, 201)

Method

Praeface [quadrature of the parabola: compare Dijksterhuis [1956/1987] 313–318]

Archimedes to Eratosthenes greeting.

I sent you on a former occasion some of the theorems discovered by me, merely writing out the statements and inviting you to discover the proofs, which at the time I did not give.

[two theorems squaring cylindrical sections] ...

Seeing moreover in you, as I say, an earnest student, a man of considerable eminence in philosophy, and an admirer of mathematical inquiry, I thought fit to write out for you and explain in detail in the same book the peculiarity of a certain method, by which it will be possible for you to get a start to enable you to investigate some of the problems in mathematics by means of mechanics. This procedure is, I am persuaded, no less useful even for the proof of the theorems themselves; for certain things first became clear to me by a mechanical method, although they had to be demonstrated by geometry afterwards because their investigation by this method fell short of demonstration. But it is of course easier, when we have previously acquired, by the method, some knowledge of the questions, to supply the proof than it is to find it without any previous knowledge.

This is a reason why, in the case of the theorems the proof of which Eudoxos was the first to discover, that the cone is a third part of the cylinder, and the pyramid of the prism, having the same base and equal height, we should give no small share of the credit to Demokritos, who was the first to make the assertion about the said figure though he did not prove it. It happens that we made the discovery of the presently published theorems just like those earlier discoveries, and I think it necessary to expound the method partly because I have already spoken of it and do not want to seem to have uttered vain words, but equally because I am persuaded that it will be of no little service to mathematics; for I suppose that some of my contemporaries or successors, once the method is established, will through it discover other theorems, which have not yet occurred to me.

(Cohen and Drabkin [1958] 69–71)

First then I will set out the very first theorem which became clear through mechanics, that any segment of a section of a right-angled cone [parabola] is four-thirds of the triangle which has the same base and equal height.

[lemmata from *Plane Equilibrium* 1.8, 1.4, 1.14, 1.10, and *Conoids and Spheroids* 1 follow in that order]

Let ABΓ be a segment bounded by the straight line AΓ and the section ABΓ of a right-angled cone, and let AΓ be bisected at Δ, and let ΔBE be drawn parallel to the axis, and let AB, BΓ be joined (Figure 2.4).

I say that the segment ABΓ is four-thirds of the triangle ABΓ. From the points A, Γ let AZ be drawn parallel to ΔBE, and let ΓZ be drawn tangent to the section, and let ΓB be produced to K, and let KΘ be placed equal to ΓK. Let ΓΘ be imagined to be a balance with mid-point K, and let MΞ be any parallel to EΔ. Then since ΓBA is a section of a right-angled cone, and ΓZ is tangent thereto, and ΓΔ is a semi-ordinate, EB = BΔ (for this is proved in the *Elements*); for this reason, and because ZA, MΞ are parallel to EΔ, then MN = NΞ and ZK = KA [Eukleidēs, *Elements* 6.4, 5.9] And since ΓA:AΞ = MΞ:ΞO [*Quadrature of the Parabola* 5; Eukleidēs, *Elements* 5.18] and ΓA:AΞ = ΓK:KN [Eukleidēs, *Elements* 6.2, 18], while ΓK = KΘ, therefore ΘK:KN = MΞ:ΞO. And since the point N is the barycenter

29

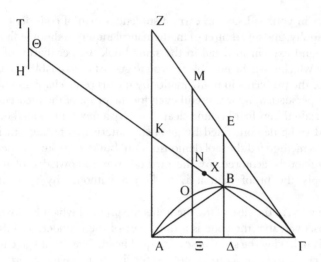

Figure 2.4

of the straight line MΞ, given that MN = NΞ [Lemma 4], if we place TH = ΞO, with Θ its barycenter, so that TΘ = ΘH [Lemma 4], then TΘH will balance MΞ in its present place, because ΘN is cut in the inverse proportion of the weights TH, MΞ, and ΘK:KN = MΞ:HT; therefore their common barycenter is K.

In the same way, as often as parallels to EΔ are drawn in the triangle ZAΓ, these parallels, remaining in the same position, will balance the parts cut off from them by the section and transferred to Θ, so that their common barycenter is K. And since the triangle ΓZA is composed of the lines in it [the "infinitesimals"], and the segment ABΓ is composed of the lines in the section formed like ΞO, therefore the triangle ZAΓ in its present place will be balanced about K by the segment of the section placed with Θ for its barycenter, so that their common barycenter is K. Now let ΓK be cut at X so that ΓK = 3·KX; then the point X will be the barycenter of the triangle AZΓ (proved in *Equilibriums* [1.5]). Then since the triangle ZAΓ in its present place is balanced about K by the segment BAΓ placed so as to have Θ for its barycenter, and since the barycenter of the triangle AZΓ is X, therefore the ratio of the triangle AZΓ to the segment ABΓ placed about Θ as its barycenter is equal to ΘK:XK. But ΘK = 3·KX; therefore triangle AZΓ = 3·segment-ABΓ. And triangle ZAΓ = 4·triangle ABΓ, because ZK = KA and AΔ = ΔΓ; therefore segment-ABΓ = four-thirds triangle ABΓ.

This indeed has not actually been demonstrated by the arguments here used, but they have given some indication that the conclusion is true; seeing therefore that the theorem is not demonstrated, but supposing that the conclusion is true, we will set out the geometrical proof which we discovered and published. [*Quadrature of the Parabola* 24 is copied out]

(Thomas [1941] 223, 225, 227, 229)

2.3 Eratosthenes of Kurênê (*ca.* −285 to −193), growing up while Kurênê enjoyed a brief autonomy from Ptolemaic rule, studied philosophy in Athens. In ca. −245 he became royal tutor to Ptolemy III ("Euergetes") and head of the Library at Alexandria. He corresponded with Archimedes and wrote on a wide variety of literary, mathematical, philosophical, and scientific topics (see Chapter 5.4). See also below, Section 8, Nikomachos 1.13, Eratosthenes' "sieve."

Duplication of the Cube [using a mechanical calculator similar to a slide-rule]

Given two straight lines, to find two means in continued proportion. Let AE and ΔΘ be given. I adjust the tablets [*pinakes*] in the instrument until points A, B, Γ, and Δ are all in a straight line (Figure 2.5). [The "tablets" AZ and ΓΘ slide along EΘ; AE and ΔΘ being fixed and different]

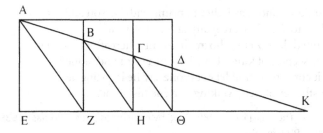

Figure 2.5

Then AK:KB = EK:KZ (since AE and BZ are parallel), and AK:KB = ZK:KH (since AZ and BH are parallel). [The longer proof also quoted by Eutokios shows that the movable tablets are all the same size and parallel, so their diagonals are parallel. Then AZ has been slid over the middle tablet, hiding its diagonal above B, and the tablet with ΓΘ has been slid beneath the middle one, hiding its own diagonal above Γ.] Therefore EK:KZ = KZ:KH; and this proportion is AE:BZ = BZ:ΓH. Similarly we will be able to show that ZB:ΓH = ΓH:ΔΘ. And therefore AE, BZ, ΓH, and ΔΘ are in continued proportion. That is, two means between the two given lines have been found.

Now if the given lines are not equal to AE and ΔΘ, by taking AE and ΔΘ proportional to the given lines we shall obtain the means between them, and then back-transfer the results to the given lines. Thus we will have done what was required. And if it is required to find more means, we will always insert one more tablet on the instrument than the number of means to be found (the proof is the same).

(Cohen and Drabkin [1958] 65–66)

Epigram

Friend, if you're thinking to render a minimal cube to its double, or
    Even whatever in shape, Formed as solid, you must
Metamorphose, this is ready for use; when you've laid out the measure of
    Deep-sunken grain-pit or byre, Hollowed-out vault of wide well,
Using this system, and once you have taken the two-fold concurrent of
    Means, they are found on the points, Topping interior rods.

Nor should you seek for the difficult way of Archútas' cut cylinders[a],
Neither for Ménaichmos' "Three" (Lines from the conics produced)[b];
Further, the line which the pious Eudoxos constructed by bending you
    Need not resort to for means: All are far harder than this.
Thus you may easily render a myriad means on this "Mid-graph,"
    Starting from origin small, Using these tablets of mine.

Fortunate man Ptolemaíos[c], as father in prime with his son and him
    All that is dear to the Muse, Royally favored as well,
Freely you granted: lord Zeus who are heavenly, hereafter let him in
    Quest for the scepter of kings Take it from Your hand alone;
All of this, let it come true; and let everyone say he is gazing at
    What Eratósthenes placed, Making Kuréné his home.

[[a]Archutas' method, see Thomas [1939] 285–289; [b]Menaichmos', see Thomas [1939] 279–283; [c]Ptolemaios is Ptolemy III]

                                                   *(Powell [1925] 66)

2.4 Apollonios of Pergê wrote around −200 ± 10 in Pergamon (and possibly Rhodes), when it was an independent city, and worked in Alexandria and Ephesos. Besides the *Konika* (which became the standard work displacing all predecessors), he wrote on polyhedra, the helix, and "pi" (all lost), among others (see also Chapter 3).

Conics

1 Preface

Apollonios to Eudemos greeting. [books 4–8 dedicated to "Attalos," the king of Pergamon, lived −268 to −196, or perhaps the astronomer and mathematician of Rhodes, *ca.* −175 ± 25]

    If you are in good health and other things are as you wish, it is well; I am pretty well too. During the time I spent with you at Pergamon, I noticed how eager you were to make acquaintance with my work in conics; I have therefore sent to you the first book, which I've revised, and I will send the remaining books when I am satisfied with them. I suppose you have not forgotten hearing me say that I took up this study at the request of Naukrates the geometer [otherwise unknown], at the time when he came to Alexandria and stayed with me, and that, when I had completed the investigation in eight books, I gave them to him rather hastily, because he was on the point of sailing, and so I was not able to correct them, but put down everything as it occurred to me, intending to make a revision at the end. Accordingly, as opportunity permits, I now publish serially whatever happens to be corrected. As certain other persons whom I have met have happened to get hold of the first and second books before they were corrected, do not be surprised if you come across them in a different form.

Of the eight books the first four form an elementary introduction. The first includes the methods of producing the three sections and the opposite branches [of the hyperbola] and their fundamental properties, which are investigated more fully and more generally than in the works of others. The second book includes the properties of the diameters and the axes of the sections as well as the asymptotes, with other things generally and necessarily used in *diorismoi* [solvability conditions]; and what I call diameters and axes you will learn from this book. The third book includes many remarkable theorems useful for the syntheses of solid loci [surfaces] and for *diorismoi*; of these most, and the most elegant, are new, and discovering them made me realize that Eukleidês had not worked out the point-locus with respect to three and four lines [products of distances to given lines determining a point-locus], but only a chance portion of it, and that not successfully; for the synthesis could not be completed without the theorems discovered by me. The fourth book investigates how many times the sections of cones can meet one another and the circumference of a circle; in addition it contains other things, none of which have been discussed by previous writers, namely, in how many points a section of a cone or a circumference of a circle can meet [the opposite branches of hyperbolas].

The remaining books are more for abundance [now extant only in Arabic]: one of them discusses fully *minima* and *maxima* [least and greatest lines to a conic from a given point], another deals with equal and similar sections of cones, another with *diorismic* theorems, and the last with *diorismic* conic problems [lost]. When they are all published it will be possible for anyone who reads them to form his own judgement. Farewell.

Proposition 8 [our "parabola" – so named by Apollonios in Proposition 11]

If a cone be cut by a plane through the axis, and it be also cut by another plane cutting the base of the cone in a line perpendicular to the base of the axial triangle [formed by the first plane], and if the diameter of the section made on the surface [drawn from the vertex of the parabola bisecting all lines parallel to the line perpendicular to the base of the triangle] be either parallel to one of the sides of the triangle or meet it beyond the vertex of the cone, and if the surface of the cone and the cutting plane be produced to infinity, the section will also increase to infinity, and a straight line can be drawn from the section of the cone parallel to the straight line in the base of the cone so as to cut off from the diameter of the section towards the vertex an intercept equal to any given straight line.

Let there be a cone whose vertex is the point A and base the circle BΓ, and let it be cut by a plane through the axis, and let the section so made be the [axial] triangle ABΓ; now let it be cut by another plane cutting the circle BΓ in the straight line ΔE perpendicular to BΓ, and let the section made on the surface be the curve ΔZE; let ZH, the diameter of the section ΔZE, be either parallel to AΓ or let it, when produced, meet AΓ beyond the point A (Figure 2.6). I say that if the surface of the cone and the cutting plane be produced to infinity, the section ΔZE will also increase to infinity.

For let the surface of the cone and the cutting plane be produced; it is clear that the straight lines, AB, AΓ, ZH are simultaneously produced. Since ZH is either parallel to AΓ or meets it, when produced, beyond the point A, therefore ZH, AΓ when produced in the directions H, Γ, will never meet. Let them be produced accordingly, and let there be taken any point Θ at random upon ZH [produced], and through the point Θ let KΘΛ be drawn parallel to BΓ, and let MΘN be drawn parallel to ΔE; the plane through KΛ, MN is therefore parallel to the plane through BΓ, ΔE [Eukleidês, *Elements* 11.15]. Therefore the plane figure KΛMN is a circle [Prop. 4]. And since the points Δ, E, M, N are in the cutting plane, and are also on the surface of the cone, they are therefore upon the common section; therefore ΔZE has increased to M, N. Therefore, when the surface of the cone and the cutting plane increase up to the circle KΛMN, the section ΔZE increases up to the points M, N. Similarly we may prove that, if the surface of the cone and the cutting plane be produced to infinity, the section MΔZEN will increase to infinity.

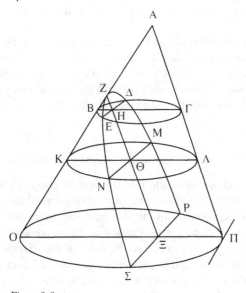

Figure 2.6

And it is clear that there can be cut off from the straight line ZΘ in the direction of the point Z an intercept equal to any given straight line. For if we place ZΞ equal to the given straight line and through Ξ draw a parallel to ΔE, it will meet the section, just as the parallel through Θ was shown to meet the section at the points M, N; therefore a straight line parallel to ΔE has been drawn to meet the section so as to cut off from ZH in the direction of the point Z an intercept equal to the given straight line.

[1.Prop.11: given ΘZ such that ΘZ/ZA = BΓ·BΓ/triangle-BAΓ, then ZH·ΘZ = ΔH*ΔH, the fundamental property of the parabola.]

(Thomas [1941] 281, 283, 285, 289, 291, 295, 297, 299, 301)

2.5 Combinatorics: no work on this subject survives, but from a passage in Plutarch we know that Xenokrates of Chalkedon (head of the Academy –338 to –314), Chrusippos of Soloi (ca. –280 to –206), and Hipparchos of Nikaia (working –146 to –126) wrote on the subject (Hipparchos also invented trigonometry; cp. Ptolemy, *Syntaxis* 1.10).

(in Plutarch, Dinner-table Talk 8.9)

Chrusippos says that the number of compound propositions that can be made from only ten simple propositions exceeds a hundred myriads [we expect only 1024]. Hipparchos, to be sure, refuted this by showing that on the affirmative side there are ten myriads plus three thousand forty-nine [103,049] compound statements, and on the negative side thirty-one myriad plus nine-hundred fifty-two [310,952; meaning unclear!]. Xenokrates asserted that the number of syllables which the letters will make in combination is myriad-and-twenty times a myriad-myriad [1,002,000,000,000; if a "syllable" can have up to any three initial and final consonants, and up to any two vowels, the number would still be no more than about one one-thousandth of this; did Xenokrates say "chiliad" for "myriad"?].

(Minar [1961] 195, 197)

2.6 Heron of Alexandria wrote a series of books on science and engineering (ca. 55 to 68; see also Chapters 5–8).

Mensurations

1.8 [area of triangle]

There is a general method for finding the area of any triangle whose three sides are given, without drawing a perpendicular. For example, let the sides of the triangle be 7, 8, and 9 monads. Add together 7, 8, and 9; that makes 24. Take half of them, which makes 12. Take away 7 monads and 5 are left. Again from the 12 take away 8, and 4 are left. And again 9 and 3 are left. Multiply 12 by 5; they make 60. Multiply these by 4; they make 240. Multiply these by 3; that makes 720. Take the square root of this, and it will be the area of the triangle.

Since 720 doesn't have a rational square root, we find the root (with a minimal error) thus. Since the square nearest 720 is 729, having a root 27, divide 27 into 720; that makes 26 and two-thirds; add 27 and that makes 53 and two-thirds. Take half of these; the result is 26 + 1/2 + 1/3. Therefore the square root of 720 will be very nearly 26 + 1/2 + 1/3. For 26 + 1/2 + 1/3 multiplied by itself gives 720 + 1/36; so that the difference is 1/36th part of a monad. If we wish to make the difference less than 1/36, instead of 729 we shall take the number now found, 720 + 1/36, and by the same method we shall find an approximation differing by much less than 1/36. [Heron then copies his proof from *Dioptra* 30]

(Thomas [1941] 471, 473)

2.7 Menelaus of Alexandria, active 96 ± 2, wrote on mathematics and materials-science (see Chapter 9), and made astronomical observations at Rome. His *Spherics* survives only in Arabic: book 1 gives triangle-theorems, book 2 augments Theodosios' *Spherics*, and book 3 covers spherical trigonometry. (Material in *italics* is not found in all manuscripts.)

Spherics

1 Definitions

1. Of the figures on the surface of the sphere, that which I call a **triangle** is enclosed by three arcs of great circles, each arc being less than a semicircle.
2. The **angle** is that which those arcs enclose *so that it forms one triangular surface.*
3. The angles enclosed by great circles which I call **equal** are those of which the arcs in the plane of their semicircles are equal *i.e., the arc of the circle which extends through both their poles and is between the two circles.*

[*Spherics* 3.1 – see Ptolemy, *Syntaxis* 1.13 in Toomer [1984b] 68–69]

*(Krause [1936] 118–119)

2.8 Nikomachos of Gerasa a neo-Pythagorean who wrote around 100, composed a handbook (*enchiridion*) on music from a Pythagorean perspective, as well this work on mathematics.

Introduction to Arithmetic

1.13 [sieve of Eratosthenes]

. . .

2–3 The method of generating these numbers is called a "sieve" by

3	5	7	3⟩9	11	13	3⟩15	17
			⟨3			⟨5	
19	3⟩21	23	5⟩25	3⟩27	29	31	3⟩33
	⟨7		⟨5	⟨9			⟨11
5⟩35	37	3⟩39	41	43	3 5⟩45	47	7⟩49
⟨7		⟨13			⟨15 9		⟨7
3⟩51	53	5⟩55	3⟩57	59	61	3 7⟩63	5⟩65
⟨17		⟨11	⟨19			⟨21 9	⟨13
67	3⟩69	71	73	3 5⟩75	7⟩77	79	9 3⟩81
	⟨23			⟨25 15	⟨11		⟨9 27
83	5⟩85	3⟩87	89	7⟩91	3⟩93	5⟩95	97
	⟨17	⟨29		⟨13	⟨31	⟨19	

Figure 2.7

Eratosthenes, since we take the odd numbers mingled and indiscriminate and we separate out of them by this method of generation, as if by some instrument or sieve, the prime and non-composite numbers by themselves, and the secondary and composite numbers by themselves, and we find separately those that are mixed [1.13.9: squares of primes]. The method of the sieve is this: I set out all the odd numbers in order, from the triad, in as long a series as possible (Figure 2.7), and then starting with the first I observe which ones it can measure, and I find that it can measure the terms two places apart, as far as we care to proceed. . . .

4–5 Then taking a fresh start from the second number I observe which ones it can measure, and find that it measures all the terms four places apart, the first by the quantity of the first in order (three times); the second by that of the second (five times); the third by that of the third (seven times); and thus in series forever. Again, as before, the third term 7, taking over the measuring, will measure terms six places apart, and the first by the quantity of 3, the first of the series, the second by that of 5, for this is the second number, and the third by that of 7, for this has the third position in the series. . . .

7–8 Now if you mark the numbers with certain signs, you will find that the terms which succeed one another in the measuring neither measure all the same number (and sometimes not even two will measure the same one), nor do absolutely all of the numbers set forth submit themselves to a measure, but some entirely avoid being measured by any number whatsoever, some are measured by one only, and some by two or even more. Now those that are not measured at all, but avoid it, are primes and non-composites, sifted out as it were by a sieve. . . .

(D'Ooge [1926] 204–205)

2.8–10 [polygonal numbers; compare Theon, *Mathematics* 1.19]

8.1 Now a triangular number is one which, when it is analyzed into monads, shapes into triangular form the equilateral placement of its parts in a plane; examples of which are 3, 6, 10, 15, 21, 28, and so on (Figure 2.8); for their regular formations expressed graphically will be at once triangular and equilateral.
. . .

8.3 The triangular number is produced from the natural series of the numbers set out in a line, and by the continued addition of successive terms one-by-one from the beginning. . . .

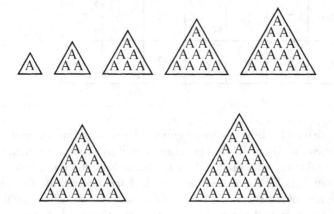

Figure 2.8

9.1 The square is the next number after this, which shows us no longer three, like the former, but four angles in its graphic representation, but is none the less equilateral (Figure 2.9). Take for example 1, 4, 9, 16, 25, 36, 49, 64, 81, 100; for the representations of these numbers are equilateral square figures as here shown (and it will be similar as far as you wish to go).
. . .

9.3 This number also is produced if the natural series is extended in a line, increasing by a monad, and no longer the successive numbers are added to the numbers in order, as was shown before, but rather all those in alternate places, that is, the odd numbers. [e.g., 1 + 3 + 5 + . . .]
. . .

10.1–2 The pentagonal number is one which likewise upon its resolution into units and depiction as a plane figure assumes the form of an equilateral pentagon such as 1, 5, 12, 22, 35, 51, 70 and analogous numbers. . . . For in a like and similar manner there are added together to produce pentagonal numbers the

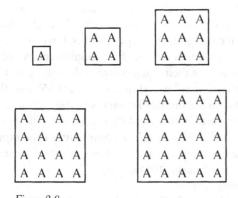

Figure 2.9

terms from the monad (to any extent whatever) that are two places apart, i.e., those which differ by three. [i.e., 1 + 4 + 7 + 10 + 13 + . . .]

(D'Ooge [1926] 241–243)

2.9 Ptolemy (Claudius Ptolemaeus) of Alexandria (*ca.* 100 to *ca.* 175) wrote on a wide variety of "mathematical" topics (he is quoted in each of Chapters 2–8). He viewed mathematics including astronomy as a surer insight to reality than were theology or physics (the *Syntaxis* was published *ca.* 150). Here he explains trigonometry, Hipparchos' invention.

Syntaxis

1.10.1–10 [Trigonometry: Toomer [1973]]

1 For the user's convenience, then, we shall subsequently set out a table of their amounts, dividing the circumference into 360 parts, and tabulating the chords subtended by the arcs at intervals of half a degree, expressing each as a number of parts in a system where the diameter is divided into 120 parts. We do this because of its arithmetical convenience, which will become apparent from the actual calculations. But first we shall show how one can undertake the calculation of their amounts by a simple and rapid method, using as few theorems as possible, the same set for all. We do this so that we may not merely have the amounts of the chords tabulated unchecked, but may also readily undertake to verify them by computing them by a strict geometrical method. In general we shall use the sexagesimal system for our arithmetical computations, because of the awkwardness of the fractional system [Knorr [1982]]. Since we always aim at a good approximation, we will carry out multiplications and divisions only as far as to achieve a result which differs from the precision achievable by the senses by a negligible amount.

[2 starting with values determined from inscribed pentagon, hexagon, and decagon]

3   [chords of supplementary angles] We can, then, consider the above chords as established individually by the above straightforward procedures. It will immediately be obvious that if any chord be given, the chord of the supplementary arc is given in a simple fashion, since the sum of their squares equals the square on the diameter. For instance, since the chord of 36° was shown to be 37;4,55^P, and the square of this is 1375;4,15^P, and the square on the diameter is 14400^P, the square on the chord of the supplementary arc (which is 144°) will be the difference, namely 13024;55,45, and so chord of 144° ≈ 114;7,37^P. Similarly for the other chords. [The superscript "P" indicates Ptolemy's units of arbitrary "parts" in terms of which he calculates trigonometric lengths: Toomer [1984b] 7–9.]

[4: auxiliary theorem; 5: chord of difference]

6   Let us now consider the problem of finding the chord of the arc which is half that of some given chord. [due to Hipparchos]

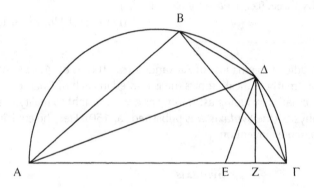

Figure 2.10

Let ABΓ be a semi-circle on diameter AΓ. Let ΓB be a given chord. Bisect arc ΓB at Δ, join AB, AΔ, BΔ, ΔΓ and drop perpendicular ΔZ from Δ on to AΓ (Figure 2.10). I say that

ZΓ = ½ (AΓ – AB).

Let AE = AB, and join ΔE. Then since [in the triangles ABΔ, AΔE]

AB = AE, and AΔ is common,

the two pairs of sides AB, AΔ, and AE, AΔ are equal.

Furthermore ∠ BAΔ = ∠ EAΔ

∴ base BΔ = base ΔE

But BΔ = ΔΓ [by construction]

therefore ΔΓ = ΔE.

So, since, in the isosceles triangle ΔEΓ, perpendicular ΔZ has been drawn from apex to base

$EZ = Z\Gamma$

But $E\Gamma = [A\Gamma - AE =] A\Gamma - AB$

$\therefore Z\Gamma = \frac{1}{2}(A\Gamma - AB)$.

Now, if the chord of arc $B\Gamma$ is given, the supplementary chord AB is immediately given.

Therefore $Z\Gamma$, which is $\frac{1}{2}(A\Gamma - AB)$, is also given.

But, since, in the right-angled triangle $A\Gamma\Delta$, the perpendicular ΔZ has been drawn,

triangle $A\Delta\Gamma$ is similar to triangle $\Delta\Gamma Z$ (both right-angled).

therefore $A\Gamma{:}\Gamma\Delta = \Gamma\Delta{:}\Gamma Z$

$\therefore A\Gamma \cdot \Gamma Z = \Gamma\Delta^2$

But $A\Gamma \cdot \Gamma Z$ is given.

Therefore, $\Gamma\Delta^2$ is given, and so chord $\Gamma\Delta$, which subtends an arc half of the arc of the given chord $B\Gamma$, is also given.

By means of this theorem too a large number of chords will be derived by halving the arcs of the previously determined chords, and notably, from the chord of 12°, the chords of 6°, 3°, 1½°, and ¾°. By calculation we find the chord of 1½° to be approximately $1;34,15^P$ where the diameter is 120^P, and the chord of ¾° to be approximately $0;47,8^P$ in the same units.

[7: chord of sum]

8   It is obvious that by so combining the chord of 1½° with all the chords we have already obtained, and then computing successive chords, we will be able to enter in the table all chords of arcs which when doubled are divisible by three [i.e., multiples of 1½°]. Then the only chords remaining to be determined will be those between the 1½° intervals, two in each interval, since our table is made at ½° intervals. If, therefore, we can find the chord of ½°, this will enable us to fill out all the remaining intermediate chords, by finding the sum or difference intervals. Now, if a chord, e.g., the chord of 1½°, is given, the chord corresponding to an arc which is one-third of the previous one cannot be found by geometrical methods [i.e., trisecting an angle is impossible: compare Pappos, *Mathematical Collection* 4.57–59]. (If this were possible, we should immediately have the chord of ½°). Therefore we shall first derive the chord of 1° from those of 1½° and ¾°, by establishing a lemma which, though it cannot in general exactly determine the sizes of chords, in the case of such very small quantities can determine them with a negligibly small error.

[9: approximating chord of ε as ε for small angles]

10   Such, then, I think, is the easiest way to undertake the calculation of the chords. But, as I said, in order that we may have the actual amounts of the chords readily available for every occasion, we shall set out tables below. They will be arranged in sections of 45 lines to achieve a symmetrical appearance. The first column will contain the arcs tabulated at intervals of ½°, the second the

41

corresponding chords in units of which the diameter contains 120, and the third the thirtieth part of the increment in the chord for each interval (so that we may have the average increment corresponding to one minute of arc, which will not be sensibly different from the true one-minute-increment). Thus we can easily calculate the amount of the chord corresponding to fractions which fall between the tabulated half-degree intervals. [The table approximates "pi" as $3;8,30 = 3\frac{17}{120} =$ 3.141666..., given in *Syntaxis* 6.7.]

It is easy to see that, if we suspect some scribal corruption in one of the values for the chord in the table, the same theorems which we have already set out will enable us to test and correct it easily, either by taking the chord of double the arc of the chord in question, or from the difference with some other given chord, or from the chord of the supplement.

(Toomer [1984b] 48–56)

2.10 Diophantos of Alexandria (150 ± 80) wrote a textbook of "algebra" in which example solutions in rational numbers are found to problems posed in the Babylonian style similar to Heron's *Metrika* (see Knorr [1993]). (He also wrote on polygonal numbers, as in Nikomachos.)

Arithmetika

1 Praeface ["Algebra"]

Knowing that you are anxious, my most esteemed Dionusios, to learn how to solve problems in numbers, I have tried, beginning from the foundations on which the subject is built, to set forth the nature and power in numbers. Perhaps the subject will appear to you rather difficult, as it is not yet common knowledge, and the souls of beginners despair of success; but it will be easy for you to grasp, with your enthusiasm and my teaching; for desire receiving instruction is swift to knowledge.

As you know, in addition to these things, that all numbers are made up of some multitude of units, it is clear that their formation has no limit. Among them are: *squares*, which are formed when any number is multiplied by itself; the number itself is called the *side of the square*; *cubes*, which are formed when squares are multiplied by their sides, *square-squares*, which are formed when squares are multiplied by themselves; *square-cubes*, which are formed when squares are multiplied by the cubes formed from the same side; *cube-cubes*, which are formed when cubes are multiplied by themselves; and it is from the addition, subtraction, or multiplication of these numbers or from the ratio which they bear one to another or to their own sides that most arithmetical problems are woven; you will be able to solve them if you follow the method shown below.

1.28 [quadratic equations]

To find two numbers such that their sum and the sum of their squares are given numbers.

It is a necessary condition that double the sum of their squares exceed the square of their sum by a square. [also, the numbers found must be rational] And the following is made up: let it be required to make their sum 20 monads and the sum of their squares 208 monads. Let their difference be $2x$, and let the greater $= x + 10$ (again adding half the sum) and the lesser $= 10 - x$. Then again their sum is 20 monads and their difference is $2x$. It remains to make the sum of their squares 208 monads. But the sum of their squares is $2x^2 + 200$. Therefore $2x^2 + 200 = 208$ monads, and $x = 2$ monads. To return to the hypothesis: the greater $= 12$ and the lesser $= 8$ monads. And these satisfy the conditions of the problem.

2.8 [$x^2 + y^2 = z^2$]

To divide a given square number into two squares. [only possible for squares]

Let it be required to divide 16 into two squares. And let the first square $= x^2$; then the other will be $16 - x^2$; it shall be required therefore to make $16 - x^2 = a$ square. I take a square of the form $(mx - 4)^2$, m being any integer and 4 the root of 16; for example, let the side be $2x - 4$, and the square itself $4x^2 + 16 - 16x$. Then $4x^2 + 16 - 16x = 16 - x^2$. Add to both sides the negative terms and take like from like. Then $5x^2 = 16x$, and $x = 16/5$. One number will therefore be $256/25$, the other $144/25$, and their sum is $400/25$ or 16 monads, and each is a square.

(Thomas [1941] 519, 521, 537, 551, 553)

2.11 Anatolios of Alexandria (250 ± 30) a professor of Aristotelian philosophy and a Christian bishop from 263 (appointed to the see of Laodikaia in 268), wrote this school text of neo-Pythagorean numerology similar to the *Mathematics* of Theon of Smurna, who paraphrased Philolaus of Kroton *On Nature*, Archutas of Taras *On the Decade*, and Speusippos *Pythagorean Numbers*. He published a computation of the date of Easter (McCarthy [1995/6]).

On the Decade

The nature of the decade and the numbers within it displays a myriad beauties to those able to perceive such things sharp-sightedly in the mind. We will say what is possible about each of the numbers. But now we preface this, that the Pythagoreans referred every number to 10, and that above 10 there is no further numeral, since the numbers increase again as we turn back to the monad after completing each decade. Further, they especially honored the *tetraktus* [first four numbers] because the decade is constituted from the tetrad.

About the monad [compare Theon of Smurna, *Mathematics* 2.40]

The monad is prior to every number: from it all numbers derive, and it derives from none. Accordingly, it's called "Gonad," being the matter of numbers. For were it abolished, there would be no number. One is indivisible, unfaultable, not changed from its nature even in multiplication. Likewise (although not actually but potentially), it is odd, even, even-odd [i.e., twice-an-odd-number; compare Philolaus fr. 5; Archutas as noted by Aristotle, *Pythagoreans* in Theon, *Mathematics* 1.5; and Eukleidês, *Elements* 7.Def.9], cube, square, and everything else. One designates a point.

The Pythagoreans called it "Mind" likening it to the "One," an intelligible god, ungenerated, ideal-beauty ideal-goodness [Plotinus 1.8.13 says these are prior to virtue], in all things, comparing the "One" especially to Phronesis [prudence] among the virtues. For the straight and concordant is one. Further, they thought it was essence, cause, truth, simplicity, paradigm, order, symphony, among greater and lesser the equal, in interval the middle, in multitude the mean, in time past and future the present moment. Further, they imagined it was Receptacle [Plato, *Timaios* 51, 53], Ship, Vehicle [Plato, *Phaidros* 246–248], Love, Life, Prosperity. In addition, they also said that around the middle of the elements lies a unitary fiery cube [Anatolios' near-contemporary, Hippolutos of Rome, records the gnostic belief in a fiery spark enveloped in dark waters: 5.21.1], whose midmost position Homer knew, saying: "As far below Hades as Heaven from Earth." [*Iliad* 8.16]

About the heptad [compare Theon of Smurna, *Mathematics* 2.46; underlined passages probably belong at the respective symbols]

The heptad alone of the numbers within the decade neither generates nor is generated by any other number except by the monad. For this reason it is called by the Pythagoreans the Motherless Virgin [Athena]. (Of the other numbers within the decade, the number 4 is generated by the dyad, and with that it generates 8; the number 6 is generated by the triad, but it does not generate; and 3 and 5 generate: 3 generates 6 and 9, and 5 generates 10.) The series from the monad to 7 makes the perfect number 28, equal to its aliquot parts [Eukleidês, *Elements* 7.Def.22, 9.Prop.36]. (*)The 28 days of the moon are completed according to heptads. From the monad a series of 7 numbers doubled make 64, the first number which is both a square and likewise a cube: 1, 2, 4, 8, 16, 32, 64. From the monad a series of 7 numbers tripled make 729, a square and a cube (square of 27, cube of 9): 1, 3, 9, 27, 81, 243, 729; and always increasing 7 such terms does the same, for in the sequence starting from 64 [as the first term], the 7th term in doubling is the cube of 16 [i.e., 4096]. Further, the heptad composed from the dimensions and the 4 limits shows body and instrument (the limits are point, line, surface, and solid; the dimensions are length, width, and depth). 7 is said to be the number of the first harmony, i.e., a fourth: 4/3 [that concord's ratio], and of the first geometric ratio: 1, 2, 4. (#)It is called *Telesphoros* [fulfillment]: for 7 months are favorable to generation. In illness the heptad is indicative of the *krisis* [see Chapter 11]. In the prototypical right triangle, 7 is the sum of the sides around the right angle (one of these sides is 4, the other is 3).

There are 7 planets. There are 7 phases of the moon: twice crescent-shaped, twice half, twice gibbous, once full. (*) The Bear has 7 stars. According to Herakleitos, "the heptad contributes to the *logos* [account] of the seasons and of the moon, and is distributed throughout the Bears [constellations] for a sign of deathless remembrance." The Pleiadês have 7 stars. The Equinoxes occur 7 months apart, as do the solstices [again counting inclusively].

(#) Apart from the *hegemonikon* [center of control; see Chapter 12], the parts of the soul divide into 7: the 5 senses and the voice and generation. There are 7 complete parts of the body: head, neck, chest, 2 feet, 2 hands. There are 7 viscera: stomach, heart, lung [singular; see Chapter 11], liver, spleen, two kidneys. Herophilos says that the intestines of a man measure 21 cubits, which is 3 heptads [fr. 100b-von Staden; Theon has "28"]. The head has 7 openings: two eyes, two ears, two nostrils, mouth.

We see 7 things: body, distance, shape, size, color, movement, position. There are 7 changes of sound: sharp-tone, deep-tone, circumflex-tone, rough, smooth, long, short. There are 7 movements: up, down, forward, backward, rightward, leftward, in a circle. There are 7 vowels: α, ε, η, ι, ο, υ, ω. The lyre has seven strings. Terpander [*ca.* –600] says of the lyre [which he calls "phorminx"; fr.5]:

having turned away from fourth-harmonized song,
we sound new odes on the seven-stringed phorminx.

From seven numbers, Plato composes the soul in the *Timaios* [35b].

Straits usually change seven times a day [Pliny 2.100]. Everything is seven-loving. From babyhood to old age there are 7 ages: child, youth, adolescent, young man, man, elder, and old man, and within seven years we pass from child to youth, from youth to adolescent, and in turn to the other ages of life [he then quotes Solon of Athens, fr.27].

About the decade

10 can be generated from an even and odd: for five times 2 is 10. 10 is the circle and limit of every number, which wrap and turn back around it (as those running the "long-course" [about 2 km] do around the turning post). Yet, it is the limit of the boundlessness of numbers. We count from the monad all the way up to ten, and stopping we then say the rest: one-ten, two-ten. Then 20 is formed by summing twice what made ten: for ten is composed of one, 2, 3, 4; whereas 20 is the sum of twice 1 and twice 2, twice 3, twice 4 (and likewise for the subsequent decades).

The decade is called "power" and "perfection," since it completes every number and encompasses all nature within it: even and odd, mutable and immutable, good and evil [Theon, *Mathematics* 2.49 says the same]. It is called the "Receiver" [*dekhas* instead of *dekas*] because of receiving all.

Perimeter equal to area of figures is found for the square 16, for the rectangle 18 [the sole rectangular figures in integer side whose area equals perimeter; Zenodoros around −180 wrote on isoperimetric figures]: the sides of these are 4 and 6 (for 4 times 4 is 16, and 6 times 3 is 18), and 4 and 6 make 10. Further, 10 is generated from the first numbers of the *tetraktus* summed: 1, 2, 3, 4.

The decade generates 55, offering a marvelous beauty: first, it is formed from the sum of doubles and triples in series, by doubling: $1 + 2 + 4 + 8 = 15$; by tripling: $1 + 3 + 9 + 27 = 40$: these added together make 55. Plato in the *Timaios* [35b] recalls these as the beginning of psychogony ("one portion from the whole" and so on). Second: 55 is the sum of the decade just as 385 is the sum of the decade in square: for if you form the products from the monad up to the decade you will add up to the indicated number 385 [$= 1 + 4 + 9 + 16 \ldots + 100$], and 385 is seven times 55. Third: 55 is a triangle number [the 10th such]. Fourth: if you reckon "one" ['εν] in letters, you find 55 according to addition [ε is 5 and ν is 50]. Fifth: the most fecund hexad [because it's the marriage-number: Anatolios 6 or Theon, *Mathematics* 2.45], multiplied by itself, generates the square 36, which has seven parts generated thus: twice 18, thrice 12, four-times 9, six-times 6, 9 times 4, 12 times 3, 18 times 2. The sum of these 7 parts makes 55 [$= 1 + 2 + 3 + 4 + 6 + 9 + 12 + 18$]. Sixth: 5 triangle numbers in sequence generate 55: $3 + 6 + 10 + 15 + 21$. Likewise, the 5 squares in sequence: 1, 4, 9, 16, 25 make 55.

According to Plato, from the triangle and the square is the genesis of all [*Timaios* 53–57]. From the equilateral triangle three figures are formed: pyramid, octahedron, icosahedron – the shape of Fire, of Air, and of Water – and from the square is the cube, this is the shape of Earth.

*(Heiberg [1901] 29–30, 35–37, 39–40)

3

ASTRONOMY

Humans have always gazed upward and wondered at the lights in the sky, telling explanatory stories. The sun, moon, and stars were depicted and studied by the earliest societies, as at Stonehenge (*ca.* −3000: Thurston [1994] 45–55) and the solar-aligned pyramids (*ca.* −2700: Neugebauer [1980]). Egyptian myths make Earth and Heaven primordial deities, and Mesopotamians saw gods in the moving stars (Marduk was Jupiter, for example). In the eighth century BCE, some Babylonians sought to forecast the changes they saw, using only mathematics; a century or more later, Greeks began to seek mundane explanations for the heavenly events.

The following chapter, on astrology, considers the Greek search for meaning in and influence from the heavens; here we examine their efforts to explain. The earliest Greek *astronomy* (the natural philosophy of the sky) is known to us only in bits and scraps (of the authors we mention only Plato and Aristotle are intact), and we confine our account to the more certain items. However, certain principles and practices persisted, such as that the *kosmos* was eternal and geocentric, and the perception that celestial events were more perfect and regular than earthly and mutable happenings. Moreover, although magnification by lens and mirror was known (see Chapter 7), no evidence of an ancient telescope has been found, and all ancient astronomical observations were made with the naked eye.

From the beginning, moon and sun were our timekeepers; Babylonians and Greeks, and surely most peoples, refined that grand clock by noting the times of year when certain bright stars (Sirius, for example, and the Pleiades) were visible, and when the sun reached his extreme northerly and southerly points (the solstices). Besides the calendric use of astronomy were three observations. Some stars are visible all year round (for example, those we name the "Big Dipper"); they were designated as lying within Greece's Arctic Circle (Greeks named the "Big Dipper" the Bear, *arktos*). The seven wandering (*planêtai*) stars did so always within a circular band, which came to be called the ecliptic (since only if both sun and moon were on its central circle could an eclipse occur). Another circular band seemingly filled with stars like dust the Greeks called "The Milk."

Our clear knowledge of Greek attempts to offer natural explanations of the sky begins with Xenophanes of Kolophon, who wrote verse (in the rhythms of

Homer and Hesiod) during the latter half of the sixth century BCE (–510 ± 30). He advocated a more abstract and less anthropomorphic concept of divinity (as unitary, transcendent, and omnipotent); correspondingly he suggested that the sun and other stars were not gods but coalesced daily from sparks (Keyser [1992b]). His *kosmos* seems to have been linear, with essential "up" and "down," like all early flat-earth myths. Around the turn of the century (–500 ± 20) the famously cryptic aristocrat Herakleitos of Ephesos wrote prose (in terms reminiscent of Zen Buddhism) arguing the inherent unity and dynamic balance of opposites, manifested in the continual and eternal changes of the *kosmos*. "Fire" (almost "energy") is the ever-living fundamental stuff of the world, as can best be seen in the sky; change occurs in measures, also clearest in the stars (modeled as bowls of fire).

Parmenides of Elea composed his difficult poem early in the next century (–470 ± 20), claiming a transcendent grasp of eternal truth and of mortal opinion, a grasp whose essence is that the only valid objects of thought are those which actually exist. He may be the first to have asserted that the *kosmos* was spherical (i.e., has a center rather than an up/down), and that the moon gets her light from the sun; he obscurely pictured some of the heavenly bodies as "rings of fire and night" (like Hesiod, *Works and Days* 282). Around mid-century (–455 ± 20), the aristocrat Empedokles of Akragas sought to explain the *kosmos* in epic verse. He believed in four primal "roots" (fire, air, earth, and water) which eternally cycled in the axial vortex of the outwardly spherical *kosmos*, driven by Love (attraction?) and Strife (repulsion?). The sun and sky and earth separated out of the vortex, and (it seems) will one day revert thereto; as the vortex slowed it tilted like a top. He may have been the first to state that lunar light is a *reflection* of solar light and that both bodies circle the earth, eclipsed when *shadowed*.

At nearly the same time (perhaps –455 ± 20), Anaxagoras of Klazomenai, working in Athens, propounded partly similar doctrines, in prose. His *kosmos* rotated, but purposively, and spun off the heavenly bodies (and earth), among which the earthlike moon was illumined by the sun, while the "Milk" was faint stars rendered visible in the shadow of the earth. Notably he argued that the sun was an incandescent stone, and that there were numerous unseen bodies whirling about the earth, one of which was the meteorite fallen at Aigospotamoi in –466 (possibly associated with Halley's comet).

Demokritos of Abdera (writing *ca.* –410 ± 30) explained the *kosmos* by positing just two fundamental entities, atoms ("individuals") and void. Atoms, of all shapes and sizes, randomly whirl and coalesce into innumerable *kosmoi*, each with a central flat earth and circumterrene fiery bodies (stars, and in some *kosmoi* a sun). In our *kosmos*, the tambourine-shaped earth, most closely orbited by an earthlike moon, hides the sun at night behind its raised rim. Beyond our *kosmos* are infinite atoms and unending void containing innumerable transient *kosmoi*, which ultimately perish and resolve into chaos, as will our own.

Eschewing speculation, the Athenian astronomers Meton and Euktemon observed solstices (one fixed date is that of –431 June 27) around Greece, and founded thereon a calendar better attuned to sun *and* moon. Meton proposed a cycle of 235 lunar months (110 of which contained 29 days, the rest 30), over 19 solar years, while Euktemon is credited with the observation that the intervals between a successive equinox and solstice are unequal.

Around the same time (–420 ± 10) Philolaus of Kroton (a follower of the teachings of Pythagoras) either introduced or popularized the notion that the earth and the *kosmos* were shaped as and harmonized by the sphere. A second radical departure was that the earth, the sun, and all the planets (including an unseen "counter-earth") orbited an unseen central "hearth" of the *kosmos*, which illumined the sun and stars. These whirling spheres resound in a harmony to which we have become deaf.

The writings of Plato (died in –347 at the age of about 75) are full of astronomical references, particularly in the dialogues *Phaidôn* 97–99, 108–110, the *Republic* books 7 and 10, the *Timaios*, and the (possibly partly posthumous) *Epinomis*. He imagines a *kosmos* (similar to that of Philolaus) which is as a whole a living being, in which the stars and planets serve as divine and musical timepieces in their orbits around the spherical earth, which is much larger than we imagine, the Mediterranean basin being a small air-filled hollow, its true surface being exposed to the *aither*. Despite the apparent irregularity of planetary motions, Plato asserted, and set it as a project for astronomers to find, their essential uniform (i.e., circular) and ordered motions. But all that he says is provisional, indirect, and tinged with irony, and for Plato astronomy was but a stage on the way to individual perfection in wisdom.

Eudoxos of Knidos wrote several works on astronomy in the mid-fourth century BCE (around –355 ± 10), three of whose titles were *Phainomena, Mirror,* and *Speeds* (possibly written in that order). The first two described the heavens, giving star positions and observations of the sun, moon, and planets; *Speeds* introduced a model in which for each planet several concentric spheres with inclined axes spun and counter-spun to produce a motion similar to the retrogradations and stations of the planets. The theory was qualitative, and cannot in fact reproduce the motion of the inner planets or Mars.

Late in the century, Kallippos of Kuzikos (*ca.* –340 ± 10) and Aristotle, in his works *On Heaven* and *Metaphysics* (book Λ), added spheres to represent more precisely (but no more accurately) the motions of the inner planets, and the inequality of the seasons (Aristotle added yet more spheres to make their combined motion strictly mechanical). Kallippos also refined the Metonic cycle by taking four of them, less one day, to produce 940 lunar months over 76 years; this cycle began in –329, was used by astronomers for several centuries to date observations, and gives a year of 365¼ days. Aristotle also argued for a fixed central spherical earth (by now the standard model), and spherical planets and stars, the moon and sun being closer to the earth than any other bodies (neither he, nor Eudoxos, had any reason to specify the order of the other planets). All were composed of *aither* (or "first body") which has a natural and eternal circular motion, but is otherwise unalterable. All motion in the finite closed universe is initiated by the outer sphere of the fixed stars, or by an external abstract "Prime Mover." His presentation is notoriously complex and inconsistent, which generations of commentators have sought to excuse or explain.

3.1 Autolukos of Pitanê around –300 wrote two works on spherical astronomy; his terminology is awkward (astronomical vocabulary was not yet fixed) and his approach is purely geometrical, very similar to Eukleidês' *Phainomena*, from the same years. He considers only the phenomena of the fixed stars, and so does not refer to Eudoxos' model (see Berggren 1991).

Moving Sphere 6

If on a sphere the great circle defining the visible and invisible part of the sphere [the horizon] is inclined to the axis, it will be tangential to two circles equal and parallel to one another [i.e., the arctic and antarctic circles]. Of these, the one towards the visible pole will always be visible, and the one towards the invisible pole will always be invisible [i.e., the arctic circle never sets, while the antarctic circle never rises].

For on a sphere let the great circle ABΓ inclined to the axis define the visible and invisible part of the sphere (Figure 3.1). I say that the circle ABΓ will be tangential to two circles equal and parallel to one another, and of these the one near the visible pole will always be visible, and the one near the invisible pole will always be invisible. For, let the visible pole of the sphere be Δ and let the great

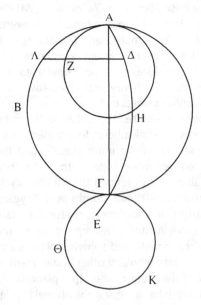

Figure 3.1

circle AΔE be drawn through Δ and the poles of the circle ABΓ, and let the arc ΓE be equal to arc AΔ [i.e., the sphere's poles are Δ and E]. And around pole Δ and with radius ΔA let circle AZH be drawn, and around pole E and with radius EΓ let circle ΓΘK be drawn. It is then clear that circle AZH is equal and parallel to circle ΓΘK, and further that circle ABΓ is tangential to circles AZH and ΓΘK.

I say that circle AZH is always visible, and that circle ΓΘK is always invisible. Because, if circle AZH is not always visible, then during the rotation of the sphere, circle AZH will intersect horizon ABΓ [at some point other than A]. Let it intersect at point Λ, and let the lines AΔ, ΔΛ, and AΓ be joined. Since on the sphere the great circle AΔΓ intersects some circle of those on the sphere, i.e., ABΓ, at its poles, it bisects it and is perpendicular to it. Therefore AΓ is a diameter of circle ABΓ and circle AΔΓ is perpendicular to circle ABΓ. Then on diameter AΓ of circle ABΓ is perpendicularly erected the segment AΔΓ of the circle, and the arc AΔΓ of the erected segment is divided unequally at Δ, arc AΔ being smaller, as is clear. Therefore line AΔ is the smallest of all lines extending from Δ onto the circle ABΓ [compare Theodosios, *Spherics* 3.1: because AΔ and ΔΓ are the only lines lying in the plane AΔΓ, and AΔ is less than ΔΓ by construction]. It follows that line AΔ is smaller than line ΔΛ. However it is also equal, because point Δ is a pole of circle AZH [so that AΔ and ΔΛ are both lines from the pole to the circle and hence are equal], which is impossible. Therefore during the motion of the sphere, the circle AZH will not set. Similarly then we can show that the circle ΓΘK will not rise. Therefore circle AZH is always visible and circle ΓΘK is always invisible.

[The work *Risings and Settings* is preserved in two books, of which the second either builds on or is a second edition of the first: Neugebauer [1975] 2.751 and Fig. 47; the following definitions open the first book: "The visible morning rising is when a star is seen to rise just before sunrise, the visible morning setting is when a star is seen to set just before sunrise; the visible evening rising is when a star is seen to rise just after sunset, and the visible evening setting is when a star is seen to set just after sunset." See Dicks [1970] 13 and Evans [1998] 190–197]

Risings and Settings

2.2–4

2.2 Of the twelve signs, the one ahead of the one in which the sun is, is seen rising at dawn, but the one following it is seen setting at sunset.

Let AB be the circle of the signs, and ΓΔ the horizon. Let a twelfth part EΔ be taken and let the sun be in its middle. And let the twelfth part ahead of the sun be ΔH and let the twelfth part following it be EΘ (Figure 3.2a). I say that the arc ΔH makes a morning rising and that EΘ makes an evening setting.

For the arc ΔH is seen to rise separated by more than half the arc of a sign, so that it makes a morning rising, but the arc ΔE is not seen rising and EΘ which rises in daytime is not seen. But as the *kosmos* turns arc ΔH makes a morning

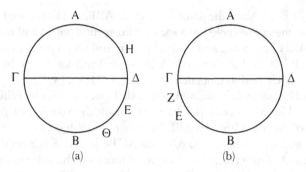

Figure 3.2

rising, while ΔE is not seen rising. But arc EΘ is seen to set separated by more than half the arc of a sign. Therefore EΘ makes an evening setting but ΔH a morning rising.

2.3 During the night time the arcs of eleven signs are seen, six that have already risen and five that are rising.

Let AB be the circle of the signs, and ΓΔ the horizon. Let the arc of a sign be taken as ΓE and the sun be in about the middle of it at Z (Figure 3.2b). Since it is supposed that the stars escape the rays of the sun which is at Z, it is clear that the star Γ makes a visible evening setting, so that the whole semicircle ΓAΔ has six signs. The remaining six signs belong in semicircle ΓBΔ, and one, ΓE, is occupied by the sun, so the remaining five are rising, so that eleven signs are seen.

2.4 Those of the fixed stars that are intersected by the zodiac towards the North or towards the South will be visible for five months from the visible morning rising to the visible evening rising.

Let the circle AB define the visible and invisible parts of the heaven, and let ΓΔ and EZ be the tropics, and HΘ be the equinox circle, and KHΛΘ be the circle of the signs. And let three stars M, Θ, and N be rising (Figure 3.3). I say that for M, Θ,

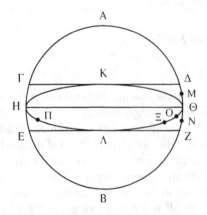

Figure 3.3

N take five months from the morning rising to the evening rising. For let an arc ΘΞ of one sign be taken and let it be bisected at O; let the sun be at O. Then surely M, Θ and N will make a morning rising. When the sun has moved to the opposite sign let it have moved through an arc of five signs, and let it be at point Π. Therefore from point O the sun will move along an arc of five signs, while from point H it is an arc of half a sign. When H is setting, stars M, Θ, and N rise; and when the sun moves from O to Π, the stars go from a morning rising to a evening rising.

(Bruin and Vondjidis [1971] 10–11, 57–59)

3.2 Klearchos of Cyprian Soloi (writing *ca.* –330 to –270), a student of Aristotle, traveled and wrote widely (only fragments survive).

(title unknown) ["Man in the moon"]

Klearchos asserts that what is called the face consists of mirrored likenesses, that is images of the great ocean reflected in the moon, for the visual ray when reflected naturally reaches from many points objects which are not directly visible and the full moon is itself in uniformity and luster the finest and clearest of all mirrors. Just as you think, then, that the reflection of the visual ray to the sun accounts for the appearance of the rainbow in a cloud where the moisture has become somewhat smooth and condensed, so Klearchos thought that the outer ocean is seen in the moon, not in the place where it is but in the place whence the visual ray has been deflected to the ocean and the reflection of the ocean to us.

(Cherniss and Helmbold [1957] 41)

3.3 Aristarchos of Samos studied under Straton of Lampsakos (between –286 and –269; he observed the summer solstice in –279), and correctly hypothesized that the earth moved around the fixed sun (published in a book whose name we do not know, but attested by his contemporaries Archimedes and Kleanthes): Ptolemy, *Syntaxis* 1.7 records the arguments that convinced Greeks of geocentricity. Before that work he had suggested that the order of the planets was Moon, Mercury, Venus, Sun, Mars, Jupiter, and Saturn. Aristarchos' method here is pre-trigonometric and would yield a result close to correct if the angle in hyp. 4 and the diameter in hyp. 6 were corrected. His work is the first actually to *calculate* the sizes and distances of any heavenly body, which had long been a subject of speculation (Aristotle, *Meteorologika* 3.2–5 evidently assumed they were not far above the clouds); Hipparchos repeated his work (Theon of Smurna, *Mathematics* 3.39; Toomer [1975])

Sizes and Distances

Hypotheses [sun and moon]

1. That the moon receives its light from the sun.
2. That the earth is in the relation of a point and center to the sphere in which the moon moves. [he acknowledges that he is neglecting lunar parallax]
3. That, when the moon appears to us halved, the great circle which divides the dark and the bright portions of the moon is in the direction of our eye. [compare pseudo-Aristotle, *Problems* 15.7]
4. That, when the moon appears to us halved, its distance from the sun is then less than a quadrant by one-thirtieth of a quadrant. [the correct value, very difficult to measure, is 1/540 of a quadrant; Aristarchos probably derived the angle from an assumed upper limit to the inequality between lunar phases of ½ day: Evans [1998] 72]
5. That the breadth of the earth's shadow is that of two moons. [a lower bound: it can be as large as 2.6 lunar diameters]
6. That the moon subtends one fifteenth part of a sign of the zodiac. [in fact it is 4 times smaller]

We are now in a position to prove the following propositions:

A. The distance of the sun from the earth is greater than eighteen times, but less than twenty times, the distance of the moon from the earth; this follows from the hypothesis about the halved moon.
B. The diameter of the sun has the same ratio to the diameter of the moon. [will be derived from Prop. 7 and the observation, made during eclipses, that their apparent angular diameters are the same]
C. The diameter of the sun has to the diameter of the earth a ratio greater than that which 19 has to 3, but less than that which 43 has to 6; this follow from the ratio thus discovered between the distances, the hypothesis about the shadow, and the hypothesis that the moon subtends one fifteenth part of a sign of the zodiac.

2

If a sphere be illuminated by a sphere greater than itself, the illuminated portion of the former sphere will be greater than a hemisphere. For let a sphere the center of which is B be illuminated by a sphere greater than itself the center of which is A (Figure 3.4). I say that the illuminated portion of the sphere the center of which is B is greater than a hemisphere.

For, since two unequal spheres are comprehended by one and the same cone which has its vertex in the direction of the lesser sphere, let the cone comprehending the spheres be drawn, and let a plane be carried through the axis; this plane will cut the spheres in circles and the cone in a triangle.

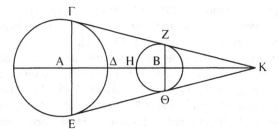

Figure 3.4

Let it cut the spheres in the circles ΓΔΕ, ΖΗΘ, and the cone in the triangle ΓΕΚ.

It is then manifest that the segment of the sphere towards the circumference ΖΗΘ, the base of which is the circle about ΖΘ as diameter, is the portion illuminated by the segment towards the circumference ΓΔΕ, the base of which is the circle about ΓΕ as diameter and at right angles to the straight line AB; for the circumference ΖΗΘ is illuminated by the circumference ΓΔΕ, since ΓΖ, ΕΘ are the extreme rays. And the center B of the sphere is within the segment ΖΗΘ; so that the illuminated portion of the sphere is greater than a hemisphere.

(Heath [1913/1966/1997] 353, 355, 359, 361)

3.4 Chrusippos of Soloi (*ca.* −280 to −206) wrote so comprehensively and authoritatively on the Stoic teachings that his interpretation became standard (however, all his writings are lost). He studied and wrote in Athens for about a quarter of a century, and then was head of the Stoic school from −231.

Providence

Book 1 [Here he is describing the Stoic teaching that the *kosmos* periodically is consumed by flame and in a way reborn; compare Plato, *Timaios* 33cd, "the *kosmos* is continuously eating and excreting itself."]

1 For since death is the separation of soul from the body, and the soul of the *kosmos* is not separated but grows continuously until it has completely used up its matter on itself, the *kosmos* must not be said to die.

2 The *kosmos* alone is said to be self-sufficient because it alone has within itself everything it needs, and it gets its nourishment and growth from itself since its different parts change into one another.

3 When the *kosmos* is fiery through and through, it is directly both its own soul and *hegemonikon*. But when, having changed into moisture and the soul which remains therein, it has in a way changed into body and soul so as to be compounded out of these, it has got a different principle.

(Long and Sedley [1987] 275–276)

3.5 Apollonios of Pergê wrote around −200 ± 10 in Pergamon (and possibly Rhodes), when it was an independent city, and originated the epicyclic theory of planetary motion, which was at first intended, like Eudoxos' theory, as qualitative but physically plausible. The model explained the well-known solar anomaly (the inequality in the lengths of the seasons) as well as the observation that Venus and Mars appear brighter and fainter (i.e., closer or farther) at different times, neither of which could happen under Eudoxos' model. Apollonios demonstrated the equivalence of epicyclic motion (in which a planet moves on a circle whose center is itself in geocentric circular motion) and eccentric motion (in which a planet moves on a circle whose center is fixed but not geocentric). This question was probably raised by the heliocentric model of Aristarchos, in which the geocentric solar motion was shown to be equivalent to a heliocentric motion of the earth.

Ptolemy quotes Apollonios' theorem that eccentric circular geocentric orbits are equivalent to epicyclic geocentric orbits:

If the synodic anomaly is represented by the epicyclic hypothesis, in which the epicycle performs the motion in longitude on the circle concentric with the ecliptic towards the rear [i.e., in the order] of the signs, and the planet performs the motion in anomaly on the epicycle [uniformly] with respect to its center, towards the rear along the arc near the apogee, and if a line is drawn from our point of view intersecting the epicycle in such a way that the ratio of half that segment of the line intercepted within the epicycle to that segment intercepted between the observer and the point where the line intersects the epicycle nearer its perigee is equal to the ratio of the speed of the epicycle to the speed of the planet, then the point on the arc of the epicycle nearer the perigee determined by the line so drawn is the boundary between forward motion and retrogradation, so that when the planet reaches that point it creates the appearance of station.

If the anomaly related to the sun is represented by the eccentric hypothesis (which is a viable hypothesis only for the three planets which can reach any elongation from the sun) [also valid for the inner planets Mercury and Venus if the speed of the mean sun is augmented by the speed of the planet's anomaly], in which the center of the eccenter moves [uniformly] about the center of the ecliptic with the speed of the [mean] sun towards the rear of the signs, while the planet moves on the eccenter in advance [i.e., in the reverse order] of the signs with a speed uniform with respect to the center of the eccenter and equal to the [mean] motion in anomaly, and if a line is drawn in the eccenter through the center of the ecliptic in such a way that the ratio of half the whole line to the smaller of the two segments of the line formed by the position of the observer is equal to the ratio of the speed of the eccenter to the speed of the planet, then when the planet arrives at the point in which the above line cuts the arc of the eccenter near the perigee, it will produce the appearance of station.

(Toomer [1984b] 555–556)

3.6 Hegesianax of Alexandria (near Troy), a friend of Antiochos "the Great," was an ambassador, and a scholar (*ca.* −195 ± 20), who wrote a hexameter poem, *Phainomena*, describing and explaining the heavens. In the two fragments preserved he seems to be versifying popular belief – or Klearchos' model (Plutarch *On the Face in the Moon* 2–3[920e–921b] quotes them together).

> Luna's whole orb is illumined by fire, and enflamed in the middle are shining far brighter than earthly material her bluish-gray pupils, her eyes, and her supple white brow: and adjacent's a face to be seen . . .
>
> . . .
>
> Or, when a great and opposing wet surge of the billowing sea becomes pictured by glimmering forth from a blazing reflector on high . . .
>
> *(Lloyd-Jones and Parsons [1983] 238)

3.7 Hupsikles of Alexandria wrote two astronomical treatises around −160 or −150 (under Ptolemy VI ("Philometor"), whose reign was troubled by civil and foreign wars); he is the first Greek we know of to have employed the Babylonian division of the circle into 360 parts. His method of determining rising times was soon superseded by the trigonometry of Hipparchos (see Chapter 2.9), but continued to be used by astrologers (see Vettius Valens 3.13). His other astronomical work, on the harmony of the spheres, is lost.

Anaphorikos

4.1–4 [rising times of zodiac signs]

With the circle of the zodiac divided into 360 equal arcs, let each part of the arcs be called local; likewise, with the time (in which the zodiac turns from any given point to the same point) divided into 360 equal times, let each part of the times be called temporal. Given these suppositions and using the theorems above we will show how, in a given locale, and knowing the ratio which the longest day has to the shortest [i.e., the latitude], for each of the signs of the zodiac, we can know in how many temporal parts it will rise. Consider the latitude (*klima*) of Alexandria near Egypt, where the longest to the shortest day has the ratio of 7 to 5: in fact we have determined this by the midday shadows at the solstices, measured with gnomons.

Let the circle of the zodiac be laid out, in which the equinoctial diameter is AH, and let the circle be divided into the signs at ABΓΔEZHΘKΛMN [iota unused], and let the point A be the start of the Ram [Aries], B the start of the Bull [Taurus], Γ the start of the Twins [Gemini], and let the following parts be assigned to the following signs (Figure 3.5).

And since the longest day is that during which the semicircle after the Crab [Cancer] rises, i.e., ΔHΛ, and the time of the shortest day during which the

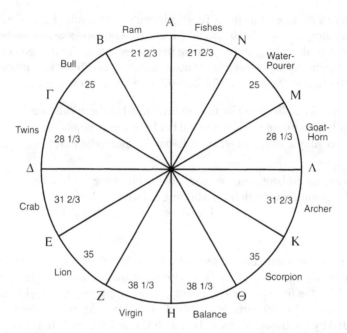

Figure 3.5

semicircle after the Goat-Horn [Capricorn] is ΛΑΔ, then the time of the rising of the semicircle ΔΗΛ to the time of the rising of the semicircle ΛΑΔ has the ratio 7 to 5. And the entire circle rises in 360 temporal parts: thus the semicircle ΔΗΛ rises in 210 temporal parts and the semicircle ΛΑΔ rises in 150. And the quadrant ΔΗ rises in the same time as the ΗΛ quadrant, and the ΛΑ quadrant rises in the same time as the ΑΔ quadrant: for they are the same distance from the equinox. Thus the quadrant ΔΗ will rise in 105 temporal parts and the ΔΑ quadrant will rise in 75: thus the rising time of the ΗΖΕΔ quadrant exceeds the rising time of the ΔΓΒΑ quadrant by 30 temporal parts.

And since the risings of the arcs ΗΖ, ΖΕ, ΕΔ, ΔΓ, ΓΒ, and ΒΑ are six successive stages in equal difference starting from the greatest at Η (for that is assumed as a hypothesis for the business of rising-times), the difference by which the sum of half the number exceeds the sum of the rest is equal to the product of their common difference times the square of half the quantity [theorem 1: in a series of common difference *D*, the sum of the *N*/2 greater values minus the sum of the *N*/2 lesser values equals *D* · (*N*/2)²]. And the difference is 30 temporal parts and the square of half the quantity is 9, and the 9th part of 30 is 3;20 [i.e., 3 + 20/60]. Then the difference of the risings in the twelve segments ΗΖ, ΖΕ, ΕΔ, ΔΓ, ΓΒ, and ΒΑ is also 3;20.

Again, since the risings of the arcs ΗΖ, ΖΕ, ΕΔ are arbitrary quantities in a sequence of constant difference (odd in number), the sum of them all is as many times the middle term as the number of them [theorem 2: in a series of common difference *D*, with an odd number of values, the sum equals *N* times the average, which is the middle term]. The rising time of all of them is 105, and there are three of them, and the third of 105 is 35. So the arc EZ, which is the arc of the Lion [Leo] will rise in 35 temporal parts; likewise the arc BΓ, which is the arc of the Bull will rise in 25 temporal parts [i.e., using the series ΔΓ, ΓB, and BA].

And the risings of the following arcs differ by 3;20 temporal parts each. Thus the Ram will rise in 21;40 temporal parts, the Bull in 25, the Twins in 28;20, the Crab in 31;40, the Lion in 35, the Virgin in 38;20, and the arc ZH in the same time as ΘH, and the arc EZ in the same time as ΘK. And those that are the same distance from the equinoctial diameter will rise in the same time: so the Balance [Libra] will rise in 38;20, the Scorpion in 35, the Archer [Sagittarius] in 31;40, the Goat-horn in 28;20, the Water-Pourer [Aquarius] in 25, and the Fishes [Pisces] in 21;40.

*(deFalco and Krause [1966] 36–37)

3.8 Hipparchos of Nikaia worked and wrote in Rhodes in the early years of its subjection to Rome. His observations date his work to −146 to −126, and we know something about seven of his astronomical works: *Sizes and Distances* (of the sun and moon), *Parallaxes* (of the moon) which he used to determine the distance of the moon, *Displacement of the Solsticial and Equinoctial Points*, *Length of the Year*, *Intercalary Months and Days*, *System of the Fixed Stars*, and *Motion of the Moon in Latitude*. His work built on Babylonian observations (which he seems to have been the first Greek astronomer to employ), the mathematics of Apollonios, and the systematic observations of fixed stars, lunar eclipses, and planetary positions (at least of Venus and the moon) by Timocharis (−294 to −271) and by Aristullos (for whom only fixed-star observations are recorded: see Goldstein and Bowen [1991]). Based on these observations he constructed an epicyclic theory of the lunar and solar motions (thus allowing the prediction of lunar eclipses), but refused to create a theory of planetary motion (compare Ptolemy, *Syntaxis* 9.2). He also discovered what we call the precession of the equinoxes.

(various works) [length of the year]

[Ptolemy records that Hipparchos assumes that the sun has a single and invariable anomaly, the period of which is the length of the year as defined by the solstices and equinoxes; that is, he determined the constant length of the year.]

For in his work *Length of the Year* he compares the summer solstice observed by Aristarchos at the end of the fiftieth year of the First Kallippic Cycle [−279] with the one which he himself had determined, again with accuracy, at the end of the forty-third year of the Third Kallippic Cycle [−134], and then says:

It is clear, then, that over 145 years the solstice occurs sooner than it would have with a 365¼ -day year by half the sum of the length of day and night.

Again, in *Intercalary Months and Days* also, after remarking that according to the school of Meton and Euktemon the length of the year comprises 365¼ + 1/76 days, but according to Kallippos only 365¼ days, he comments, in his own words, as follows:

As for us, we find the number of whole months comprised in 19 years to be the same as they found, but we find the year to be even less than ¼ [day beyond 365], by approximately 1/300 of a day. Thus in 300 years its deficit is 5 days compared with Meton's figure, and 1 day compared with Kallippos.

And when he more or less sums up his opinions in his *List of his own Writings*, he says:

I have also composed a work on the length of the year in one book, in which I show that the solar year (by which I mean the time in which the sun goes from a solstice back to the same solstice, or from an equinox back to the same equinox) contains 365 days, plus a fraction which is less than ¼ by about 1/300 of the sum of one day and night, and not, as the mathematicians suppose, exactly ¼-day beyond the above-mentioned number of days.
[compare Jones [1991]]

(Toomer [1984b] 139)

Displacement of the Solsticial and Equinoctial Points

[Hipparchos records alignments of stars within the zodiac to stars outside the zodiac, in order to allow later astronomers to confirm that the precession affects stars outside the zodiac: Hipparchos had at first thought that only zodiacal stars precessed, in which case they would be very long-period planets]

Stars in Cancer. Hipparchos records that the star in the southern claw of Cancer [α Cnc], the bright star which is in advance of the latter and of the head of Hydra [β Cnc], and the bright star in Procyon [α CMi] lie almost on a straight line. For the one in the middle lies 1½ digits [the "digit" is a Babylonian unit equal to 5' of arc] to the north and east of the straight line joining the two end ones, and the distances from it to each of them are equal.

Stars in Leo. He records that the easternmost two [μ, ε Leo] of the four stars in the head of Leo [μ, ε, κ, λ], and the star in the place where the neck joins the head of Hydra [ω Hya], lie on a straight line. Also, that the line drawn through the tail of Leo [β] and the star in the end of the tail of Ursa Major [η UMa] cuts off the bright star under the tail of Ursa Major [α CVn] 1 digit to the West [i.e., passes 1 digit to the east of it]. Similarly, he records that the line through the star under the tail of Ursa Major and the tail of Leo passes through the more advanced of the stars in Coma [Berenices; maybe 15 and 7 Com].

[Hipparchos continues through all 12 signs of the zodiac; Ptolemy adds his own list of star-alignments.]

(Toomer [1984b] 322)

But the sphere of the fixed stars also performs a motion of its own in the opposite direction to the revolution of the universe, that is, the motion of the great circle through both poles, that of the equator, and that of the ecliptic. We can see this mainly from the fact that the same stars do not maintain the same distances with respect to the solsticial and equinoctial points in our times as they had in former times: rather, the distance of a given star towards the rear with respect to those same points is found to be greater in proportion as the time is later.

For Hipparchos too, in his work *Displacement of the Solsticial and Equinoctial Points*, adducing lunar eclipses from among those accurately observed by himself, and from those observed earlier by Timocharis, computes that the distance by which Spica is in advance of the autumnal equinoctial point is about 6° in his own time, but was about 8° in Timocharis' time. For his final conclusion is expressed as follows:

> If then 'Wheat-ear' [Spica, α Virgo] *for example, was formerly 8°, in zodiacal longitude, in advance of the autumnal equinoctial point, but is now 6° in advance,*

and so forth. [Hipparchos was tentative because he realized that the measurements of Timocharis were not perfect: compare Ptolemy, *Syntaxis* 7.3]. Furthermore he shows that in the case of almost all the other fixed stars for which he carried out the comparison, the rearward motion was of the same amount.

. . .

But in the 50th year of the Third Kallippic Cycle [–128/7], as Hipparchos records from his own observations, that star had a distance to the rear of the summer solstice of 29⅚°. Therefore the star on the heart of Leo has moved 2⅔° towards the rear along the ecliptic in the 265 or so years from the observation of Hipparchos to the beginning [of the reign] of Antoninus [137/8], which was when we made the majority of our observations of the positions of the fixed stars. From this we find that 1° rearward motion takes place in approximately 100 years [the modern value is 1° in 72 years], as Hipparchos too seems to have suspected, according to the following quotation from his work *Length of the Year*:

> For if the solstices and equinoxes were moving, from that cause, not less than 1/100 of a degree in advance [i.e., in the reverse order] of the signs, in the 300 years they should have moved not less than 3°.

In the same way we took sightings of Spica and the brightest among those stars near the ecliptic, from the moon, and then were in a better position to use those stars to take sightings of the rest. We thus find that their distances relative to each other are, again, very nearly the same as those observed by Hipparchos, but their individual distances from the solsticial or equinoctial points are in each case about 2⅔° farther to the rear than those derivable from what Hipparchos recorded.

(Toomer [1984b] 327–329)

3.9 Theodosios of Bithunia writing around –100 updated Autolukos and Eukleidês in his *Spherics*; his other works are *On Habitations* (describing the heavens as seen from various latitudes), and *On Days and Nights* (describing the changes in the length of daylight over the year).

Spherics

1 Definitions

1. A *sphere* is a solid figure enclosed by one surface, such that all the straight lines falling onto it from one point (of those lying in the interior of the figure) are equal to one another.
2. The *center* of the sphere is that point.
3. The *diameter* of the sphere is a straight line drawn through the center and terminated at each end by the surface of the sphere, and around which (while it remains straight) the sphere is turned.
4. The *poles* of the sphere are the termini of this axis.
5. The *pole* of a circle on the sphere is the point on the surface of the sphere from which the straight lines falling onto the circumference of the circle are equal to one another.
6. Two planes are said to have the same angle between them as do two others whenever the perpendicular straight lines drawn to the intersection of the planes at the same points enclose equal angles at those points.

[prop. 1.2: find the center, given a sphere; prop. 1.6: the circle on the sphere whose center is the center of the sphere is the largest such circle, and more remote circles parallel to it are smaller]

1.11

In a sphere the great circles bisect one another. [i.e., the equator and the ecliptic bisect one another] Let there be in a sphere two great circles AB and ΓΔ which divide one another at the points E and Z (Figure 3.6). I say that the circles AB and ΓΔ bisect one another. Let the center of the circles be found and let it be the point H, and it will also be the center of the sphere [prop. 1.2], and let the lines EH and

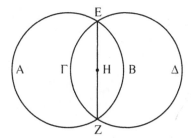

Figure 3.6

HZ be joined. And since the points E, H, Z are in the plane AB, and in the plane ΔΓ, the points E, H, Z are in the planes of both the circles AB and ΓΔ: and thus they are in their intersection. Now the intersection of any two planes is a straight line, and thus the line EHZ is straight. And since the point H is the center of the circle ΓΔ, the line EZ is its diameter; thus each of the lines EΓZ and EΔZ is a semicircle. Thus the circles AB and ΓΔ bisect one another.

1.17

Given a circle on a sphere, and given that a line to it from its pole is equal to the side of a quadrilateral inscribed in the great circle, then the given circle is a great circle. [proof omitted]

1.20

Through two given points on the surface of a sphere, to draw the great circle. [finding the great-circle route between two points] Let the two given points on the surface of the sphere be A and B; it is required to draw the great circle through the points A and B (Figure 3.7). Now if the points A and B are on a diameter of the sphere, it is clear that infinitely many great circles may be drawn through A and B. Let the points A and B not be on a diameter of the sphere, and from A as a pole and with the radius being the side of the quadrilateral inscribed in the great circle, let a circle ΓΔE be drawn – then the circle ΓΔE is a great circle, for the line from its pole is equal to the side of the quadrilateral inscribed in the great circle [prop. 1.17]. Again from B as a pole and with the radius being the side of the quadrilateral inscribed in the great circle, let a circle ZEH be drawn – then the circle ZEH is a great circle, for the line from its pole is equal to the side of the quadrilateral inscribed in the great circle [prop. 1.17]. And let the straight lines EA and EB be drawn from point E to A and B. Thus each of the lines AE and EB is equal to the side of the quadrilateral inscribed in the great circle; thus EA is equal to EB. Thus a circle drawn from pole E with a radius EB will also pass through A, because EA is equal to EB. Let it do so, and let the circle be BAΘ – then the circle BAΘ will be a great circle, for the line from its pole is equal to the side of the

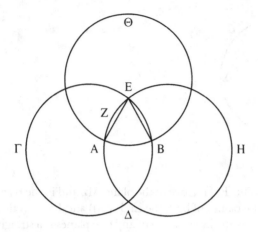

Figure 3.7

quadrilateral inscribed in the great circle [prop. 1.17]. Thus through the two given points A and B on the surface of the sphere the great circle BAΘ is drawn.

2.1–2

The parallel circles on a sphere have the same poles. [proof omitted]
The circles on a sphere having the same poles are parallel. [proof omitted]

2.6–7

If the great circle on a sphere is tangent to one of the circles on the sphere, it will also be tangent to another one, equal and parallel. [proof omitted]

Given two equal and parallel circles, the great-circle tangent to one of them is also tangent to the other. [proof omitted; compare Autolukos, *Moving Sphere* 6]

*(Heiberg [1927] 2, 20, 22, 30, 36, 42, 48, 50)

3.10 Poseidonios of Apamea (in Syria) became a citizen of the "free city" of Rhodes (a Roman protectorate), where he wrote and taught between about –100 and –50. He was widely traveled and well-connected, but was not very expert in astronomy (Neugebauer [1975] 2.652; Keyser [1998] 247). From Seneca (*Natural Questions*, 7) we know that (probably between Hipparchos and Poseidonios) three authors, Epigenes, Apollonios of Mundos, and Artemidoros of Parion, wrote on cometary theory (Keyser [1994b]). Epigenes combined the old theory of Anaxagoras (that comets were created when two planets approached one another) with Aristotle's ideas on earthly vapors rising up to form comets. Apollonios seems to be the "Pythagorean" who theorized that comets were long-period planets with unusual orbits. Artemidoros imagined multiple usually unseen planets, illuminated or even ignited by their approach to ordinary stars, plus a spherical shell of fire around the *kosmos* in which "windows" opened letting out gouts of flame, i.e., comets.

(title unknown)

fr. 131b E–K [origin of comets]

You should know that comets come from dry exhalation [a vapor hypothesized by Aristotle to account for a wide variety of terrestrial phenomena; see Chapter 9]; the exhalation, set easily alight by its own heat, sends something like rays upwards from its underlying material as it is the nature of fire to move upwards. It is these rays from any star people think are comet-tails.

The Pythagoreans [Apollonios of Mundos?] count comets with the planets appearing at long intervals of temporal revolutions and in varied places, and one of them they proposed was even seen to the north beyond the zodiacal circle. And that is precisely an objection to their theory; for the zodiacal circle is the place defined for the planets; and what appears outside the zodiacal circle could not be a planet.

[This objection may be from Artemidoros:] And if it had been one of the planets, expert scientists would certainly have observed its revolutions as they do with planets. Anyway, this is not a single phenomenon, but many; nor are they seen in one form [or "place"?], but some are called "comets": they appear above the stars and have a tail; others are called "beardies" and appear below, while others, "swords," in between; and there are other variations.

Poseidonios says that the generic principle of comets is when a denser bit of air is shot under pressure into the *aither* and is bound in the whirling revolution of the *aither*, and then as their basic material [or "sustenance"] flows in they seem to increase to spin faster; and so they are seen to expand and decrease, as at one point they grow greater as the material increases, at another, as it fails, they contract. That is why they form above all in northern regions, where the air is thicker and condensed. Changes of weather coincide with their conflagration and destruction; droughts or violent rain storms, the opposite, occur at their dissolution, as their formation occurs in the atmosphere. This is what Aratos meant when he said they provided tokens of drought. [Elsewhere Poseidonios suggested that comets dwelt near the sun and become visible as they recede, when "they flee from the rays of the sun."]

(Kidd [1999] 185–186)

3.11 Alexander of Ephesos, writing around –60, versifies a contemporary description of the heavens and the harmony of the spheres (Strabo 14.1.25). His hexameter poem is preserved in two quotations by Theon of Smurna (3.15). The order of the planets he gives was by now the standard one. The harmony of the spheres was asserted probably first in Philolaus and certainly in Plato, *Republic* 10(617a–c); Aristotle, *On Heaven* 2.9(290b12–291a27) rejects the idea; later writers accepted it: Nikomachos, *Enchiridion* 3.3; Ptolemy, *Harmonics* 3.13–16.

[harmony of the spheres]

(a) Sphere upon sphere in the heavens is rising up higher and higher:
Nearest to Earth is the sphere of Selénê, divinity lunar,
Second in order is Stilbôn of Hermes who's strumming the lyre and
Next is the brightest in splendor: Phosphóros of Lady Kuthéry[a],
Fourth of the spheres up above us, Helíos, is borne on his horses,
Fifth in this order's Puróeis of murderous Ares of Thrace, then
Sixth in this order's a glorious star, the Phaëthôn of Zeus, then
Seventh in order's the Phaínôn of Kronos who's next to the stars; and
All of these sound out together the sevenfold tones of the lyre, thus
Forming a harmony based on their intervals, each upon each.

(b) Earth being set in the middle produces the deep tone *hupátê*;
Stars that are fixed in their sphere are producing the *nétê sunêmmenon*;
Midmost of wandering stars, Helíos, is sounding in *mésê*,
Making the tetrachord concord with stars in their circle of coldness;
Slackening Phaínon is raising a note just a half-tone below, with
Intervals equal 'tween Phaínôn, Phaëthôn and terrible Ares:
Sun who is pleasant to mortals is yielding one tone just below these;
Lower a fourth less a note[b] is Kuthéry from dazzling sun, and
Stilbôn of Hermes is carried resounding a half tone below her:
Likewise below him's the orb of curvaceously hueful Selénê;
Earth takes her place in the center while sounding a fifth from the Sun, and is
Girdled by zones five in number[c], from misty to burning with fire, thus
Equally fitted with rays that are fiery and frosts that are freezing.
Heaven fulfills and completes this in making an octave of six tones.
Hermes the son of lord Zeus has become as a singing Seirên, his
Seven-string lyre's depicting the *kosmos* divinely devised.

[[a]"Kuthéry": Aphrodite; [b]"fourth less a note": a "trihemitone" or minor third; [c]five "zones": see Chapter 5.]

[Greek music allowed more scales than our major and minor. The succession of the notes in the two-octave ("greater perfect") system of their scales is: (lowest) *proslambanomenos, hupatê hupaton, parhupatê hupaton, lichanos hupaton, hupatê meson, parhupatê meson, lichanos meson, mesê, paramesê, tritê diezeugmenon, paranetê diezeugmenon, netê diezeugmenon, tritê huperbolaion, paranetê huperbolaion,* and *netê huperbolaion* (highest). In the octave-plus-tetrachord ("lesser perfect") system, the notes above *mesê* are *tritê sunêmmenon, paranetê sunêmmenon,* and *netê sunêmmenon*.]

*(Dupuis [1892/1966] 228, 230)

3.12 Anonymous This anonymous work (traditionally ascribed to Aristotle) seems to have been written around –50; it was influenced by Stoicism, and gives a textbook-like summary of astronomy.

Kosmos

2 (392a6–31) [aither: star material]

The substance of heaven and stars we call *aither*, not, as some think, because being fiery it burns (they err about its capacity, which is quite different from that of fire), but because it always runs (*aeí theín*) in its circular orbit; it is an element different from the four elements, pure and divine (*theíon*). Now, of the stars which are encompassed in it, some are fixed and move with the whole heaven always having the same seats; in the middle of these the "animal-bearing" circle, as it is called, set obliquely through the tropics, passes round like a belt, divided into the twelve regions of the signs. The others, being planets ["wanderers"], naturally do not move at speeds like the fixed stars nor each other, but each in a different circle, so that one is nearer the earth and another higher. The number of the fixed stars is undiscoverable by people, although they all move on one visible surface of the whole heaven; but the number of planets is summed up in seven units, arranged in the same number of circles in a series, so that the higher is always greater than the lower, and all the seven, though contained one within another, are nevertheless encompassed by the sphere of the fixed stars. The circle which is always in the position next to this sphere is that which is called the circle of Phainôn or Kronos [Saturn]; then comes the circle of Phaëthôn or Zeus [Jupiter]; next Puroeis, named after Herakles and Ares [Mars]; next Stilbôn which some dedicate to Hermes [Mercury], some to Apollo; after this is the circle of Phosphoros, which some call after Aphrodite [Venus] and others after Hera; then the circle of the sun; and the last, the circle of the moon, is bounded by the earth's sphere. [This is the order proposed by Plato, *Timaios* 38d; Aristotle, *On Heaven* 2.10 (291a29–b10) leaves the question to the "mathematicians"] The *aither*, then, encompasses the divine bodies and the order of their orbits.

(Furley [1955] 349, 351, 353)

3.13 Xenarchos of Kilikian Seleukeia (*ca.* –75 to 17) taught the geographer Strabo, and was a friend of the emperor Augustus (Strabo 14.5.4). Only Simplicius preserves his work against the existence of Aristotle's *aither* (or "first body," which by Xenarchos' time had become known as the fifth element).

Against the Fifth Element

1–8

1 [arguing against the thesis that there are two simple magnitudes only, the straight

and the circular] For there is also a simple line on the cylindrical helix, because every part of it matches every like part. And if there is a simple magnitude besides the two, there should be a simple motion/change besides the two, and another simple body besides the five, one having that motion. [Earth, Air, Fire, and Water have linear motion, *aither* has circular]

2 [admitting that the helix is composed of straight and circular] Let there be a quadrilateral and let this be moved about in a circle, with one side remaining fixed, the axis of the cylinder. And let a point be carried on the parallel side, the one being rotated, and in the same amount of time let this point run along that line, and let the parallelogram return again to the same position whence it started moving. In this way the parallelogram makes a cylinder, and the point carried on the straight line makes a helix, and this is simple, therefore homoiomerous. [simple = non-decomposible; homoiomerous = having all parts alike]

3 [questioning the simplicity of forced motions of simple bodies] So rectilinear motion is not really natural to any of the four elements, but only when they are becoming. But becoming is not simple, but is between being and not-being, just as is moving, which is mediate between the attained place and the prior-occupied place, and becoming [*genesis*] is similar to motion which is itself a kind of change.

Thus, the fire said to be borne upwards we do not properly call fire, but fire-becoming, and once it has come to its proper place and floated up above the other elements and has come to rest, then it is properly said to be fire. For it becomes a species in so far as it is buoyant in this place. And earth is properly earth whenever it lies beneath the other elements and reaches the middle place; and as for water and air, water is properly so when it floats atop earth and stands beneath air, and air is properly so when it floats atop water and stands beneath fire.

It is false that the natural motion of a simple body is simple: for it is capable of motion when it is not really the body, but its motion is contingent upon the body coming into being. So then if it is necessary to attribute some motion and a simple one to already existing elements, one should attribute *circular* motion, if indeed these are the only two simple motions, circular and linear, and linear motion belongs to the four elements when becoming, but not being. For one would not improperly attribute circular motion to fire and to the other three stationarity.

4 [compound objects and compound motions] So there are two simple lines, the circle's-arc and the straight, and each of the four (earth, water, air, fire) whenever it properly exists has its natural motion as linear. But what prevents the circular motion of one or some or even all of these substrates [elements] from existing by nature? But it is not possible to hypothesize only one motion, for that is patently false. For each of the elements in the middle has two natural motions: water is naturally borne up out of the earth and downward out of air and fire, while air is borne downward out of fire and upward out of water.

5 [non-simplicity of circular motion] It is impossible that the circular motion of a simple body be natural, since in simple homoiomerous bodies all parts move at

the same speed; but in the circle, the parts at the center are always slower than those at the periphery, since in the same time they move a shorter distance; but also in the sphere, the circles near the poles move more slowly than those farther off, and fastest of all is the greatest of the parallel circles [equator].

6 [dragging of air by upper fire] We must study whether circular motion in the dragged air is natural or not: and if someone should say it is unnatural, then some upward motion will be natural for it. Well, since downward motion is unnatural for air, there will be two contraries to one motion [but that is impossible] – therefore air and fire are circularly dragged by nature.

7 [pairs of opposites to one entity] We say, even in the works about ethics, that there are two contraries to each of the virtues, for example rashness [*panourgia*] and simplicity [*euetheia*] to wisdom [*phronesis*], and recklessness [*thrasutes*] and cowardice [*deilia*] to bravery, and similarly for the others. . . . But if these claims are true, it is not necessary that heaven be made of some fifth body because of their not being two things opposed to one, the circular motion of fire [caused by the rotating heaven] and its downward motion both opposed to its upward motion: for the upward motion is opposed to the downward as excess to defect, while the motion common to both, the rectilinear, is opposed to the circular as inequality to equality.

8 [the *kosmos* and the boundless] Yet if the *kosmos* is in an infinite void, what is the cause of its remaining where it is? For no body having weight is able to rest without an external restraint, if not in its proper place. But in the void there is no distinction to make one place proper for a body and another not. But if the *kosmos* does not rest but is moved, given that the void is uniform, why is it moved in one direction rather than another? But if it were moved in every direction, it would expand, for no god could reasonably oppose himself to the expansion, for being fire himself he would have the motion of fire. . . .

But to say that the *kosmos* located in the void is contained and controlled by pervasive *pneuma*, is to babble. For it would also be necessary that each of the other bodies is itself also contained at rest [what is true of the *kosmos* must be true of each of its parts]. Then the containment *pneuma* would cooperate to prevent the parts expanding, yet would no longer also be *pneuma* to maintain the *kosmos* at rest, in virtue of being itself moved and not impeding the natural impulse [*rhope*] of any body [i.e., the hypothesis is contradicted]; in addition to which, also saying that each of the bodies is contained by some *pneuma* is false, as we have shown elsewhere [we lack that proof].

*(Heiberg [1894] 13–14, 21–22, 23–24, 50, 55–56, 286; for Sections 1–5 and 7 see Sambursky [1962/1987] 127–130)

3.14 Geminus produced his introductory textbook of astronomy about mid-first century CE (perhaps in Rhodes); here his material is similar to the introduction of Eukleidês' *Phainomena*.

Phainomena

1.13–17 [solstices and equinoxes]

The periods between the solstices and the equinoxes are divided as follows [the inequality of the seasons discovered by Euktemon]. From the vernal equinox to the summer solstice there are 94½ days [Hipparchos made it ¼ day less]. For in this number of days the sun traverses the Ram, the Bull, and the Twins, and, arriving at the first degree of the Crab, brings about the summer solstice. From the summer solstice to the autumnal equinox there are 92½ days [Hipparchos had the same value], for in this number of days the sun traverses the Crab, the Lion, and the Virgin, and, arriving at the first degree of the Scales, brings about the autumnal equinox. From the autumnal equinox to the winter solstice there are 88⅛ days, for in that number of days the sun traverses the Scales, the Scorpion, and the Archer, and, arriving at the first degree of the Goat-Horn, produces the winter solstice. From the winter solstice to the vernal equinox there are 90⅛ days, for in that number of days the sun traverses the remaining three signs, the Goat-Horn, the Water-Pourer, and the Fishes. The days forming these periods, when all added together, make up 365¼ days, which, as we saw, was the number of days in the year.

5.54–61, 68–69 [horizon]

The horizon is the circle which from our point of view divides the visible from the invisible part of the *kosmos* and halves the whole sphere of the *kosmos*, so that it cuts off a hemisphere above the earth and a hemisphere below the earth. There are two horizons, one perceived and the other seen by reason. The perceived horizon is circumscribed by our eye at the limit of sight, and has no greater a diameter than 2000 stades [about 300 km, valid for elevations of up to about 1200 m]. The horizon seen by reason is the one reaching the sphere of the fixed stars and halving the whole *kosmos*.

The horizon is not the same in every place and city. But as far as perception goes it is almost the same horizon out to 400 stades [about 60 km], so far as the length of days, the latitude, and all the phenomena remain the same. If the distance increases, as the habitation alters the horizon becomes different, changing according to the latitude, and all the phenomena modify. Well, it is required that the change of habitation of more than 400 stades be taken in a North–South direction. For to those dwelling on the same parallel even at a distance of myriads of stades [thousands of kilometers], the horizon differs but the latitude is the same and all the phenomena are similar. In the exact theory, with the smallest displacement in any part of the *kosmos*, the horizon and the latitude change and all the phenomena differ.

[62–67: similar argument about the meridian lines]

Another oblique circle is the circle of the Milk; it is rather broad and oblique to the tropic circle. It is composed of fine-grained cloudlike stuff and is the only visible circle in the *kosmos*. Its width is not constant, but it is wider in some parts and narrower in others. (Thus in most globes, the circle of the Milk is not inscribed.) [Aristotle, *Meteorologika* 1.8 (345a11–6b15) recounts various theories about the "Milk": a burnt track where the sun once ran; stars in the shadow of the earth (Anaxagoras); a reflection of the sun on the heavenly sphere; he advocates the theory that the haze is caused by the larger quantity of bright stars near it which partly ignite exhalations rising from the earth.]

*(Aujac [1975] 4–5, 30–31, 33)

3.15 Aristokles of Messênê, a follower of Aristotle (*ca.* 50 ± 100), wrote a work entitled *Philosophy* summarizing the teachings of various philosophers.

[Reporting the Stoic doctrine of universal conflagration]

At certain fated times the entire *kosmos* is subject to conflagration, and then is reconstituted afresh. But the primary fire is, as it were, a sperm which possesses the principles of all things and the causes of past, present, and future events. The nexus and succession of these is fate, knowledge, truth, and an inevitable and inescapable law of what exists. In this way everything in the world is excellently organized as in a perfectly ordered society.

(Long and Sedley [1987] 276)

3.16 Apollinarius (of Aizanoi?), working around 80 ± 50, constructed lunar tables based on the Babylonian 248-day period of the moon's motion and phases (a linear zigzag function widely used by Greek astronomers from *ca.* −150 to *ca.* 150), and wrote on solar eclipses and on astrology.

(title unknown) [length of month]

"Month" is the interval resulting from the combined motion of the sun and moon. "Restitution of latitude" is the name of the interval from when the lunar center coincides with the ecliptic to when it has revolved through the latitudinal limits and is returned to the plane of the ecliptic [either the tropical month of 27.3216 days or else the draconitic month of 27.2122 days]. "Restitution of depth" is the name of the interval in which the exact apogee of the surface of the star's sphere, starting from the exact apogee of its own motion, is returned again to its exact apogee [i.e., anomalistic month, given in Babylonian and Greek texts as 27;33,20 days; modern value is 27;33,16,33, . . .]. "Restitution of longitude" is the name of the period in which the center of any star, having set out from some plane of one of the circles drawn through the poles of the zodiac, and having revolved around the zodiac, is returned to that same plane from which it began revolving [i.e., sidereal month, compare Section 3.8 Hipparchos]. Besides, a restitution is called mean or not mean [i.e., calculated or observed]. Now the length of a month has been determined as being a composite of the motion of the sun and moon; for the moon, after starting from its conjunction with the sun and revolving around its own circle, takes up additionally as much arc as the sun has traversed during the intervening time until the catching up [i.e., the synodic month, given in Babylonian and Greek texts as 29;31,50,8,20 days].

(Jones [1990] 39, 41)

3.17 Plutarch of Chaironeia (*ca.* 50 to *ca.* 120) was a Platonist teacher of philosophy writing under the Flavian emperors, Trajan, and in the early years of Hadrian; much of his copious output is in the form of moralizing essays or dialogues. The dialogue quoted here discusses the moon, starting with the question of the "face" in the moon (which seems to contradict the notion that the moon is made of immutable and homogenous *aither*).

The Face in the Moon

8 *(924d–f)* [the moon could be "lunar" material in *its* proper place]

Yet if all heavy body converges to the same point and is compressed in all its parts upon its own center, it is not as center of all but as a whole that the earth would appropriate heavy bodies as parts of herself; and the downward tendency of falling bodies is evidence not of the earth's centrality but of the affinity and cohesion to earth of those bodies which when thrust away fall back again. For as the sun attracts to itself the parts of which it consists [as suggested by Xenophanes: see Keyser [1992b]], so the earth too accepts as her own the stone that has its own downward tendency. Consequently every such thing eventually unites and coheres with her. But given some body not originally allotted to the earth or detached from it, but having somewhere independently its own constitution and nature, as those men would say of the moon, what is to hinder it from being permanently separate in its own place, compressed and bound together in its own parts? For it has not been proved that the earth is the center of all, and the way in which things here press together and concentrate upon the earth suggests how in all probability things there fall towards upon the moon and remain there. The man who drives together into one place all earthy and heavy things and makes them parts of a single body, I do not see why he does not apply the same necessity to light objects, but allows so many separate concentrations of fire and, since he does not collect all the stars together, clearly does not think that there must also be a body common to all things that are fiery and move upward.

25 *(940a–e)* [life on the moon?]

[The moon causes tides which shows it is not a fiery body, and so may support life.] For it's likely, my friend, that the moon's nature is contrary [*antipathê*] to that of the sun, if she not only naturally softens and loosens all that he condenses and dries, but even liquefies and cools the heat that he casts upon her and imbues her with. They err then who believe the moon to be a fiery and glowing body; and those who think that animals there are equipped like those here for generation, nourishment, and livelihood, seem blind to the diversities of nature, among which one can discover more and greater differences and dissimilarities between animals than between them and inanimate objects.

[Certain mythical beings may live without eating.] It is plausible that those on the moon, if they exist, are slight of body and capable of being nourished by what they get. After all, they say that the moon herself, like the sun which is an animal of fire many times as large as the earth, is nourished by the moisture on the earth [referring to Aristotle's theory of exhalations arising from the earth: see Chapter 9], as are the innumerable other stars too; so light and frugal of requirements do they suppose the creatures that the upper region supports. We have no comprehension of these beings, however, nor of the fact that a different place and nature and climate are suitable to them. Just as, assuming that we were unable to approach the sea or touch it, but only saw it from afar and learned that it is bitter and unpotable and salty water, if someone said that it supports in its depths many large animals of all sorts of shapes and is full of beasts that use water just as we use air, he would seem to relate myths and marvels; such appears to be our experience about the moon when we disbelieve that any people dwell there. Those men, I think, would be much more amazed at the earth, when they look out at the sediment and dregs of the universe, as it were, obscurely visible in moisture and mists and clouds as a lightless and low and motionless spot, to think that it engenders and nourishes animals which partake of motion, breath, and warmth.

(Cherniss and Helmbold [1957] 67, 69, 71, 175, 177, 179)

3.18 Theon of Smurna a Platonist of the era of Hadrian (*ca.* 130 ± 10) wrote a book whose full title well describes it: *Aspects of Mathematics Useful for the Reading of Plato* (book 3 is on astronomy); much of it may be from his contemporary Adrastos of Aphrodisias the follower of Aristotle (whom he often quotes explicitly).

Mathematics

3.33 [a quasi-heliocentric theory from *ca.* –90 ± 50,
known also to Vitruuius 9.1.6]

For the sun, Phosphoros, and Stilbôn it may be that there are two proper spheres for each, and that the hollow spheres of these three stars have similar speeds and in the same time go round the sphere of the fixed stars, in the opposite direction, while their solid spheres have their centers on the same line, with the size of the sphere of the sun being the lesser, the sphere of Stilbôn larger than that, and that of Phosphoros still larger. It may also be that there is one hollow sphere common to all, the solid spheres of the three being in its interior around the same center as one another, with the smallest (and actually solid) sphere being that of the sun, around that one the sphere of Stilbôn, then enveloping both and filling the whole interior of the common hollow sphere is that of Phosphoros.

It is for this reason that these three make a retardation or equal-speed retrogression along the signs, and other dissimilar motions, and are always seen overtaking and being overtaken and eclipsing one another. But the star of Hermes is separated by at most about 20 degrees on either side of the sun, at evening or at dawn, while the star of Aphrodite is separated by at most 50 degrees [the exact values are 28° and 47°; see Dicks [1970] 25]. One might suspect that this is the truest position and order, so that this place would be the place of ensoulment of the *kosmos*, as *kosmos* and lifeform, as if the very hot sun were the place of the heart of the whole being, because of his motion and size and the common path of the stars around him.

For among animals the center of being (i.e., of the animal as animal) is different from the center of size. For example, as we said, for us as people and as animals our center of ensoulment is at the heart, always very hot and in motion, and therefore being the source of all the power in our soul [i.e., of all the capabilities which the soul has], such as life itself and locomotion, desire and imagination and understanding; but our center of size is different, i.e., about the navel. [see Chapter 12 for more about the heart as center]

So likewise, to evaluate the greatest and most worthy and divine matters based on small and chancy and mortal things, the center of size of the whole *kosmos* is at the chilled and motionless earth. But the center of ensoulment of the *kosmos* (as *kosmos* and living being) is at the sun, being a kind of heart of the whole, wherefrom they say its soul begins and moves through the whole body out to the extremities.

*(Dupuis [1892/1966] 300, 302)

3.19 Ptolemy (Claudius Ptolemaeus) of Alexandria (*ca.* 100 to ca. 175) wrote on a wide variety of "mathematical" topics (he is quoted in each of Chapters 2–8). His astronomical writings recapitulated and in a way completed the program enunciated by Plato; his *Syntaxis* was published *ca.* 150 (serving to render astrology more reliable), and the *Planetary Hypotheses* a bit later – his model was so effective that it prevented alternatives and became standard for over a millennium. He viewed mathematics including astronomy as a surer insight to reality than were theology or physics, and studied astronomy to enable astrology (see Section 4.13).

Syntaxis

7.4 *On the method used to record the positions of the fixed stars*

Thus, from our observations and comparisons of the above stars, from similar observations and comparisons of the other bright stars, and from the fact that we found the distances of the other stars with respect to the bright stars which we had established to be in agreement with prior results, we have confirmed that the sphere of the fixed stars, too, has a movement towards the rear with respect to the solsticial and equinoctial points of the amount determined (in so far as the available elapsed time allows); furthermore, we have confirmed that this motion of theirs takes place about the poles of the ecliptic, and not those of the equator (i.e., the poles of the first motion). So we thought it appropriate, in making our observations and records of each of the above fixed stars, and of the others too, to give their positions, as observed in our time, in terms of longitude and latitude, not with respect to the equator, but with respect to the ecliptic (as determined by the great circle drawn through the poles of the ecliptic and each individual star). In this way, in accordance with the hypothesis of their motion established above, their positions in latitude with respect to the ecliptic must necessarily remain the same, while their positions in longitude must always traverse equal arcs towards the rear in equal times.

7.5

2   [The constellation of the Great Bear – Ursa Major]

Table 3.1 Constellation of the Great Bear – Ursa Major

	Description	Longitude in degrees	Latitude in degrees	Magnitude	Modern label
1	The star on the end of the snout	Π25⅓	+39⅚	4	o UMa
2	The more advanced of the stars in the two eyes	Π 25⅙	+43	5	2(A) UMa
3	The one to the rear	Π 26⅓	+43	5	π² UMa
4	The more advanced of the two stars in the forehead	Π 26⅙	+47⅙	5	ρ UMa
5	The one to the rear	Π 26⅔?	+47	5	σ² UMa
6	The star on the tip of the advance ear	Π 28⅙	+50½	5	24(d) UMa
7	The more advanced of the two stars in the neck	♋0½	+43⅚	4	τ UMa
8	The one to the rear	♋2½	+44⅓	4	23(h) UMa
9	The northernmost of the two stars in the chest	♋9	+42	4	υ UMa
10	The southernmost of them	♋11	+44?	<4	φ UMa
11	The star on the left knee	♋10⅔	+35	3	θ UMa
12	The northernmost of the two in the front left paw	♋5½	+29⅓	3	ι UMa

Table 3.1 (continued)

	Description	Longitude in degrees	Latitude in degrees	Magnitude	Modern label
13	The southernmost of them	♋6⅓	+28⅓	3	κ UMa
14	The star above the right knee	♋5⅔	+36	4	18(e) UMa
15	The star below the right knee	♋5⅚	+33	4	15(f) UMa
16	The stars in the quadrilateral: the one on the back	♋17⅔	+49	2	α UMa
17	. . . the one on the flank	♋22⅙	+44½	2	β UMa
18	. . . the one where the tail joins the body	♌3⅙	+51	3	δ UMa
19	. . . the remaining one, on the left hind thigh	♌3	+46½	2	γ UMa
20	The more advanced of the two in the left hind paw	♋22⅔	+29⅓	3	λ UMa
21	The one to the rear of it	♋24⅙	+28¼	3	μ UMa
22	The star on the left knee-bend	♌1⅔	+35¼	>4	ψ UMa
23	The northernmost of the two in the right hind paw	♌9⅚	+25⅚	3	ν UMa
24	The southernmost of them	♌10⅓	+25	3	ξ UMa
25	The first of the three stars on the tail next to where it joins the body	♌12⅙	+53½	2	ε UMa
26	The middle one	♌18	+55⅔	2	ζ UMa
27	The third, on the end of the tail	♌29⅚	+54	2	η UMa

9.2 [planetary theory is much more difficult]

So much, then, for the arrangements of the spheres. Now it is our purpose to demonstrate for the five planets, just as we did for the sun and moon, that all their apparent anomalies can be represented by uniform circular motions, since these are proper to the nature of divine beings, while disorder and non-uniformity are alien to them. Then it is right that we should think success in such a purpose a great thing, and truly the proper end of the mathematical part of theoretical philosophy. But, on many grounds, we must think that it is difficult, and that there is good reason why no-one before us has yet succeeded in it. For, firstly in investigations of the periodic motions of a planet, the possible inaccuracy resulting from comparison of two observations (at each of which the observer may have committed a small observational error) will, when accumulated over a continuous period, produce a noticeable difference sooner when the interval over which the examination is made is shorter, and less soon when it is longer. But we have records of planetary observations only from a time which is recent in comparison with such a vast enterprise: this makes prediction over much greater time insecure. Moreover, in investigation of the anomalies, considerable confusion stems from the fact that it is apparent that each planet exhibits two anomalies, which are moreover unequal both in their amount and in the periods of their return: one return is observed to be related to the sun, the other to the position in the ecliptic; but both anomalies are continuously combined, whence it is difficult to distinguish the characteristics of each individually. It is also confusing that most of the ancient planetary observations have been recorded in a way which is difficult to evaluate, and crude. For the more continuous series of observations concern stations and phases [i.e., first and last visibilities]. But detection of both of these particular phenomena is fraught with uncertainty: stations cannot be fixed at an exact moment, since the local motion of the planet for several days both before and after the actual station is too small to be observable; in the case of the phases, not only do the places [in the sky] immediately become invisible together with the bodies which are undergoing their first or last visibility, but the times too can be in error, both because of atmospheric differences and because of differences in the sharpness of vision of the observers. And in general, observations of planets with respect to one of the fixed stars, when taken over a comparatively great distance, involve difficult computations and an element of guesswork in the quantity measured, unless one carries them out in a manner which is thoroughly competent and knowledgeable. This is not only because the lines joining the observed stars do not always form right angles with the ecliptic, but may form an angle of any size (hence one may expect considerable error in determining the position in latitude and longitude, due to the varying inclination of the ecliptic); but also because the same interval between star and planet appears to the observer as greater near the horizon, and less near mid-heaven; hence, obviously, the interval in question can be measured as at one time greater, at another less than it is in reality. [compare *Optics* 3.59]

Hence it was, I think, that Hipparchos, being a great lover of truth, for all the above reasons, and especially because he did not yet have in his possession such a groundwork of resources in the form of accurate observations from earlier times as he himself has provided to us, although he investigated the theories of the sun and moon, and, to the best of his ability, demonstrated with every means at his command that they are represented by uniform circular motions, did not even make a beginning in establishing theories for the five planets, not at least in his writings which have come down to us. All that he did was to make a compilation of the planetary observations arranged in a more useful way, and to show by means of these that the phenomena were not in agreement with the hypotheses of the astronomers of that time.

(Toomer [1984b] 339, 342–343, 420–421)

Planetary Hypotheses

1.1.8 [model of the sun: see Murschel [1995]]

Now let the eccentric circle of the solar sphere be supposed like this in the plane of the zodiac [in the reconstructed diagram (Figure 3.8), the circle ABΓΔ], so that its radius to the line between its center [E] and the center of the zodiac [Z] has the ratio 60 to 2½, and let the straight line extending through both of these centers

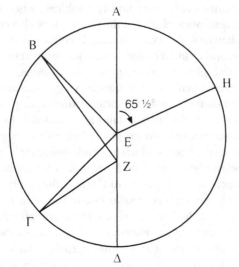

65 ½°

Figure 3.8

and the apogee of the eccentric circle be supposed to intercept the zodiac at a constant angle from the spring equinox, in the direction following the rotation of the *kosmos*, of 65½°. Let the center of the sun move along the indicated eccentric circle, from west to east, about its center at a constant speed, so that in 150 Egyptian years [365 days] and 37 entire [24-hour] days it will make 150 returns, being observed at the apogee of the eccentric circle, and let the sphere of the fixed stars be moved about the center of the zodiac and its poles, towards the east, at a constant speed in the given time 1½° (where the zodiac has 360°) [i.e., taking into account the precession of the equinoxes].

Now then, in the first year of the death of Alexander the Founder [–322], on the first day of Egyptian Thoth [Nov. 12] at noon in Alexandria, the sun was 162° 10′ in the direction of the rotation of the *kosmos* from the apogee of the eccentric circle, and the star on the heart of the Lion [Regulus, α Leo] was 117° 54′ along the zodiac from the spring equinox, also in the direction of the rotation of the *kosmos*.

*(Heiberg [1907] 80)

[1.1.9–11: corresponding, more complex, models for the moon, Mercury and Venus; in 1.1.12–14 for Mars, Jupiter, and Saturn]

1.2.2 [arrangement of heavenly bodies]

The relative arrangement of the spheres has been a subject of some doubt up to this time. The sphere of the Moon is the closest sphere to the earth; the sphere of Mercury closer to the earth than the sphere of Venus; the sphere of Venus closer to the earth than the sphere of Mars; the sphere of Mars than the sphere of Jupiter; the sphere of Jupiter than the sphere of Saturn; and the sphere of Saturn than the sphere of the fixed stars. It is clear from the course of the planets that this sphere is closer to the earth and that sphere further away, if along a straight line from the eye. But with respect to the Sun, there are three possibilities: either all five planetary spheres lie above the sphere of the Sun just as they all lie above the sphere of the Moon; or they all lie below the sphere of the Sun; or some lie above, and some below the sphere of the Sun, and we cannot decide this matter with certainty.

The distances of the five planets are not as easy to determine as those of the two luminaries, for the distances of the two luminaries were determined, mostly, on the basis of combinations of eclipses. A similar proof cannot be invoked for the five planets, because no phenomenon allows us reliably to fix their parallax. Moreover, up to this time we have not seen an occultation of the Sun, and therefore it is possible for one to assert that all five planetary spheres lie above the sphere of the Sun [transits of Venus were predictable on Ptolemy's theory, Neugebauer [1975] 1.227–230; may have been observed by Babylonians, Johnson [1882]; and are retrodicted for 23 May of –183 and 60, and 23 November of –305, –62, and 181: Meeus [1989] 47]. But to one whose intention is to know the truth this doesn't serve as proof, for I say, firstly, if a body of such small size were to occult a body of such large size and with so much light, it would necessarily be imperceptible, because of the smallness of the occulting body and the state of the parts of the Sun's body which remain uncovered; for when the Moon eclipses part of the Sun equal to, or greater than, the diameter of one of the planets, the eclipse is not perceptible. Moreover, such events could only take place at long intervals, for this happens when a planet is closest to the Sun, and this is when it is at the apogee and perigee of its epicycle; but the planet is found in the plane of the ecliptic only twice in every revolution on the epicycle, when it passes from the north to the south, and when it passes from the south to the north. And no doubt if the epicycles are at the nodes, and the planet is also at that node, and if the planet is at that node while being at the apogee or perigee of its epicycle, then the eclipse happens. Those who study observations and examine them carefully, agree that a long time must elapse before both returns coincide, i.e., the return of the epicycle and the return of the planets, and it only happens if the conjunctions coincide above the earth. With these conditions, it is clear that one cannot judge with certainty for the two inner planets, nor even for the planets on which it is agreed that they lie above the sphere of the Sun, i.e., Mars, Jupiter and Saturn.

(Goldstein [1967] 6–7 corrected from Toomer [1984b] 419, n.2)

2.3 [the stars are divine and self-willed bodies]

A physical observation now leads us to say that the aitherial bodies allow no influence and don't alter (since they are distinct forever) due to what is proper to their wonderful essence, and to their similarity to the power of the stars, in that their rays clearly penetrate through all the things dispersed around them, unhindered and uninfluenced (and similarly penetrating is what in us is like those rays, i.e., sight and understanding).

Also, it brings us to the observation that the aitherial bodies do not alter, as we already said, I mean that their forms are spherical and their activities are those of objects which resemble each other in their parts. For each of these movements different in quantity and quality, there is a body which moves around a pole in a time and space that are proper to it in its own motion according to the power of each individual star, from which power the beginning of motion takes place, which motion arises from the principal powers (which are like the powers in us), and which move bodies similar to themselves and similar to the parts of the "Whole Animal" [kosmos], according to the measure of the ratios proper to each one of them, and this happens to them without force or external coercion, since there is nothing stronger than that which allows no influence to affect it. Moreover, this is not like what occurs in those bodies which in the circumstances of their natural motion rise and fall [for example, earth or fire], because of the condition of a natural weight and a non-independent motion.

For in the first place, these motions are not by nature proper to bodies which are moved in accordance with the motions, but each of them stops and rests when it enters something related to itself, and if it passes over to something dissimilar and unrelated to itself, and if any oppositions are removed, then it tends to its proper place.

[*Planetary Hypotheses* 2.6–8,12 are partly translated by Sambursky [1962/1987] 142–145]

*(Heiberg [1907] 111–113)

3.20 Sosigenes the Peripatetic (*ca.* 165) wrote on astronomy; one extract arguing against Eudoxos' theory is preserved (he also calculated a very long common period, "perfect year," of the planets: about 65×10^{16} years).

On the Counteracting Spheres [varying distances from us of the planets]

Nevertheless the theories of Eudoxos and his followers fail to save the phenomena, and not only those which were first noticed at a later date, but even those which were before known and actually accepted by the authors themselves. What need is there for me to mention the generality of these, some of which, after Eudoxos had failed to account for them, Kallippos tried to save – if indeed we can regard him as so far successful? I confine myself to one fact which is actually evident to the eye; this fact no one before Autolukos of Pitane even tried to explain by means of hypotheses and not even Autolukos was able to do so, as clearly appears from his controversy with Aristotheros [the work is lost; this is all we know of Aristotheros, though Aratos had a mathematics teacher by the same name]. I refer to the fact that the planets appear at times to be near to us and at times to have receded. This is indeed obvious to our eyes in the case of some of them; for the star called after Aphrodite and also the star of Ares seem, in the middle of their retrogradations, to be many times as large, so much so that the star of Aphrodite actually makes bodies cast shadows on moonless nights. The moon also, even in the perception of our eye, is clearly not always at the same distance from us, because it does not always seem to be the same size under the same conditions as to medium [i.e., of air]. The same fact is moreover confirmed if we observe the moon by means of an instrument [a dioptra]; for it is at one time a disc of eleven finger-breadths, and again at another time a disc of twelve finger-breadths, which when placed at the same distance from the observer hides the moon so that his eye does not see it. In addition to this, there is evidence for the truth of what I have stated in the observed facts with regard to total eclipses of the sun; for when the center of the sun, the center of the moon, and our eye happen to be in a straight line, what is seen is not always alike; but at one time the cone which comprehends the moon and has its vertex at our eye comprehends the sun itself at the same time, and the sun even remains invisible to us for a certain time, while again at another time this is so far from being the case that a rim of a certain breadth on the outside edge is left visible all round it at the middle of the duration of the eclipse [dated to 04 September 164: Neugebauer [1975] 1.104,n.4]. Hence we must conclude that the apparent difference in the size of the two bodies observed under the same atmospheric conditions is due to the inequality of their distances at different times.

...

But indeed this inequality in the distances of each star at different times cannot even be said to have been unknown to the authors of the concentric theory themselves. For Polemarchos of Kuzikos [the teacher of Kallippos] appears to be aware of it but to minimize it as being imperceptible, because he preferred the theory which placed the spheres themselves about the very center in the universe. Aristotle too, shows that he is conscious of it when, in the *Physical Problems* [pseudo-Aristotle, *Problems* 15.4, 15.8 – or a lost work], he discusses objections to the hypotheses of astronomers arising from the fact that even the sizes of the planets do not appear to be the same always. In this respect Aristotle was not altogether satisfied with the revolving spheres, although the supposition that, being concentric with the universe, they move about its center, attracted him, Again, it is clear from what he says in Book Lambda of the *Metaphysics* [1073b17–4a14] that he thought that the facts about the movements of the planets had not been efficiently explained by the astronomers who came before him or were contemporary with him.

(Heath [1913/1966/1997] 221–223)

4

ASTROLOGY

Humans have always sought causes and patterns in causal sequences, but we often impute purpose and mind to origins of events, especially those which affect us. The "anthropo-psychic" view of causation directed at the sky led people in all cultures to attribute the seasons to the heat- and light-giving sun, and menstruation to the moon. Religious rituals are very often seasonal, tying heavenly events or entities to gods; and early ways of life naturally exploit the constant and visible relations between earthly and celestial occurrences (as agriculture is seasonal, etc.). There was widespread fear of eclipses as ominous events interrupting the expected course of life. Some events are clearly presaged (fall presages winter); and so in early Mesopotamia and China celestial omens were sought, which (like seasons) were believed to affect the whole land and people. The Egyptian sun was divine and all moving lights in the Mesopotamian sky were gods; these same peoples perceived designs of familiar things in the scattered unmoving stars (like seeing animals in clouds).

Around −3000, Sirius' visible rising coincided with the mysterious life-giving rise of the Nile, so Egyptians made him a divine and heavenly sign. In Homer's Greece, Sirius' prominence during feverish autumn earned the epithet "baleful" (*Iliad* 22.26–31). Hesiod's versified farmer's almanac correlated human deeds with astral risings and settings (harvest is best at the Pleiades' rising), sometimes apparently conferring agency on stars (*Works and Days* 571–572, 618–621), and perceiving lunar phases as ominous for good or ill (770–828). And Greeks were partly fatalistic, awestruck at the seeming inevitability of everything that follows birth (*Iliad* 6.487–489).

In the late fifth century BCE, some of these notions were given a theoretical foundation. Herodotos (−435 ± 10) describes the sun heading south as "repelled by the winds" (2.24–26), showing how tightly-coupled our *kosmos* seemed. Egyptians whom he met related that each month and day was ruled by a given god (2.82), suggesting that one might estimate the fortunes of a day by understanding its divine ruler. Demokritos of Abdera (−410 ± 30) postulated a harmony or sympathy (*sumpatheia*) between each human and the universe as a whole, both being a kind of ordered organism, a *kosmos*; he also seems to have granted a special status to the divine Babylonian planetary triad sun–Venus–

moon (fr. A86). Yet Aristophanes in his comedy *Peace* (–420) asserted that Greeks differ from foreigners in *not* worshipping sun and moon (lines 406–413; archaeology concurs). The medical library attributed to Hippokrates (–400 ± 30) explains connections between astral events and medical effects. The changes of the seasons, delineated by certain risings and settings, create analogous transitional states of health and disease (*Airs, Waters, Places* 2), and great danger attends the "turnings" of the sun (solstices), the equinoxes, the risings of Sirius and Arktouros, and the setting of the Pleiades (11). Moreover, at least in dreams, the appearance of the sun, moon, and stars is portentous of individual health, the moon corresponding to the innermost parts, the stars to the outermost, while the sun has a dual binding (*Regimen* 4.89). The physician Ktesias returned home to Knidos from service in Persia, and published (*ca.* –400) tales of Babylonian astral forecasts affecting politics (Diodoros of Sicily 2.24). Thoukudides ("Thucydides," in –395 ± 5) records a lunar eclipse which had effected a recent Greek military collapse (7.50).

Two Platonic dialogues (–365 ± 20) attribute godly power and activity to the planets. The *Timaios* tells a tale that divine stars participated in our creation (41a), that their ongoing phenomena portend earthly events (40c–d), and that whenever their dance wheels back around to its start, one great year of the *kosmos* is fulfilled (22c and 39d). The *Epinomis* offers a theory of the ties that bind heaven and earth (977–985): the worship-worthy sun, moon, and stars dance to accomplish and regulate environmental change here below; they are animate fire (we are ensouled Earth), and in between dwell invisible animals compounded from *aither* and air and water, who participate in this divine work. The two stars nearest the sun are tied to Greek gods, Hermes and Aphrodite (986–987).

The treatises of Aristotle (–335 ± 10) refer to the outer planets as the stars of the gods Kronos, Zeus, and Ares respectively, and speak of the manifold changes wrought on Earth by the teleological action of moon, sun, and stars. The sphere of fixed stars easily attains the perfection towards which it, a living and initiatory being, strives, and so has a simple motion, while the planets can only attain their perfection through complex exertions, and the sun, moon, and earth, furthest from perfection, simply resign themselves to whatever bits of goodness they (and we) can achieve through relatively simple motions (*On Heaven* 2.12 [291b24–292b25]). The uniform rotation of the sphere of the fixed stars cannot cause growth and decay, creation and destruction, which comes from the varying and continuous motion along the zodiac; the approach of any of the bodies (especially the sun) causes growth and creation, while retreats elicit destruction; and these cycles determine how long things live (*Generation and Corruption* 2.10 [336a32–b24]). This generalization from the sun to all planets is elaborated for comets, which are fiery (formed close to Earth from a terrestrial "exhalation") and in sufficient number presage winds and drought, a correlation often observed (*Meteorologika* 1.7 [344b20–345a5], citing the comets of –466, –372, and –340). His student and successor Theophrastos recorded the effects of the moon's heat on plant decay (*Plant Etiology* 3.22.2, 4.14.3).

The Stoics portrayed an interdependent *kosmos* unified and harmonized in sympathy (*sumpatheia*) by the all-pervading *pneuma*. All sorts of prognosis were thus possible, since the fated causes of events were "woven together": e.g., Chrusippos (*ca.* −280 to −206) remarked "The predictions of the diviners could not be true if Fate were not all-embracing." Yet two influential Stoics rejected astrology: Diogenes of Babylon (−240 to −153) and Panaitios of Lindos (−185 to −95). Diogenes believed that stars provided predictions not of events but only of one's character and abilities. Panaitios followed the arguments of the Platonist Karneades (−213 to −128) that imprecise observations rendered astrology impracticable, while coincident births yielding dissimilar lives, similar fates attending the separately-born, animals remaining unaffected, and races and peoples greatly varying, all show the stars' lack of effect.

This chapter considers the Greek search for meaning in and influence from the heavens. Throughout the period of this book, Greeks used the same word, *astrología*, for this and for what we call astronomy. The distinction was between a science of unvarying and hence mathematical nature (the stars themselves) and a science of mutable and hence uncertain nature (earthly response to the effects of the stars).

On terminology see Sextus Empiricus, *Against the Mathematicians* 5.1–22 and Neugebauer and van Hoesen [1959] 2–13. Note especially: Kronos (Saturn), Zeus (Jupiter), Ares (Mars), Aphrodite (Venus), Hermes (Mercury); and Ram (Aries), Bull (Taurus), Twins (Gemini), Crab (Cancer), Lion (Leo), Virgin (Virgo), Scales (Libra), Scorpion (Scorpio), Archer (Sagittarius), Goat-Horn (Capricorn), Water-Pourer (Aquarius), and Fishes (Pisces).

4.1 Berôsos of Babylon wrote for Antiochos I (*ca.* −270 ± 10) a history of Babylonia from the creation, in which he also described the *kosmos*; he is credited with introducing Babylonian astrology to the Greeks.

Babyloniaka

Book 1 (fr. 4) [the moon]

The moon is a ball, one half luminous and the rest of a blue color.

When in the course of her orbit she has passed below the disc of the sun, she is attracted by his rays and great heat, and turns thither her luminous side, on account of the sympathy between light and light. Being thus summoned by the sun's disc and facing upward, her lower half, as it is not luminous, is invisible on account of its likeness to the air. When she is perpendicular to the sun's rays, all her light is confined to her upper surface, and she is then called the new moon. As she moves on, passing by to the east, the effect of the sun upon her relaxes, and the outer edge of the luminous side sheds its light upon the earth in an exceedingly thin line. This is called the second day of the moon. Day by day she is further relieved and turns, and thus are numbered the third, fourth, and following days. On the seventh day, the sun being in the west and the moon in the middle of the firmament between east and west, she is half the extent of the firmament distant from the sun, and therefore half of the luminous side is turned toward the earth.

But when the sun and moon are separated by the entire extent of the firmament, and the moon is in the east with the sun over her in the west, she is completely relieved by her still greater distance from his rays, and so, on the fourteenth day, she is at the full, and her entire disc emits light. On the succeeding days, up to the end of the month, she wanes daily as she turns in her course, being recalled by the sun until she comes under his disc and rays, thus completing the count of the days in the month.

(Burstein [1978] 16)

4.2 Aratos of Soloi (*ca.* −315 to −240) studied Stoicism under Zenon and became court poet for the King of Macedon, where (*ca.* −270) he versified Eudoxos' almost century-old description of the fixed stars; much of what he says about Zeus is Stoic; some of the meteorological meaning of the stars is from Hesiod; the astro-meteorology depends partly on Aristotle.

Phainomena

19–44 [stars and constellations: the Bears]

Stars in their myriads, scattered in ev'ry direction *above us*,
[20] Sweep across heaven for ever in uniform motion unceasingly:
Th'axis however inclines not a whit and forever remains, always
Fixed in its place, and is holding the Earth in entirety, evenly
Balanced at center, besides which it whirls *all* the sky in its circling.
Dual and opposing, two terminal poles are affixed at both ends, of which
[25] One is invisible, always unseen, and the other up North is
Raised over Ocean. Around it *you see*, as they're wheeling in unison,
Bears in a brace, and so therefore they're also called Wagons or Carts. These
Heavenly beasts are maintaining position, with heads in direction of
Other one's thighs, and they always are carried with shoulders held
    foremost,
[30] Inversely turning symmetrical shoulders. The tale may be valid: from
Crete they ascended up heavenward, *granted this boon* by the will of lord
Zeus the majestic, because in his childhood *while dwelling in Crete*, in the
Sweet-smelling vale known as Lúktos, quite near to mount Ida *the
    wooded*,[a] they
Gently deposited Zeus in a cave where they nourished him yearlong, while

[35] Kronos got tricked and deceived by Koúretês (sprung from Diktaí).
One of the bears, she is given an epithet, "Tail of the Dog,"[b] and the
Other one's nickname is "Whirler."[c] Now Whirler's the one that
     Achaían[d] men
Use as a witness to steer by when guiding their ships *on the sea*, but the
Phoínikes cross the wide Ocean relying on Tail of the Dog. Now the
[40] First one is clear in the sky and is readily marked out *for plotting*, the
Whirler in grandeur and brightness appearing as soon as the night does. But
Tail of the Dog is the slighter yet better for sailors *to guide on*, for
She is revolving her whole self in circles much smaller *and poleward*, and
Thus the Sidónians sail in their ships on the course that is straightest.

[[a] "Ida" means wooded; [b] "Kunosoura"; [c] "Helikê"; [d] "Achaían": Greek]

63–70 [Herakles]

Next to the Dragon there circles an image, resembling a man in his
Struggles. But no-one has knowledge to say in the open[a] its nature,
[65] Nor can they tell you the work he's intent on, and all they *can say is*
Call him the "Man on his Knees."[b] Now, in laboring bent upon knees, he
Seems to be crouching, while upward from both of his shoulders, *the might
     of* his
Arms he's uplifting and stretching apart in opposing directions,
Fully a fathom extended; moreover, the tip of his right foot he's
[70] Holding just over the midpoint of tortuous Dragony head.

[[a] "in the open": or, "with conviction"; [b] Herakles; to Mesopotamians this was Gilgamesh victoriously hoisting aloft his opponent: Wagman [1992]]

254–267 [Pleiades]

Hard by the left knee of mighty Perséus[a] do all in a cluster the
[255] Pleíades[b] move in their course, and the space that is holding them all is
Not very large, and besides which each one of them's faint to observe.
     They're
Seven in number accounted in learning and lore of all people,[c] but
Six and no more can be seen by our eyes *when we cast them on high*.
Not even one of the stars has been lost from our knowledge by Zeus since
[260] Ever our oral tradition began, and yet nevertheless this is
Said to be true. Now the names of the seven are given as follows:
First Alkuónê, Merópê is next, Kelainó and Eléktrê and
Then is Sterópê the next, then Teugétê[d] and honored dame Maia. Now
All and alike they are tiny and dimly perceived, yet famous their
[265] Turnings and windings at morning and evening with Zeus as the
     cause, since he

Set them as authorized signals thus marking the start of the summer[c] and
Winter as seasons *in heaven* and onset of season of ploughing.

[[a] ε Per; [b] in Greek, four syllables: Plê-i-ádes, and not considered part of the Bull until
Ptolemy; [c] they are seven in most cultures; [d] in Greek, four syllables: Tê-ugétê; [e] summer
began at the visible morning rising of the Pleiades (Hesiod, *Works and Days* 383, 572),
around the second week of May, and winter, when ploughing started, began at their
visible morning setting (Hesiod, *Works and Days* 384, 615), around the first week of
November: Dicks [1970] 36. The modern use of the equinoxes and solstices to mark not
the middles but the ends of the seasons is mere confusion.]

(Kidd [1997] 73, 75, 77, 91, 93: versified)

4.3 Eratosthenes of Kurênê (*ca.* −285 to −193) was royal tutor to Ptolemy III
("Euergetes") and head of the Library at Alexandria from *ca.* −245. He wrote on a
wide variety of literary and philosophical topics (see Chapters 2 and 5). Some
judge the work here quoted, *Katasterismoi* (*Constellations*) spurious, but even if
so, it is quite likely to be closely based upon Eratosthenes' work. (He uses "right"
and "left" as from our position, inside the sphere.)

Constellations

12 [Leo: like Scorpion and Twins, a Babylonian constellation;
compare Ptolemy, *Syntaxis* 7.5.26]

The Lion is one of the bright constellations. It is believed this sign was honored
by Zeus because of ruling the quadrupeds. Some say the constellation bears
witness to the first labor of Herakles, who, seeking fame, killed this animal
without a weapon, strangling it with his hands. Peisander of Rhodes tells the
story [in his lost two-book epic, *ca.* −600]. Herakles thereafter wore the lion's skin,
as having performed a notable feat. This is the lion he slew at Nemea.

The Lion has three stars on its head [ε, μ, λ]; one on the chest [α: Regulus]; two
below the chest [31, ν]; one bright star on the right foot [ξ]; one on the middle of
the belly [46 or 52]; one under the belly [53]; one on the haunch [δ]; one on the
back knee [σ]; one bright star on the edge of the foot [τ or υ: "lower legs" and "hind
claws"]; two on the neck [γ, ζ]; three on the spine [41, 54, η]; one in the middle of
the tail [θ or ι: "buttocks" or "hind thighs"]; one bright star at the end of the tail
[β]. [Four 4th-magnitude stars are omitted: κ on the nostrils and o, π, and ρ on the left
front leg; in the earliest Greek horoscope known, the coronation-horoscope of Antiochos I
of Kommagênê of −61, there are also 19 stars in Leo: Neugebauer and vanHoesen
[1959] 14.]

Above the Lion's tail seven faint stars are visible in the shape of a triangle;
these are called the Lock of Berenikê Euergetis [Coma Berenices 7, 15, 23, etc.: first
identified as a constellation by the astronomer and mathematician Konon of Samos
ca. −245].

43–44 [Planets, and the "Milk"]

43   On the five stars called "wanderers," because they have a motion peculiar to themselves: they are associated with five gods. The first star is a large one, Phainôn [shining], associated with Zeus. The second star is not large: it is called Phaëthôn [radiant] and is named for Helios. The third star is the star of Ares. It is called Puroeides [fiery]; it is not large and its color is similar to the star in the Eagle [to which only four are assigned]. The fourth star is Phosphoros [light-bearer], white in color, the star of Aphrodite. This is the largest of all stars and is called by two names, Phosphoros and Hesperos. The fifth is the star of Hermes, Stilbôn, small and bright. It was given to Hermes because he ordained the arrangement of the heavens, the positions of the stars, the calculation of the seasons, and the appearance of weather-signs. This star is called Stilbôn [gleaming] because of its appearance.

44   This is one of the heavenly circles, known as the "Milky" [*galaxía*]. Now, it was not possible for the sons of Zeus to share in heavenly honor before one of them had been nursed at the breast of Hera. And so, it is reported, Hermes brought Herakles shortly after his birth and placed him at Hera's breast, and he nursed. When Hera noticed, she pushed Herakles away and thus the remaining milk was spilled, forming the Milky Circle.

(Condos [1997] 125, 167, 109)

4.4 Petosiris dedicated to king Nechepso (*ca.* −140 ± 5) a lengthy astrological poem (15 books are cited) propounding prognoses based on celestial events (Keyser [1994b] 641–642). "Nechepso" was supposedly a pharaoh of the seventh century BCE (based on the historical Necho [*nk3w*] the II, −609 to −593), and "Petosiris" ("Osiridoros") a priest of the late fourth century BCE. Although most foreign astrological ideas came into Greek thought from Mesopotamia, the Greeks themselves believed that astrology was Egyptian (e.g., Diodoros of Sicily 1.81, 2.30–31) – this book may be why. (These two extracts are preserved by the astrologer Hephaistion of Thebes, around 380.)

fr. 7 [signs from eclipses; compare Dorotheos 1.1.4–8]

If during an eclipse the stars dart across into the remnant of the moon, they reveal a battle of enemies according to the zone assigned by the sign; if there is a storm during a solar or lunar eclipse, this foretells mortal diseases.

   If the whole sun is eclipsed in Ram or becomes shadowy like a mirror or as if having the light of the moon without its beams [partial eclipse], it indicates the destruction of great and highly esteemed men in Syria. If in Bull the same sort of thing appears for two days, the destruction will be very intense and fall on many. If it becomes red through the whole day so that the ground reflects a bloody red color, it reveals that there will be a destruction of crops and men and there will be slaughter in many places. If in Twins a blood-red color occurs from sunrise to sunset, it indicates many evil things and the barrenness of crops in Libya and Kilikia. If a blood-red color occurs in Crab, it will stir up trouble for the realm of the Indians and Syrians and Egyptians. [Lion: missing] If in Virgin, it foretells that there will be destruction and rebellion of crowds from the ruler. In Scales the occurrence of the color red indicates that mobs revolt from the ruler of Libya. The sun growing dim as it rises means there will be war and poverty in Libya and Kilikia and Italy and Phoenicia and other settlements in the west. In Scorpion it foretells evil for Libya and Aithiopia; in Archer the sun's rising as a red comet indicates war for a leader of Asia. If it extends its tail towards the north, it predicts distress and barrenness of orchards. An eclipse in Goat-Horn indicates dreadful things for the leader of Egypt and those under him – storms and tumult. In Water-Pourer, the color red foretells that an encamped army will be marshaled by allies of the king from the south. If a comet occurs in Fishes and some other sacred star [comet] is seen out of the zodiac, they say that in Egypt and Syria there will be slaughter and butchery and many incredible, undignified, portentous, things and instability for a long time.

If when the stars in Ram have been seen by day [total solar eclipse], this indicates insurrection and slaughter. [missing: all from Bull to Scorpion] If by day a star is seen in Archer, it foretells war in Egypt and Asia. If in Goat-Horn the sacred star has made a transit, casting a great light, it is a sign of distress and calamity for maritime cities to the south. The stars shining by day in Water-Pourer foretell war in Egypt and alliances for injustice. If in Fishes a star is seen during the day somewhere, it shows distress for those men and the increase of evil and deceits.

[this paragraph may be Hephaistion's summary:] Total eclipses occurring beneath the earth are the causes of earthquakes, as they have usually foretold. The kingly triangle, through Ram, Lion, and Archer is for kings and for the royal court. A solar eclipse means death for tyrants or rulers in the East and Asia, while a lunar eclipse means death for tyrants in Europe and the West. Especially in Water-Pourer and Lion a solar eclipse indicates barrenness and the desiccation of rivers and other bodies of water: in Water-Pourer waters withdraw in the North; in Lion, the waters in Egypt and the South dry up. And the retreats of the Ocean at the risings and settings, the flood-tides and ebb-tides of the Atlantic Sea and Indian Ocean which happen each day-and-night due to the risings and settings of the moon, it [lunar eclipse in Leo] predicts to be roused and profoundly moved. [compare perhaps Poseidonios in Chapter 5]

fr. 10 [signs from comets; fr. 9 is a longer list of comet types: see Keyser [1994b]]

Of the comets, the "horseman" is also called the sacred star of Aphrodite. It is the size of the whole full moon, swift in motion, tail flickering and dissipating at the end. It is carried in the same direction as the *kosmos* [east to west] through the

12 signs of the zodiac. When it appears it brings about falls for kings and tyrants, and the changes in those lands from which it sends forth its tail.

The "sword" is also called the comet of Hermes, and it appears strong and rather green in color and has fairly long beams around it. When it appears towards the East, for the king of the Persians and Egyptians it indicates a plot and potions issuing from governors [satraps: provincial governors in the Seleukid empire]. Appearing to the West it foretells likewise to those living to the West.

The "torch" of Ares, in contrast, is longer and fiery, nearly resembling darting torches. Appearing turned back towards the East it indicates for Persia and Syria drought, lightning, destruction of crops' and burning of royal courts. Turned back towards the South it indicates similar outcomes as well as a throng of shields for those in Libya and Egypt. Turned back towards the West, it reveals the same things for those inhabitants.

The natural comet, called Zeus', appears calmer and flickers only its tail and appears silvery and brandishes its tail far away, so that it does not have an opposite side [?]. It has a male face, as is suitable for a god, whence, where it might rise and set, it indicates good things when Zeus is in Crab or Scorpion or Fishes.

The "discus," called Kronos, round and colored like Kronos, both being golden [electron]. Around its circumference it pours forth its beams. It has only one face and is disposed similarly to every klima. It sets in motion all sorts of wars in every land and the death of a great king. It reduces them in rule and repute.

There is another comet, rosy and large, rather round, which is called "Eilethuia" [goddess of childbirth], having the appearance of a maiden [lunar?], having golden beams in a circle round its head, pleasant in appearance, and similar in color to a mixture of silver and gold. It is a sign of cutting evil men to pieces and a change of circumstance for the better and of the divorce of lovers.

There is another comet, Titan [solar?], which is called Tuphôn, exceedingly painful and fiery, misshapen and slow-moving. Its tail extends rather far behind. Customarily it follows the sun along the boundaries of the arctic pole. Its appearance is the cause of many evils, the destruction of crops and kings in the East and West.

The "planks" and "beards" occur after the others beyond the zodiac in the arctic region.

*(Riess [1891/1893] 341–342, 347–348)

4.5 Hipparchos of Nikaia worked and wrote in Rhodes in the early years of its subjection to Rome. His astronomical observations date his work to −146 to −126. This extract is from his only extant book, a commentary on Aratos' descriptive poem (above) – writing commentaries was a popular activity of scholars from the third century BCE on, and after about 300 CE became the primary mode of scientific writing; this is the earliest scientific commentary to survive. Another book, briefly and simply explaining the origin of the zodiac signs, is found in astrological manuscripts attributed to Hipparchos. See also Chapters 3, 5, 6.

Commentary on Aratos' "Phainomena"

3.5.1–6 [rising times of the fixed stars:
employed especially in horoscopes]

In addition to the theory about co-risings and co-settings of stars, I think it is also useful for us to trace carefully some of the fixed stars separated from one another in sequence by one equinoctial hour. This is useful for us both to reckon accurately the time of night and also to observe carefully the times of lunar eclipses and the many other observations of *astrología*.

The star at the tip of the tail of the Dog [η CMa, magnitude 2.5] lies on the circle through the poles and the tropic points [Crab and Goat-Horn], on the semicircle having the winter tropic point. From this star separated by one hour is the star on the spine of the neck of the Hydra [ζ Hya, magnitude 4] and very nearly the bright star of the Bear on its fore-knees [θ UMa, magnitude 3].

The second hour-long interval is delimited at the beginning of the Lion by the small star of those in the Lion [ν Leo, magnitude 5, at 0° of Lion], which precedes the bright star in the chest [α Leo, magnitude 1, at 2½° of Lion] by a little more than a "cubit" [2° or 2½°: see Toomer [1984b] 322, n.5]. This starlet does not precede by as much as a *daktul* [5'] the circle through the poles which delimits the second hour-long interval.

The third hour-long interval is delimited at about the middle of the Lion by the more southerly of the two stars [θ Leo, magnitude 3, at 16⅓° of Lion] lying on both sides of the bright star lying on the loins [δ Leo, magnitude 2].

The fourth hour-long interval is delimited at about the beginning of the Virgin by the star lying on the right angle of the triangle under the Mixing Bowl [δ Crt, magnitude 4, at 0° of Virgin]; and the bright star on the loins of the Great Bear is less than 1/20 part of an hour short of the circle delimiting the fourth hour [δ UMa, magnitude 3].

The fifth hour-long interval is delimited at about the middle of the Virgin by no star precisely, but the Vintage-harbinger [ε Vir, magnitude 3.5, at 12⅙° of Virgin] lies more than 1/10 of an hour short thereof, as does the bright star on the right shoulder of the Virgin [δ Vir at 14⅓° of Virgin, or γ Vir at 13⅙° of Virgin: both magnitude 3].

The sixth hour-long interval is delimited at the circle through the equinoctial points, which is very close to the star of Centaur which is one of the bright ones in the southern part of the thursos which are mutually separated by about half a "cubit" [1°], which lies at about the middle of the chest of Centaur [μ, ν, and φ Cen, all magnitude 4.5; φ Cen at 15⅙° of Scales is meant], and the midmost of the stars of Boötes, left foot [τ Boo, magnitude 4] falls short by about 1/20 of an hour from the circle delimiting the sixth hour.

[Hipparchos continues for the other 18 hours]

*(Manitius [1894] 270, 272)

4.6 Imbrasios of Ephesos, about whom nothing else is known, wrote this work on forecasting medical prognoses sometime in the first century BCE or first century CE (25 ± 75?) (see Weinstock [1948]). Similar "lunaria" are attributed to Serapion of Antiocheia, an astrologer from about –80 and to Melampous (around –250; see Chapter 12).

Predictions about the Sick

2 [lunar effects on illness]

Before all you must look at the moon in its motion (adding or subtracting by numbers not by lights as some have wrongly done [i.e., "look up the moon in tables, as we do for horoscopes"]), so when someone is abed look to see in which of the twelve-fold signs the moon is beginning to move with its proper numbers, coming to the same degree (the point of same degree is measured at the diameter), it will provide the best indicator. And especially if there is a node, when the moon leaves the node towards the full moon and is proceeding towards the greater quantity of its course, and is waning at about 180°. And if someone becomes ill when the moon is waning, the disease is forecast until the point of same degree, and after the point of same degree the moon will turn the disease to health.

Consider also the motions or aspects of the stars. For if when someone becomes ill the moon is configurate [i.e., in trine or quartile or sextile] and waxing, it will increase the disease and cause danger. If the moon is increasing by lights and by numbers around opposition with Ares, it causes danger unalterably. If the moon should happen to be somehow configurate with light-bringing [*phaëthôn*] Zeus, it will cause the diseases to be risk-free. If in quartile it's forecasting, it will recycle them safely until opposition. And if she is configurate with blazing Kronos [unusually "*phlegethôn*" not "*phainôn*"], and decreasing by numbers, she will bring about unstable and dangerous diseases around 90°. And if she is leaving a node towards the lesser part, around 180°, she will cause death unfailingly. And if the sun or opposition . . . [there is a gap in the text]. And if she is configurate with Aphrodite or Hermes at the beginning of illness, and decreasing in numbers or even abating in light, the disease will be reduced around 90° and she's predicting it will turn to health. If she's leaving a node while waning, she establishes health at 180°.

You will need to consider the rest of the stars, how they are placed in signs. As the moon proceeds through the 12 signs, the causes of diseases are recognized there. Before all one must know that if anyone is careless about working the pebbles [using the abacus], nothing will turn out correctly for him, and if he fails the method is not at fault. And the numbers are to be observed as for a birth, for from these the truth will be known.

13 [moon in Water-Pourer]

If someone gets ill when the moon is in Water-Pourer, and is waxing in numbers and in lights, and Kronos is with her in quartile or opposition, the origin of the disease will be from work or sleeplessness or travel. The disease will be irregular in intensity, and he will be safe until opposition, especially if a benefic star is in aspect to the moon.

If someone gets ill when the moon is in Water-Pourer, and is waning in numbers and in lights, and Ares is with her in opposition or quartile, the disease will be from a pre-existing cause, swollen groin, the shin, or trouble in the privates. Fever will be especially sharp, and there will be thirsts and desire for cold and frequent vomitings.

If the moon is rolling along at the full and is configurate, or Ares is added, until the quartile she will disturb the mind. Bloodlettings will be useful, as will be both not reducing his desire for cold and sleeping in the light.

If she is rolling along toward waxing and is in quartile [with whom?] and the disease is not at all lessening, the patient will die around opposition, but if given the same situation Zeus or Aphrodite or even both are in aspect and the moon herself was rolling along toward waning when he fell sick, he will be safe after being in danger until opposition.

If she is waning in light and in numbers, and is configurate with Kronos or Hermes added, the sufferings will be the same, but he will suffer triple fluxes around the limbs, and after the sufferings have become chronic he will turn dropsical. If she is moving from mid-position toward waning, he will die by the same suffering. If the moon runs the mid course at the time of falling sick, the pre-existing state will be the cause of illness, but having endured a long time in the disease, he is saved.

*(Kühn v. 19 [1833] 531–533, 565–567)

4.7 Dorotheos of Sidon (*ca.* 25 to 75) wrote an astrological poem in Homeric style for his son Hermes: an augmented paraphrase in Arabic provides most of the extracts, and there are a few Greek bits (from which these verses are rendered).

Judgments from the Stars about Nativities

1.1

1 I shall relate to you, my son, and I shall explain to you so that you may depend on and be confident in your heart about what I shall show you of my work and words about the stars which indicate for men what will pertain to them from the time of a child's birth till his leaving the world, if God wills. I have traveled, my son, to many places, and I have seen wondrous things in Egypt and in Babylon on the Euphrates. I collected the best sayings from my foremost predecessors, like the bees which gather honey from the trees and all kinds of plants, for from them comes good honey.

2–3 [planetary rulers]

Curly-haired Ram, and the Lion *so mighty*, and Drawer of Bow do by
Day to Helíos belong, and to Zeus in the Night-time, each getting the rule
   in suc-
cession, while Kronos the awefull's allotted the third share of these three.
Taurus and Virgin and Horn of the Goat: over them there doth rule in the
Day-time the Lady foam-born[a], and by night it is goddess Selénê;
Third after these comes the lord of the wars, who is wielding his sceptre,
   now
Taking the boy-child of Maia[b] as helper in sign of the Virgin.
Ruling in Twins and in Libra and chilling Aquarius, during the
Day-time is Phainôn, but nightly it's he who did slaughter old Argos, while
Kronos receives as his portion the rank which is third after these two.
Kúpris' allotment in day-time is *feminine*[c] Scorpion, Crab and the
Last-coming Fishes, and during the night it's Puróeis who gets it, and
Darting-eyed queenly Selénê holds sway after these do.

[a "foam-born" is Aphroditê; b Hermes was born of Maia and slew Argos; c for "feminine"
see Ptolemy, *Tetrabiblos* 1.12]

8 [mansions; upliftings and abasements] I know, my son, to which of the seven
planets each sign belongs, and I know which of the signs are oblique in rising and
straight in rising. Know the houses of the planets: Crab is the house of the Moon,
Lion the house of the Sun, Goat-Horn and Water-Pourer the houses of Kronos,
Archer and Fishes the houses of Zeus, Ram and Scorpion the houses of Ares,
Bull and Scales the houses of Aphrodite, and Twins and Virgin the houses of
Hermes.

1.1.9–1.2.2 [on "upliftings" or "exaltations" and "abasements" or
"humiliations" see Theon of Smurna, *Mathematics* 3.12 (northward
position of planet) and Vettius Valens 2.19, 3.4]

Such are the houses, and Kronos rejoices the more in Aquarius,
Zeus in the Archer is gladdened, and Ares is happy in Scorpion,
Kúpris in Taurus is glad in her mind, and now Hermes is pleased with the
Virginal: one of the mansions for each of the heavenly lights t'have.
Starry upliftings: Helíos at nineteenth part of the Ram, while
Luna achieves her most height at the third of the parts of the Bull; but
Kronos in twenty and first of the Scales, and the Bearer of Aigis[a] in
Fifteen degrees of the Crab, and lord Ares at fourth of the Goat-Horn's
Hebdomads; three times nine in the Fishes is greatest for Kúpris.
Each has abasement, and they are exactly on opposite sides of these.

[a Zeus is the bearer of the fearsome Aigis.]

1.6 The power of the seven planets [contrast Ptolemy, *Tetrabiblos* 1.4–5:
Sun and Ares are hot and dry; Kronos cool and dry; moon moist; Zeus
and Aphrodite temperate; and Hermes mutably wet and dry]

Now I will tell you the power of the seven planets. Each planet is benefic when it is in its house or in its triplicity or its exaltation so that what it indicates of good is strong and increases. The same is also true for a malefic, if it is in its own place, its evil becomes lighter and decreases. Say how much Kronos harms one who is born by day and Ares one who is born at night – especially if Ares is in a feminine sign and Kronos in a masculine sign. It is best if they are in one of their dignities. And as for the planets, if they are under the rays of the Sun towards the west, their power disappears and they have no power. If it is retrograde in motion, there is difficulty and misfortune in births and other things.

1.24 [fortune and property, horoscopes 2, 3, and 4]

This nativity was diurnal, and Ram was becoming visible at that hour in the east from out of the depths of the sea, and the lord of the sun's triplicity was Kronos and Hermes (Figure 4.1). Kronos was in what follows the cardine of the west and Hermes in what follows the cardine of Midheaven, which is the place of fortune, so that the native should be wealthy, rich, powerful in business affairs, great in property, seizing eminence and fortune and increasing in both. [26 January 13: perhaps M. Antonius Pallas, d. 62, or C. Iulius Callistus, d. 52?].

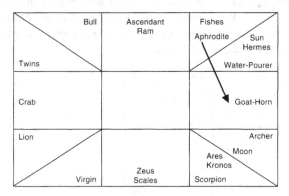

13 ce, January 26
(Aphrodite on wrong side of Sun)

Figure 4.1

This nativity was diurnal (Figure 4.2), and the first lord of the [Sun's] triplicity was the Sun, the second Zeus, and both of these are in cardines in their own exaltations, so that the newly born will be praised with the praise of kings and nobles and wealthy men. Because Kronos is the third lord of the triplicity and is cadent from a cardine and therefore he is in Zeus' house and in aspect of trine, so for that reason he will be praised with the praise of kings. [30 March 22: perhaps C. Licinius Mucianus, cos. suff. *ca.* 64, or M. Antonius Primus?].

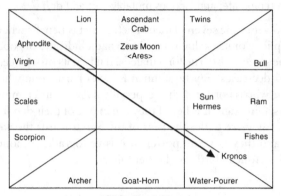

22 CE, March 30
(Ares omitted: Aphrodite placed in opposition)

Figure 4.2

Another nativity whose ascendant rising from the earth was Scorpion, and the nativity is nocturnal, and the positions of the planets are according to what is in this diagram (Figure 4.3). The lord of the moon's triplicity is Ares and Aphrodite and the moon. Because the three of them are in cardines this man is mighty in eminence, powerful in leadership so that crowns of gold and silver are placed on him and he is praised. [2 April 36: perhaps (a) C. Iulius Antiochus Epiphanes of Kommagênê, son of Antiochos IV of Kommagênê; (b) Sex. Iulius Frontinus, cos. 72 or 73 and writer on aqueducts; (c) C. Calpetanus Rantius Quirinalis Valerius Festus, cos. 71; or (d) M. Ulpius Traianus, cos. 70 and father of the emperor?].

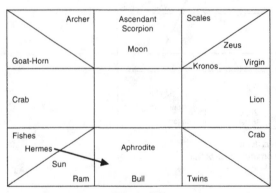

36 CE, April 02
(Hermes on wrong side of Sun)

Figure 4.3

(paraphrase: Pingree [1976] 161–162, 164–165, 185–187)
*(verses: Pingree [1976] 323–324)

4.8 Geminus produced his introductory textbook of astronomy about mid-first century CE (perhaps in Rhodes).

Phainomena

17.1–25 [arguments against astrology: weather signs]

1 The theory concerning weather-signs amongst laymen proffers the absurd opinion that alterations in the weather [lit.: "air"] come about from the risings and settings of the stars. Mathematics and physics holds another opinion.

2 First, one must grasp that warning signs of rain-storms and winds occur around the earth, and do not reach very high. For they are variable and irregular exhalations from the earth, so that they cannot reach the sphere of fixed stars, and even the clouds don't quite reach 10 stades high.

[3–11: Sacrifices left on high mountains are unaffected by weather, and one can look down from there on the clouds; lists of weather signs are simply empirical not predictive.]

12 Having observed from the beginning and having set in order the astronomical calendar, they determined the parts of the circle of the zodiac, in which changes in weather generally occur; they then inquired which of the stars were rising or setting during such times, and they employed their risings and settings as signs to predict the changes of weather.

[13–14: That is why we use visible phases and not the true phases; and when Hesiod, *Works and Days* 383–384 speaks of the effects of the Pleiades he does not mean it literally.]

15 Whether stars are fiery or *aitherial*, as some suppose, they all share the same substance and power and have no sympathy (*sumpatheia*) with events on earth.

16 The entire earth has the ratio of center-point with respect to the sphere of fixed stars, and no efflux or emanation from the fixed stars penetrates to the earth. How can one suppose that those stars, from which no force radiates towards us, are causes of rain-storms and winds?

17 Now from the sun and the moon a force does penetrate to the earth during their changes when they are waxing and waning. Thus it is reasonable that there is sympathy (*sumpatheia*) in their cases according to each one's power. But the risings and settings of the fixed stars have the value of a sign, just as we have already said.

18 Thus we must not surmise that the same weather-signs are generated by the same stars, but according to the different *klimata* [latitudes], the risings and settings of the stars change their significance.

19 And it is necessary to have proper signs for the change of the weather in each different "horizon" [region]. For the same calendar cannot be correct in Rome and in Pontos and in Rhodes and in Alexandria, but it is necessary that the observations in different "horizons" differ, and that in each city different stars be taken to establish the weather-signs.

20 From this it is evident that the risings and settings of the stars have not naturally generated the effects on the weather, but in each "horizon" different observations have occurred as well as different changes of the weather.

21 Thus all weather-signs conveyed in the calendars are not always correct. Sometimes they are wholly wrong, and the greatest winter storms arise when the risings and settings signify fair weather. Other times fair weather occurs in town, rain-storms in the country.

22 Often the prediction is valid three or four days after the rising or setting of a star; other times the event anticipated the sign by four days. Thus those who failed in the predictions by those weather-signs have the excuse that they anticipated the weather-sign or the event happened late.

23 From all this it is clear that only generalities based on weather-signs have been inscribed in the calendars, established not by some skill or necessity, but inscribed from continuous observation. Consequently, errors are frequent. Thus one must not blame astronomers if they fail in weather-signs.

24 But if someone errs in predicting an eclipse or the rising of a star, one would rightly approve both the pursuit and whoever administers blame. For all things established by skill ought to have an infallible outcome.

25 Matters of weather-signs carry neither complete approval if they succeed, nor censure if they fail. For this part of astronomy is empirical and is not worthy of rebuke.

[Geminus discusses Sirius: 17.26–45]

*(Aujac [1975] 83, 85–88)

4.9 Balbillos (Ti. Claudius Balbillus) served as tribune in the invasion of Britain (43), and as prefect of Egypt (55–59); he was the astrologer royal of Claudius, Nero, and perhaps Vespasian (i.e., *ca.* 60 ± 20) (see Cramer [1954] 112–140). This extract from his chapter on "method concerning the length of life from starter and destroyer" is preserved by Hephaistion of Thebes (*ca.* 380) who tells us it gives horoscopes computed for a father (–71 to –21) and son (–42 to 28).

Astrologoumena [concerning the length of life from starter and destroyer]

The sign rising before the "starter" is called "destructive." So if the destroyer star is found in the same sign (even if in aspect with a benefic star) it always destroys. However, if the destroyer is found in a different sign, if Kronos is the destroyer, when Aphrodite is within 8° either by figure or in *epanaphora*, or when Zeus is within 12°, by uniting with the destroyer they prevent the crisis. But if Ares is the destroyer, unless there are two benefic stars, they don't release the destroyer. When Zeus and Aphrodite fall in a Center, the first adds 12 years since the birth is diurnal, and the other adds 8 since the birth is nocturnal.

He also says that when the sun is starter and is found in the first half of Ram, Bull, Lion, or Archer, it adds the years of the sign's ascension-time [in hours]; but

98

if it's found in the second half of the same signs, only its cycle adds years, i.e., 19 [the eclipse cycle]. The same thing is true, when the moon is starter, for the same signs plus Crab [the house of the moon]. He also says that there are 4 destroyers: sun, moon, Kronos, Ares.

*(Boudreaux [1922] 235–236)

He also gives the following example for starter and destroyer:

Table 4.1 Balbillos, Horoscope 1

Horoskopos	Goat-Horn	Sun 9°ᵃ Kronos 5° Aphrodite 11° Ares 12°
2	Water-Pourer	
3	Fishes	Zeus 20° [or 8° ?]
4	Ram	ascending node 4° [wrongly copied in 'Bull']
5	Bull	
6	Twins	
7	Crab	setting 9°
8	Lion	
9	Virgin	
10	Scales	midheaven 3°ᵇ
11	Scorpion	descending node 4°
12	Archer	Hermes 17° Moon 19°

ᵃ [therefore the Lot of Fortune will be wherever the moon is].
ᵇ [reckoned from the Lot of Fortune, not from the Horoskopos].

And he says that the sun is the starter of the birth, whereas Ares could not become destroyer, because Aphrodite is rising after the sun within 8°, although Ares is in his own exaltation [Goat-Horn]. And so one comes to the destructive sign, which is the one rising before the starter, i.e., Archer. And finding there the moon [in Archer 19°], he says that it becomes the destroyer. Taking the orbit of the sun up to the quartile of the degree of the moon, i.e., Fishes 19, then, he says, comes the destruction [so he is predicting 70 degrees, i.e., 70 years of life]. Zeus, being in Fishes, 20°, was not able to assist, because the moon is located in the destructive sign [Archer; the birthdate is 27 December –42].

And again he says that for another birth:

Table 4.2 Balbillos, Horoscope 2

Horoskopos	Twins	
2	Crab	
3	Lion	
4	Virgin	Zeus 14°
5	Scales	
6	Scorpion	Moon 4°
7	Archer	
8	Goat-Horn	Kronos 4° Sun 22°
9	Water-Pourer	Ares 14° Hermes 12°
10	Fishes	Aphrodite 25°
11	Ram	
12	Bull	

And since the luminaries [sun and moon] did not fall in a Center, he went to the *epanaphorai*, and he did not take the Horoskopos as starter, nor the sun, which was in *epanaphora* of the setting point, but he took Kronos in Goat-Horn as starter. [Hephaistion: "And this, I think, because Kronos had the greater claim in the birth and was his own house" (Goat-Horn).] And he says that Ares in Water-Pourer is the destroyer and he computed the distance from Ram to Ares [i.e., from Ram 4° at the quartile of Kronos, so 50°, meaning 50 years], and so long, he said, would be the length of life. [the birthdate is 21 January –71]

(Neugebauer and vanHoesen [1959] 77–78)

4.10 Thessalos of Tralles (writing 65 ± 15) "founded" the Methodist medical sect (see Chapter 11) and probably composed this work of "astro-botany" dedicated to Claudius or Nero (unless it was by his student Krinas of Massalia: compare Pliny 29.9).

Remedies

Praeface 27–28

[He followed Nechepso's book of remedies "using stones and plants based on the zodiac," but failed; in a séance Asklepios' spirit explains:]

> King Nechepso, although a most sound-minded man and adorned with all virtues, obtained nothing of what you seek to learn from my divine voice; relying on his own talent he worked out the *sumpatheias* of stones and plants, but he didn't know the seasons and places where the plants must be gathered. For "all fruits of the season" [*Odyssey* 9.131] wax and wane in the effluences of the stars, since that true, subtle, and divine *pneuma* pervades every place and especially those where the effluences of the stars were at the birth of the *kosmos*.

["Asklepios" then gives the example of hemlock, *Conium maculatum*; for plants gathered at certain times, see Dioskourides, in Section 11.13.]

Book 1, chapter: Twins

Twins, 3rd sign: from the twenty-fifth of Pachôn, which is Artemisios 20, and according to the Roman calendar from the 19th of May. The third herb is holy vervain [*Verbena supina*].

This plant's juice is recovered at that time, then a drug is prepared active against cancers, callosities, and what are called "figs." It is prepared like this:

12 drachmas of saffron
8 drachmas of dried roses

14 drachmas of Pontic melilot [*Trigonella graeca*]
8 drachmas of chopped alum

Grind them all together and pour in 4 *sextarii* [total 1.8 l] of the plant-juice, and boil in a new pot until it has the consistency of honey, and use. It heals all bodily ills.

Book 2, chapter: Kronos

The herb of Kronos is called "evergreen" [*Sempervivum tectorum*, house-leek]. It is extracted only by boiling and there's no need to explain the manner of its extraction to you, since it is necessary that it have the consistency of honey. Its power is active for gout: there are two kinds of gout, hot and cold. The cold is recognized by the patient enjoying the application of heat, and the hot kind by the opposite effect. When you've diagnosed the suffering, employ the juice for hot gout mixed with rose oil and rubbed on the feet; and for the cold, do the same mixed with Syrian perfume [compare Dioskourides 1.1–68 on aromatics: from Syria: 1.4 *kuperos*, 1.7–9 *nardos*, 1.16 *kostos*, 1.20 *aspalathos*, 1.27 *krokomagma*, 1.31 *elaiomeli*, and 1.66 *sturax*]. If the condition is severe, for cold power add *castoreum* [beaver musk: widely used] and for hot power add the juice of wormwood [*Artemisia arborescens*, compare Disokourides 3.113].

*(Friedrich [1968] 55, 58, 93–94, 215)

4.11 Pitenius Dozens of horoscopes by unknown astrologers are preserved on papyrus in the sands of Egypt. Titus Pitenius computed this one for Hermon (born on 31st March 81), probably in the early second century CE (the name Pitenius is unparalleled, but members of the gens Petillius are known around this time). He seems to have used Ram 5° as his vernal equinox; some of his terminology is unique; on the decans, see Neugebauer [1955].

[Hermon's horoscope]

The Egyptian men of old who had faithfully studied the heavenly bodies and had learned the motions of the seven gods, compiled and arranged everything in perpetual tables [compare Ptolemy, *Syntaxis* 9.2: chapter 3.19] and generously left to us their knowledge of these things. From these I have accurately calculated and arranged for each one of the seven according to degree and minute, aspect and phase, and, simply, not to waste time in enumerating each item, whatever concerns its investigation. For thus the way of astrological prediction is made straight, unambiguous, that is, consistent. Farewell, dearest Hermon.

Time of the [equinoctial] "tropic" of the third year of the Divine Titus, the sixth day of Pharmouthi, the third hour of the night; as the Romans reckon, the Kalends of April; and according to the old [365-day] calendar, Pachon the first to the second.

Accordingly, the sun, the mightiest and ruler of all, moving from the spring equinox had attained in Ram fourteen degrees and six minutes, that is, the tenth part of a degree; in the sign of Ares; in the terms of Hermes; in the *stoicheion* [degree?] of Zeus; in exaltation in a male and northerly sign; shining upon the Flank of Ram; in the second decan, called Sentachor [Egyptian: ḫntw hry, "lower jugs"]; the dodekatemorion was shining upon the first joint of the tail of Scorpion [ε Sco].

And the divine and light-bringing moon, waxing in crescent had advanced in Bull thirteen degrees, and a thousandth part of a degree; in the sign of Aphrodite; in its own exaltation; in the terms of Hermes; in a female and solid sign; like gold; mounting the back of Bull; in the second decan called Arôth [Egyptian: ᶜryt]; its dodekatemorion again was shining on about the same place in Scorpion.

And Phainôn, star of Kronos, had completed six degrees in Fishes, lacking a sixtieth of a degree; in the sign of Zeus; in the terms and the exaltation of Aphrodite, at its morning rising; descending from the Swallow-fish [the northern of the two fish]; of magnitude four thirds [modern computation gives 1.3].

Phaëthôn, the star of Zeus, traversing its exaltation, in Crab, had attained six degrees and ten sixtieths of the third order (which are one twenty-one-thousand-six-hundredth part of a degree); in the sign of the moon; in the terms and depression of Ares; at its second station; two *daktuls* [10′] north of the Bright Star at the Back [ζ Can?]; of the magnitude three halves [modern computation gives −1.8] and commanding.

And Puroeis, the star of Ares, has mounted in Water-Pourer sixteen degrees and a twentieth; and has mounted the sign of Kronos; the triangle of Hermes; the term of Zeus; the Star in the Cloak, called Ganymede, homonymous with the whole constellation, far to the east.

And Phosphoros, the star of Aphrodite, had completed in Fishes sixteen degrees and four minutes, which is the fifteenth part of a degree; in the sign of Zeus; in its own exaltation; rising at dawn; at the Southern Fish; like crystal; in the terms of Hermes; distant two lunar diameters [1°] from the Star in the Connecting Cords [d Psc?].

And Stilbôn, the star of Hermes, had mounted in Ram ten full degrees; in perigee [Ptolemy, *Syntaxis* 9.7 gives Ram 10° as the perigee of the eccenter]; having completed its phase [as morning star] before the seventh day of the month; wherefore it will dominate the theme of the nativity.

And the rudder of them all, the Horoskopos, has cut off eighteen degrees in Scorpion; the terms of Hermes; the sign of Ares; the triangle of the moon; the decan Thoumouth [Egyptian: ṭms n ḫntt].

And the meridian at right angles to this had struck the Back of the Lion.

And the Lot of Fortune, counting inversely [the moon is 29° from the sun], will be in the sign [i.e., house: Archer] of Zeus and in its triangle; this Lot some will ignorantly assign to Scales [by subtracting instead of adding]; thus the Lot would be back in the house of Aphrodite.

Good luck! Titus Pitenius computed it as is set forth.

1   Hermes
2   Sun [i.e., both in Bull]
1   Kronos
2   Aphrodite [i.e., both in Fishes]
in order that you may not make a mistake in the arrangement of the pairs [?]

Computed in Hermopolis [modern El-Ashmunein], where the horizon has the ratio [of longest to shortest day] seven to five. The time of pregnancy: 276 days. With good fortune.

(Neugebauer and vanHoesen [1959] 23–24)

4.12 Manethon (the name suggests Egyptian wisdom), whose own horoscope dates his birth to 27 or 28 May 80, wrote an astrology text in epic verse, around 120 ± 20. (The books labeled 2, 3, and 6 in the manuscripts were his 1, 2, and 3 – the numbers we use; the books labeled 1, 4, and 5 are later additions.)

Apotelesmatika

1(2).399–445 [sun and moon signs]

But it's the *súnkrasis* (that is, commixture) which every truth in the
[400] Life of a mortal determines, the stars with their shining supporting in
Aspect the various places above which they found their position, and
That which they do when appearing conjoint with Helíos I'll tell now.
All of them greatly rejoice when they're found at their rising at morning, which
Happens because they are happy in places Helíos *made* royal, and
[405] When they're approaching the time of their rise (like a time of
    renewal), it's
Then they appear very strong and accomplish perfection for mortals;
But their advance *on their paths* is more slow at the hour of the evening, or
Late after noon, when they cower beneath his unstable effulgence, and,
Being then strengthless and impotent, dim and make blunter his strength.
[410] Foremost is Phainôn who joined with Helíos in any and each of the
Signs of the zodiac, whether benefic the sign or malefic, is
Found to be stronger in daytime and *known* to be harmful at nighttime, for
Phaínôn, he brings it about that *all* fathers are famous and prosper, while
Mothers receive an allotment of swift and parturient death, and he
[415] Shrinks and diminishes th'ancestral household and what they possess,
    yet
Not so their poverty's utter and lacking all means of support.
[417–435: Phaëthôn, Ares, Aphrodite, and Helíos]
Thus far the stars have causation, alone in the sky with Helíos when
Joinèd, and similar deeds they decree when appearing on opposite

Sides of Helíos. The moon in appearing together with planets by
Fatal conjunctions, in lives that are wretched and woeful, accomplishes
[440] Deeds through ambiguous effluence: now I will tell you about that.
Luna encounters the star that is Phaínôn's and forms a conjunction: then
Stronger she grows in augmenting her rays and for mothers it's horrible
Illness and spasms with chilling she offers, and adds on diseases,
Keeping all ailments from making much progress, and thus they become
     most
[445] Fearful destructive diseases.

[Phaëthôn, Ares, and Aphrodite follow: lines 446–464]

2(3).399–428 [length of life; Titan and Huperíon both mean the sun]

Since in my mind and my heart is the urge in the first place to sing about
[400] Matters *decisive and* crucial, pay heed: I'll explain very clearly
How you may tell to each one at their birth who's the ruler of anyone's
Length or duration of life, and moreover tell when it will finish.
Seemingly not for all mortals does *longness of life* in the same way get
Found, nor alone from the locus is number of years to be taken:
[405] Various quadrants are vital in reckoning different birth-times.
Birthings *of people* in which you will find that lord Titan at dawn has
Made his appearance while entering Center, then counting degrees of the
Heaven beginning with that one, enumerate years of his lifetime; for
Birthings in nighttime enumerate starting from Lunar degree. When-
[410] ever the sun and Selénê away from a Center have entered the
Cadent, or running ahead have arrived at descending degrees in their
Course round the pole, let the numbering start with whichever of stars is the
Star that is ruling the birth and possesses the maximum power.
Should you perceive that the cadent is entered by powerful star, then
[415] Reckon *and tell* from that moment the quadrant that's vital for
     counting.
Further, whenever you're seeking to find the commencement of life, you'll
Closely inspect its duration of rising, and from the horizon you'll
Count the degrees of ascent to be sure that you know ev'ry one of them:
Thus you'll impart from the count of degrees the whole number *of years,*
     and how
[420] Much of a lifetime of mortals that's wretched, by Fate has been taken.
Middle degrees of this path which determine the life-course of men, you'll
Sharply inspect to be sure there's no ray *from a star,* either quartile or
Even a star who's opponent, to Krónos, or Ares, destroyers, or
Further, conjunction as well may be trimming his life to a shortness.
[425] Often has bright Huperíon, far-shining, abstracted from *one who was*
Born in the nighttime the breath of his life by *the shine of* those rays. The

104

Side of the quartile *in heaven* determines each quadrant that's vital, for
Thus to the Fates was this maximal finish of mortals made pleasing.

[end of book 2]

3(6).1–34 [children]

Come and approach me, dear Muse who composed and arranged all my
    earlier
Clear-sounding song, and *lord* Phoíbos[a] and Hermes whose sandals are
    golden, and
Grant if you please to my ode I now sing your desirable witness:
Burdened my soul is with wretched fatigue *and I'm wearied*, but still I will
[5] Hasten my efforts until I attain the far limits of song; for
Failing to follow it through to the end would be wholly disgraceful.
Granted to me to be sung in this last book are matters like these ones:
Raising and nourishing infants, and pain from the loss of your children,
    and
Which ones the parents themselves will reject and expel by their own
    hands to
[10] Ending destructive and fated but hastened by *deeds of* the parents, and
Marriage belovèd I'll tell of in song, and of brethren belovèd, and
Children in lineage running and which banes on each of us mortals is
Fated to fall *and afflict us* by stars as they march in their motion, and
Skills and the practice of trades I will sing of, and what sort of deeds the
[15] Stars are allotting to misery-suffering mortals *on earth*, or
How by misfortune men slip from a lifestyle that's filled up with wealth and,
Worn out by poverty, suffer in pain and distress *all their days*, and
However could one explain the begetting of someone in slavery.
Destiny evil is snatching the onset of nurture and raising from
[20] Children on whom at the moment they leapt from the womb
    of their mother,
Ares and Phaínôn have *both of them* entered the sign that's ascendant,
Gazing in aspect on bright shining Moon who is drawing to setting, and
Neither Phosphóros nor Zeus (the two granters of life) is in aspect to
Place of Selénê, and neither's in aspect to sign that's ascendant.
[25] Thus is it also for all who at birth by the wasting of Phaínôn in
Sign that's ascending are greeted; the same goes for those on whom Ares is
Gazing at setting, and neither the lively nor holy *fair* light of the
Moon is in to glimmer of Zeus the bestower of life. And
Even the more so whenever the fair-horned Puróeis directly is
[30] Gazing across at the rays of the Moon, or if Phaínôn is standing to
Right of Selénê, and also on side of the quadrant that's rising, and
Holding a place that's afar off from glimmering star, Huperíon, then

Truly those bursting in that hour the bonds of the womb and its labors will
[35] Straightaway turn out as spoil and as booty for birds *of the air.*

[[a] Phoibos is Apollo, god of sun, prophecy, and disease]

*(Koechly [1858] 16–18, 34–37)

3(6).738–750 [Manethon's nativity]

Let me turn in my verse to a *topic and* goal-post that's novel, and
Seek in my memory which were the stars *in the sky* at my birth, in
[740] Which of the seasons and underneath which of the stars did it happen
    that
Often-entreated Eileíthuia made me appear from the womb, with the
Purpose of teaching and proving to people of ages *to come* in the
Future that Fate as a gift to me granted the teaching of all of the
Knowledge of stars *in the sky,* both their wisdom and beautiful poetry. The
[745] Time when Helíos was found in the Twins, and together with that
    one was
Kupris the fair, and belovèd Phaëthôn and Hermes the golden, and
Pourer of Water was holding Selênê and Phaínôn *at once* in that moment, and
Ares was found in Crab, who has manifold feet, while around about
Middle of Heaven[c] the Centaur was turning while trailing his weapon.
[750] Thus did the Fates my nativity witness, determine and settle.

[[a] Kupris is Aphrodite; [b] Centaur is the same longitude as Libra: cp. Hipparchos, above; [c] the
location of the sun and the Midheaven shows that it was two hours after sunset.]

(Neugebauer and vanHoesen [1959] 92: versified)

4.13 Ptolemy (Claudius Ptolemaeus) of Alexandria (*ca.* 100 to *ca.* 175) wrote
on a wide variety of "mathematical" topics (see Chapters 2–8). In the *Predictions*
or *Tetrabiblos* ("four-book") he explains how astrology, which is simply applied
astronomy, works.

Tetrabiblos

1.1–2 [how astrology works]

1 Of the means of prediction through astronomy, Suros, two are the most
important and valid. One, which is first both in order and in effectiveness, is that
whereby we apprehend the aspects of the movements of sun, moon, and stars in
relation to each other and to the earth, at any given time; the second is that in
which by means of the natural character of these aspects themselves we
investigate the changes which they bring about in that which they enclose. The
first of these, which has its own science [*theôría*: "contemplation"], desirable in

itself even though it does not attain the result given by its combination with the second, has been expounded to you as best we could by the method of demonstration in its own treatise [the *Syntaxis*]. We shall now give an account of the second and less self-sufficient method in a properly philosophical way, so that a lover of truth might never compare its perceptions with the sureness of the first, unvarying science, ascribing to it the weakness and unpredictability of material qualities found in individual things, nor yet refrain from such investigation as is within the bounds of possibility, when it is so evident that most events of a general nature have as causes the Enclosing Heaven [Ptolemy refers to the entire cosmic environment as the "enclosing," *peri-echon*]. But since everything that is hard to attain is easily assailed by most people, and of the two before-mentioned disciplines, the allegations against the one could be made only by the blind, while there are specious grounds for those leveled at the second [until well into the Christian period, all Greeks used one word for both: *astrología*] (for its difficulty in parts has made them think it completely incomprehensible, or the difficulty of escaping what is known [i.e., fate] has disparaged even its goal as useless) we shall try to examine briefly the measure of both the possibility and the usefulness of such prognosis before the detailed instruction on the subject. First as to its possibility.

2   A few things would make it apparent to all that a certain power emanating from the eternal aithereal substance is dispensed onto, and permeates, the whole region about the earth, which throughout is subject to change, since, of the primary sub-lunar elements, fire and air are enclosed and changed by the motions in the aither, and in turn enclose and change all else, earth and water and the plants and animals therein. For the sun, together with the Enclosing, is always in some way affecting everything on the earth, not only by the changes at the annual seasons bringing about the generation of animals, the fruiting of plants and the flowing of waters and the changes of bodies, but also by its daily revolutions furnishing heat and moisture and dryness and cold [the four Aristotelian "principles": see Chapter 9], in regular order and in correspondence with its positions relative to the zenith. The moon, too, as the heavenly body nearest the earth, bestows her effluence [*aporrhoiai*] most abundantly upon mundane things, for most animate and inanimate things are sympathic and co-mutate with her; the streams of rivers wax and wane with her light, the seas turn their own tides with her rising and setting, and plants and animals wholly or partly wax and wane with her. Moreover, the passages of the fixed stars and the planets through the sky often signify hot and windy and snowy conditions of the enclosing air, and mundane things are affected accordingly. Then, too, their aspects to one another, by the meeting and mingling of their dispensations, bring about many varied changes. For though the sun's power prevails in the general ordering of quality, the other heavenly bodies cooperate or oppose it in particular details, the moon more obviously and continuously, as when it is at half or full phase, and the stars at greater intervals and more obscurely, as in their appearances and occultations and approaches.

2.2 ["national" characteristics – like many earlier Greeks,
Ptolemy regards the Mediterranean basin as the
most moderate clime]

The demarcation of national characteristics is established in part by entire parallels and angles, through their position relative to the ecliptic and the sun. For while the region which we inhabit is in one of the northern quarters, the people who live under the more southern parallels, I mean, those from the equator to the summer tropic, since they have the sun over their heads and are burned by it, have black bodies and thick and woolly hair, are contracted in form and reduced in size, are warm of nature, and in habits are for the most part wild because their homes are continually oppressed by heat; we call them by the general name Aithíops. Not only do we see them in this condition, but we likewise observe that their enclosing climate and other animals and plants of their region plainly give evidence of this baking.

Those who live under the more northern parallels, I mean, those who have the Bears over their heads, since they are far removed from the zodiac and the heat of the sun, are therefore cooled; but because they have a richer share of moisture, which is most nourishing and is not there exhausted by heat, they are white in color, straight in hair, tall in body, well-nourished, and somewhat cold in nature; these too are wild in their habits because their dwelling-places are continually cold. The wintry character of the air enclosing them, the size of their plants, and the untameability of their animals are in accord with these qualities. We call these people, too, by a general name, Skúthai.

The inhabitants of the region between the summer tropic and the Bears, however, since the sun is neither directly over their heads nor far distant at its noon-day transits, share in the equable temperature of the air, which varies but has no extreme changes from heat to cold. They are therefore medium in color, of moderate stature, in nature balanced, live close together, and are tame in their habits. The southernmost of them are in general more shrewd and inventive, and better versed in the knowledge of things divine because their zenith is close to the zodiac and to the planets revolving about it. Through this affinity the men themselves are characterized by an activity of the soul which is sagacious, investigative, and fitted for pursuing what are properly called mathematics [i.e., Ptolemy himself, from Alexandria]. Of them, again, the eastern group are more masculine, vigorous of soul, and frank in all things, because one would reasonably assume that the orient partakes of the nature of the sun. This region therefore is diurnal, masculine, and right-handed, even as we observe that among the animals too their right-hand parts are better fitted for strength and vigor. Those to the west are more feminine, softer of soul, and secretive, because again this region is lunar, for it is always in the west that the moon emerges and makes its appearance after conjunction. For this reason it appears to be a nocturnal clime and feminine and left-handed, in contrast with the orient.

3.1 [conception and birth as proper moments for a horoscope]

Since the chronological starting-point of human nativities is naturally the very time of conception, but potentially and accidentally the moment of birth, in cases in which the very time of conception is known either by chance or by observation, it is more fitting that we should follow it in determining the special nature of body and soul, examining the active power of the configuration of the stars at that time. For to the seed is given once and for all at the beginning such and such qualities by the dispensation of the Enclosing; and even though it may change as the body subsequently grows, since naturally [*phusikôs*] it mingles with itself in the process of growth only matter which is akin to itself, thus it resembles even more closely the proper character of its initial quality.

But if they do not know the time of conception, which is usually the case, we must follow the starting-point furnished by the moment of birth and give to this our attention, for it too is of great importance and falls short of the former only in this respect, that by the former it is possible to have foreknowledge also of events preceding birth. For if one should call the one "source" [*archê*] and the other, as it were, "beginning" [*katarchê*], its importance in time is indeed secondary, but it is equal or even more complete in potentiality, and with reasonable propriety would the former be called the genesis of the human seed and the latter the genesis of a human. For the child at birth and its bodily form take on many additional attributes which were not previously present, when it was in the womb, those very ones which belong to human nature alone; and even if it seems that the Enclosing at the time of birth contributes nothing toward the child's quality, at least its very coming forth into the light under the proper position of the Enclosing contributes, since nature, after the child is completely formed, gives the impulse to its birth under a configuration of similar type to that which governed the child's formation in detail in the first place [i.e., the stars enforce that birth-date conforms to conception-date]. Accordingly one may with good reason believe that the position of the stars at the time of birth is significant of things of this sort, not, however, for the reason that it is active in the full sense, but that necessarily and naturally it has potentially a very similar active power.

(Robbins [1940] odd pp. 3–9, 121–127, 223–227)

4.14 Antigonos of Nikaia summarized the work of earlier astrologers, perhaps around 180 ± 40. He preserves the horoscopes of Hadrian, and possibly his father (a third horoscope, for a birth-date in 113, could be Hadrian's grand-nephew, Pedanius Fuscus).

[Hadrian's horoscope: 24 January 76]

There was a person [Hadrian] having the sun in Water-Pourer 8 degrees, the moon and Zeus and the Horoskopos, the three together at the first degree of the same sign, namely Water-Pourer; Kronos in Goat-Horn 5 degrees and Hermes with it at 12 degrees; Aphrodite in Fishes 12 degrees and Ares with it at 22 degrees; Midheaven in Scorpion 22. In this configuration the house ruler of the moon, Kronos, being in its own house [Goat-Horn], allots its maximum number of years of life, 56; and since Aphrodite [in Fishes] is in aspect to it [Kronos in Goat-Horn], it allots additional 8 years, so that the total of years is 64. After 61 years 10 months the degrees of the Horoskopos and the moon [in Water-Pourer 1] come into quartile to Kronos, which, however, is not destructive because Aphrodite is in aspect to it the second time.

Such a person was adopted by a certain emperor [Trajan], akin to him, and having lived with him 2 years, became emperor about his 42nd year and was wise and educated, so that he was honored by shrines and temples; and he was married to one wife from maidenhood and was childless; and he had one sister [Domitia Paulina]. And he was at discord and conflict with his own relatives. When he had reached about his 63rd year he died, a victim to dropsy and asthma [July 138].

And why it happened in this way is explained as follows. He became emperor because the two luminaries [sun and moon] were with the Horoskopos and especially because the moon was of the same sect and in conjunction to the degree with the Horoskopos [both in Water-Pourer 1] and with Zeus which was also due to make its morning phase after 7 days. And the moon's attending stars themselves were found in favorable positions, Aphrodite in her own exaltation [Fishes], Ares in its own triangle [Fishes Crab Scorpion] and located in its own degrees, both in their own domains and in *epanaphora* with respect to the moon. And besides the *kosmos*-ruling sun was the moon's attendant in the subsequent degrees [Water-Pourer 8°] and had as attending stars Kronos in its own house [Goat-Horn] and Hermes, both at their morning rising. It is also significant that the moon was about to come into conjunction with a certain bright fixed star which is at the 20th degree [of Water-Pourer: one of φ, χ, or ψ Aqu, all of fourth magnitude]. For it is necessary to look at the conjunctions of the moon not only with the planets but also with the fixed stars.

(Neugebauer and vanHoesen [1959] 90–91)

He was big and manly and gracious because the two luminaries were with the Horoskopos and in a humanoid and male sign; he was wise and educated and profound because Hermes happened to be with Kronos in morning visibility, in the 12th sign and flanking the sun. And it was predicted he'd be like that from youth because of the phase (for morning risings cause things in youth while evening ones indicate deeds over the course of life). And one must also examine the ruler of the Midheaven locus, if it's well-placed and in aspect to the locus: for if it's rising it makes people notable and active and hard to master, as also in the evening (if it's well-placed), as in the birth discussed: Ares the ruler of Scorpion being most harmless in the Fishes, in its proper trigon, and in its own degrees, and in aspect to the Midheaven locus. But if the ruler of the Midheaven locus is poorly-placed it has the opposite effects. If the lords of the inoperative loci happen to be in the operative loci, they indicate modest circumstances of life. Zeus predicts originality and greatness of spirit and generosity and practicality, since he is in the Horoskopos and flanking the sun and is in the same Center with the moon. And the cause of the native having had many opponents and betrayers came from the two luminaries providing power but being encompassed by the two malefics, Kronos being at morning rising and flanking, and Ares at evening.

[Antigonos continues in the same vein for three or four more paragaphs of similar length]

*(Kroll [1903] 69)

4.15 Vettius Valens of Antioch composed his lengthy treatise around 180 ± 5; in the first-quoted passage he is showing how astrology forecasts the common fate (in 154) of six men who were born at different times; we also have an attempt to explain an infant death as fated (173).

Anthologies

7.6 127–160 [common fate of six men in a boat]

For the wonder of nature and that nothing happens outside fate, we will show from a brief example that those having anything happen in war and battle, or in fire and shipwreck, or in any other cause, are uniformly coerced by fate.

Sun and Hermes in Lion, moon in Scales, Kronos in Ram, Zeus in Bull, Ares and Aphrodite in Virgin, Horoskopos in Goat-Horn, klima 2 [i.e., 26 July 114]. In the 40th year he had a crisis: of moon 25, of Ram 15, making 40. Or of its diameter, Ram 20, of Scales 40, two-thirds of which is 40. Thus the crisis was double. And at the same time I found Goat-Horn, the Horoskopos, 28 and of Zeus in trine 12, making 40. And again of Goat-Horn 30 and of Bull 22 and of period 8, making 60, two-thirds of which is 40.

Sun and Hermes in Water-Pourer, moon in Scorpion, Kronos in Crab, Zeus in Scales, Aphrodite in Goat-Horn, Ares and Horoskopos in Virgin; klima 7 [i.e., 8 February 120]. In the 35th year he had a critical period, because the period of Ares was operative: 15 years and 20 for Virgin [the house of Hermes, whose period is 20] makes 35; but also the 8 of Aphrodite and the rising time 27 of Goat-Horn makes 35. And again 30 for Kronos opposite to Aphrodite and 32;30 for Crab and 8 of Aphrodite makes 70 years 6 months, the half of which makes 35 years 3 months. Besides also Zeus and Kronos shared the period for Scales 42 years 6 months, for Crab 27 years 6 months makes 70, of which half is 35.

Sun, Aphrodite and Ares in Archer, Moon in Scales, Kronos in Twins, Zeus in Virgin, Hermes in Scorpion, Horoskopos in Goat-Horn; klima 6 [i.e., 26 Nov 118]. The 36th year midway was critical. Of Twins 27 years 6 months and of Aphrodite 8 years makes 35 years 6 months. Of the sun 19 years and of Archer 35 years 6 months makes 54 years 6 months, two-thirds of which makes 36 years 4 months. There the beneficent stars shared in power.

Sun, Hermes and Aphrodite in Crab, Moon in Ram, Zeus and Horoskopos in Twins, Kronos in Scales, Ares in Lion, klima 1 [i.e., 18 July 127]. In the 27th year he had a crisis. Of the sun 19 and of Scales 8, making 27; further also of Twins 28 years 4 months, and of Zeus 12, making 40 years 4 months, two-thirds of which is approximately 27. And again of Lion 19 [house of the sun, whose period is 19] and of Crab 31 years 8 months, of Kronos 30, making 80 years 8 months, the third of which is approximately 27.

Sun in Water-Pourer, moon in Ram, Kronos in Lion, Zeus in Archer, Ares in Scales, Aphrodite and Hermes in Goat-Horn, Horoskopos in Fishes, klima 6 [i.e., 30 January 122]. In the 33rd year he had a crisis. Of moon 25 and of Scales 8 [a house of Aphrodite, whose period is 8] makes 33. And of Kronos 30 and of sun 19 makes 49, two-thirds of which is 32 years 8 months. And also the rising time of Archer was operative, Zeus being located there, that is 33.

Sun, Hermes, Aphrodite and moon in Bull, Kronos in Archer, Zeus in Scorpion, Ares in Lion, Horoskopos in Fishes, klima 2 [i.e., 24 April 133]. In the 22nd year he had a crisis: of Lion 19 and of moon 25, making 44, half of which is 22. In addition, 36 of Scorpion and 8 of Bull [a house of Venus, whose period is 8] are 44, the half of which is 22.

These six men on a voyage, with many others, encountered a violent storm and, the rudder being lost, were in danger of death by drowning as the ship took in water. But by the draught of the blowing wind and the steersman's management of the sails they escaped; and they encountered other dangers at the same time from a roving pirate ship.

7.4 11–15 [infant death]

For example let sun and Aphrodite be in Water-Pourer, moon and Zeus at the beginning of Ram, Kronos in Ram, Ares in Archer, Hermes in Goat-Horn, Horoskopos in Scorpion, Lot of Fortune in Scales, klima 6 [i.e., 17 February 173]. The apportioning stars were: Aphrodite because of Scales [where the Lot of Fortune is and a house of Aphrodite], Kronos because Aphrodite was in Water-Pourer [a house of Kronos], Ares because Kronos was in Ram [a house of Ares]. Then I reckoned with respect to periods and to rising times: first hours, then days, then months. Thus: I took for Scales 8 days and 8 hours and again for the rising time of Scales, klima 6, 43 hours, and, since Aphrodite is in Water-Pourer, I took for Kronos 57 hours and again since Kronos is in Ram, I took for Ares 15 hours, summing up to 8 days and 123 hours, which are 5 days 3 hours, making altogether 13 days 3 hours. He lived for 13 days 3 hours.

(Neugebauer and vanHoesen [1959] 110, 115–117, 119, 123, 125, 130)

5

GEOGRAPHY

Humans have always been wanderers, storytellers, and traders too – doubtless we built our first world-views on the tales of travelers. In the Mediterranean basin and elsewhere, adventurers carried obsidian and flint many hundreds of kilometers, and their accounts are the lost origins of geography. All early peoples assume the Earth is flat, and Homer (*Iliad* 18) and Babylonians sketched their world as an ocean-girdled disk; for Egyptians the sky was stretched upon remote pillars like a table above us. After their invention of the sail (*ca.* −3100: Casson [1986] 12, 22–23), longer oceanic voyages became more regular, and the Sumerians traded with the remote and wonderful land of Dilmun (Bahrain), while the Egyptians themselves sought treasure in Punt (perhaps Ethiopia or Yemen). Longer voyages – and larger realms – led to geographical studies in service of politics (territory to be ruled thoroughly must be known thoroughly), first by Persians and then by Greeks and others.

Greeks' geography, their attempt to describe and explain the world in which they lived, begins with such stories and ideas, and probably also with their extensive "colonization" of the Mediterranean and Black seas (*ca.* −800 to −500). Many Greek ports, especially Corinth and Miletos, planted semi-autonomous city-states along shores where they traded, as self-sufficient bases of operations (the Phoenicians of Tyre and Sidon did much the same: Negbi [1992]). The fantastic wanderings of Odysseus reflect this activity, or at least an interest in remote and marvelous places and peoples; characteristic is his introduction as a journeyer knowing many minds and cities.

The earliest Greek geography is preserved in the usual bits and scraps (the few works to survive intact are noted below), and we restrict our account to the more certain matters. Exploration and theory intertwine throughout Greek geography, and while the flat-Earth theory was replaced quite early, notions of symmetry, geographical determinism, and Greece as the ideal "center" persisted. The first explorer known who may be historical is Kolaios of Samos, who around −650 by chance reached far west to return with marvelous profit of stories (the "Pillars" of Herakles) and silver. Eratosthenes claimed that Anaximander of Miletos composed a map of the Earth (−580 ± 20), perhaps resembling the Homeric and Babylonian pictures. Two explorers from the western colony of Massalia

(Marseilles) each reported their coastal voyage (*periplous*): an anonymous voyager, who traveled along Spain to fabled Tartessos (the tale is partly preserved in the late Latin writer Auienius); and Euthumenes, who recorded his journey along the west coast of Africa, both around –530 (Euthumenes reports crocodile rivers and sweet seawater). A decade or more later shah Dareios annexed the Indus to his empire and sent a Greek captain, Skulax of Karuanda, to explore and report, from there to Egypt.

By around –500, Hekataios of Miletos had traveled to Egypt and elsewhere, and put out his *Tour of the Earth*, describing sites and peoples in Europe and in Asia (which included Africa), located on a circular map with peripheral Ocean; Aristagoras, the tyrant of Miletos in –498, used a similar map in his futile search for allies against Persia (Herodotos 5.49). Xenophanes of Kolophon in the latter half of the sixth century BCE (–510 ± 30) described a flat Earth "reaching down forever" (evading the issue of what lay beneath), and speculated that fossil seashells showed the periodic interchange of Earth and Sea. A bit later (–500 ± 20), Herakleitos of Ephesos seems to have believed in the same cycle, while (shortly after the failed Persian annexation of Greece) Empedokles of Akragas (–455 ± 20) conceived a four-step cycle, also involving Fire and Air.

The earliest extant Greek geography is found in the history of Herodotos (–435 ± 10), in whom geography is explicitly political – and also marvelous and capacious. He traveled widely and reports more widely, and manifestly makes ethnography an integral part of his task. But ethnographers find what they expect, and their informants collaborate. He believes in the circumnavigations of Africa under pharaoh Necho and by admiral Sataspes, withholds judgment on a hyperborean ocean, knows the Caspian is a lake, makes the Danube symmetric with the Nile, and endlessly and broadmindedly entertains with ethnographic marvels. Geographical determinism makes its first known showing: for Herodotos climate creates character, and so Greece is the best of lands midway between the cold harsh North and the hot soft South. The extremities of his flat Earth produced oddities, and its eastern lands were hotter at morning when the sun was so near.

Aristotle and others assign to various writers (the more likely ones roughly contemporary with Herodotos) a theory that the sun sets by moving behind tall mountains which ring the Earth; moreover the sun's seasonal motion (lower in the winter sky) is partly caused by Earth's disk having tilted toward the South (Anaximenes of Miletos may have suggested this –530 ± 20; more likely it was one or more of Anaxagoras of Klazomenai –455 ± 20; Leukippos of Miletos –440 ± 20; or Diogenes of Apollonia –430 ± 10: Dicks [1970] 58–59). Euthumenes had already guessed that Atlantic water poured into an upper arm of the Nile, thus explaining its mysterious mid-summer rise (Egyptians simply postulated that Nûn the primeval Ocean was subterranean); Anaxagoras opined that melting snows did it (as in northern rivers), while Demokritos of Abdera (–410 ± 30) held out for heavy rains dropped by summer winds. He seems also to have translated Hesiod's myth about Golden Men of an Edenic Age into philosophy as "noble savages."

114

Auienius (*Maritime Ports* 350–369) alleges the Athenian astronomer Eukte-mon (*ca.* –430) claimed the sea grew shallow and weedy outside the Pillars of Herakles. Also in the era of the Peloponnesian War, Philolaus of Kroton (–420 ± 20), based on Pythagorean notions that the sphere was the best shape, guessed that the Earth herself was spherical; his radical notion slowly became standard. Medical writers of this era (especially the author of the extant *Airs, Waters, Places*) adopt geographical determinism to explain regional differences in health and disease: Asia is mild and fertile, being close to the dawn, but its people are indolent and submissive; the northern Skúthai are chilled, watery, and almost barren like their wintry land; variable climate accompanies variable terrain, and produces peoples of more changeable character and livelier and freer minds.

In the aftermath of the Greek civil war, Thoukudides described the situation of Sicily ("Thucydides," in his history, –400 ± 5), stating that it was far too large to have been conquered by Athens, and far too mutable to remain long conquered, being rich and tumultuous, and full of manifold immigrants (6.1–5). The physi-cian Ktesias served in Persia, returned home to Knidos, and published accounts (*ca.* –400) out-marveling Herodotos' eastern lands (a ten-times larger sun in India). In two of his dialogues Plato wrote on geography (–365 ± 20). In the *Phaidôn* (97–99, 108–111), he describes the large spherical Earth as dimpled, its surface compared to leather balls sewn in many sections; each dimple is deep, ringed with giant mountains, and filled with dense air (above is the clear pure *aither*). Long-ago Greeks had a more fertile land, but its soil and forests have eroded away, leaving only a bald and rocky skeleton (*Kritias* 111); they defeated Atlantis, a mythical continent founded on vaguely-storied Atlantic islands, Greek subsidences, and Persian imperialism (*Kritias* 113–121).

From the same years (–360 ± 10) comes the extant *Periplous* attributed to Skulax, but containing a counter-clockwise description of the Mediterranean, Black Sea, and Atlantic coast of Africa. The Adriatic supplies amber and an outlet for the Danube; Rome is mentioned for the first time; Sardinia is huge; the Atlantic shallow and muddy, though an African port there, Kernê, traded ivory with Aithíops. Like Herodotos he attempts a general survey, though leavened with less theory but more data. Ephoros of Kumê (–350 ± 20) wrote the first "universal" history (40 books, paraphrased by Diodoros of Sicily), including a geographical survey of the known world. Giant pillars stand at its western, northern, eastern, and southern extremes (Pillars of Herakles; one near Kelts; Pillars of Bacchos by the Ganges mouth; and at the mouth of the Red Sea), which may mark the limits of our rhomboidal landmass on the flat Earth.

Aristotle's extant treatises include none devoted to geography, but he presents a brief excursus in *On Heaven* and in the *Meteorologika* (–335 ± 10). By now the spherical-Earth theory is standard, and he gives the first extant estimate of its size; he describes the theory of the five zones ("belts") around the Earth (one tropical, two temperate, and two polar), within which climate, and therefore plants, animals, and people, will be similar: often attributed to Parmenides but perhaps actually by Eudoxos (see Chapter 3). The spherical Earth is fixed at the true Center of the Kosmos, toward which all heavy things (earth and water) naturally tend; its surface may be dimpled, at any rate mighty mountains tower all around, and the center of the Mediterranean is the deepest of seas; moreover, the surface is subject to slow changes in the distribution of water and earth. These treatises were written (and Aristotle died) before much of the new data from Alexander's conquests could reach Greece. Four authors (all –305 ± 10) supplied rich material about the magical land of India: a history of Alexander by one of his officers, Aristoboulos of Kassandreia, offered much biological and geographical data (extensively mined by Strabo); Megasthenes, ambassador from Alexander and Seleukos the First to raj Chandragupta (whose capital he well depicts), composed a history and geography of India; admiral Nearchos of Crete described his sailing of Alexander's fleet from the rain-swollen Indus to the Tigris River (partly preserved in Arrian's *Indika*); as did his lieutenant Onêsikritos of Astupalaia, also supplying data on remote Taprobanê (Sri Lanka).

5.1 Hanno of Carthage traveled far down the west coast of Africa, around –500, and this report (translated from Punic) was published probably in the late fourth century BCE. What we have is an abbreviation (Arrian, *Indika* 43.11–12 records that Hanno sailed 35 days to the east, and then turned south, evidently at Mt. Cameroun, before he gave up), possibly derived from the lost *Periplous* of Xenophon of Lampsakos (–100 ± 35): compare Mela 3.93, 99, Pliny 6.199–200, and Solinus 56.10–12.

Periplous [voyage down the west coast of Africa]

The Voyage of Hanno, King of the Carthaginians, to the Libyan regions of the earth beyond the Pillars of Herakles, which he dedicated also in the Temple of Ba'al, affixing this:

It pleased the Carthaginians that Hanno should voyage outside the Pillars of Herakles, and found cities of the Libuphoinicians. And he set forth with sixty ships of fifty oars, and a multitude of men and women, to the number of thirty thousand, and with wheat and other provisions. After passing through the Pillars we went on and sailed for two days' journey beyond, where we founded the first city, which we called Thumiaterion [Mehedia, 34° 20′ N]; it lay in the midst of a great plain. Sailing thence toward the west we came to Solois, a promontory of Libya, bristling with trees [Cape Cantin, 32° 30′ N]. Having set up an altar here to Poseidon, we proceeded again, going toward the east for half the day, until we reached a marsh lying no great way from the sea, thickly grown with tall reeds [Cape Safi 32° 20′ N]. Here also were elephants and other wild beasts feeding, in great numbers. Going beyond the marsh a day's journey, we settled cities by the sea, which we called Karikos Wall, Gutta, Akra, Melitta and Arambus [one of these is Mogador, 31° 30′ N]. Sailing thence we came to the Lixos [Wadi Draa, 28° 30′ N], a great river flowing from Libya. By it a wandering people, the

Lixitae, were pasturing their flocks; with whom we remained some time, becoming friends.

Above these folk lived unfriendly Aithiopians, dwelling in a land full of wild beasts, and shut off by great mountains, from which they say the Lixos flows, and on the mountains live men of various shapes, cave-dwellers, who, so the Lixitae say, are fleeter of foot than horses. Taking interpreters from them, we sailed twelve days toward the south along a desert, turning thence toward the east one day's sail. There, within the recess of a bay we found a small island, having a circuit of fifteen stades; which we settled, and called it Kernê [perhaps Herne Is., 23° 45′ N]. From our journey we judged it to be situated opposite Carthage [assuming the west coast of Africa went south-east from the Pillars]; for the voyage from Carthage to the Pillars and thence to Kernê was the same. Thence, sailing by a great river whose name was Chretes [St. Jean river, 19° 25′ N], we came to a lake, which had three islands, larger than Kernê. Running a day's sail beyond these, we came to the end of the lake, above which rose great mountains, peopled by savage men wearing skins of wild beasts, who threw stones at us and prevented us from landing from our ships.

Sailing thence, we came to another river, very great and broad, which was full of crocodiles and hippopotami [Senegal river, 16° 30′ N]. And then we turned about and went back to Kernê. Thence we sailed toward the south twelve days, following the shore, which was peopled by Aithiopians who fled from us and would not wait. And their speech the Lixitae who were with us could not understand. But on the last day we came to great wooded mountains. The wood of the trees was fragrant, and of various kinds. Sailing around these mountains for two days, we came to an immense opening of the sea [Gambia river, 13° 30′ N], from either side of which there was level ground inland; from which at night we saw fire leaping up on every side at intervals, now greater, now less [burning grasslands?].

Having taken in water there, we sailed along the shore for five days, until we came to a great bay, which our interpreters said was called Horn of the West [the western edge of their world?]. In it there was a large island, and within the island a lake of the sea, in which there was another island. Landing there during the day we saw nothing but forests, but by night many burning fires, and we heard the sound of pipes and cymbals, and the noise of drums and a great uproar. Then fear possessed us, and the soothsayers commanded us to leave the island.

And then quickly sailing forth, we passed by a burning country full of fragrance, from which great torrents of fire [lava] flowed down to the sea. But the land could not be approached for the heat. And we sailed along with all speed, being stricken by fear. After a journey of four days, we saw the land at night covered with flames. And in the midst there was one lofty fire, greater than the rest, which seemed to touch the stars. By day this was seen to be a very high mountain, called Chariot of the Gods [probably Mt. Cameroun, the only active volcano on the west African coast]. Thence, sailing along by the fiery torrents for three days, we came to a bay, called Horn of the South [the southern edge of their world?].

In the recess of this bay there was an island, like the former one, having a lake, in which there was another island, full of savage men. There were women, too, in even greater number. They had hairy bodies, and the interpreters called them *Gorillae*. When we pursued them we were unable to take any of the men; for they all escaped, by climbing the steep places and defending themselves with stones; but we took three of the women, who bit and scratched their leaders, and would not follow us. So we killed them and flayed them, and brought their skins to Carthage. For we did not voyage further, provisions failing us.

(Schoff [1912] 3–5)

5.2 Putheas of Massalia in the late fourth century BCE traveled to Britain and Ireland and beyond to "Thoulê" (Shetlands? Faeroes? Iceland?), at 62° North. His observations reported in *On the Ocean* bent the paradigm, and were hence neglected or rejected (Strabo 4.5.5 even accused him of data-falsification), so that few quotations survive.

On the Ocean

(in Strabo 2.4.1) [the island of Thoulê]

Putheas, after asserting that he traveled over the whole of Britain that was accessible by foot, reported that the coast-line of the island was more than forty thousand stadia [*ca.* 7700 km, roughly correct], and added his story about Thoulê and about those regions in which there was no longer either earth by itself, or sea, or air, but a kind of substance concreted from all these elements, resembling a sea-lungs – a thing in which, he says, the earth, the sea, and all the elements are held in suspension; and this is a sort of bond to hold everything together, which you can neither walk nor sail upon. . . . He adds that on his return from those regions he visited the whole coast-line of Europe from Gadeira [southwestern Spain] to the Tanaïs [the Don river, emptying into the Sea of Azov].

(Jones [1917] 399)

(in Geminus 6.9) ["the sun's bedroom": one of the few verbal quotations]

The foreigners pointed out to us where the sun sleeps. For it happens in these places that the night is extremely short: two hours in some, three hours in others, so that after sunset, although only a short interval has passed, the sun immediately rises again.

*(Aujac [1975] 34–35)

5.3 Straton of Lampsakos was the third head of Aristotle's school (from –286 to –269), and wrote on a wide variety of scientific topics (so that he was called "the naturalist"). Aristotle *Meteorologica* 1.14 (352b15–3a8) suggested that low-lying Ammon (modern Siwa in the western desert of Egypt) was a dried-up lake; and in *Meteorologica* 2.1 (354a5–35) attributed Mediterranean currents to the progressively increasing depths of the shallow Maiotis (Azov), the Pontos, the Aegean sea, the Sicilian sea, and (deepest of all) the Sardinian and Turrhenic seas.

On Heaven

(fr. 91W paraphrased in Strabo 1.3.4) [seas]

The Euxine Sea did not at first have its outlet at Buzantion, but the rivers emptying into the Euxine forced and opened a passage, and then the water poured into the Propontis and the Hellespont. The same thing happened in "Our" Sea [Mediterranean] also; for in this case the passage at the Pillars was broken through when the sea had been filled by the rivers, and at the time of the outrush of the water the former shallows were left dry. The cause of this, first, is that the beds of the Atlantic and "Our" Sea are unequal, and, secondly, that at the Pillars even at the present day a submarine ridge stretches across from Europe to Libya, indicating that the inside and the outside could not have been one and the same formerly. The seas of the Pontos region are very shallow, but the Cretan, the Sicilian, and the Sardinian Seas are very deep; for since the rivers that flow from the north and east are very numerous and very large, the seas there are being filled with mud, while the others remain deep; and this is why the Pontos is sweetest, and why its outflows are toward the inclination of its bed [compare Herodotos 4.82 and pseudo-Aristotle, *Problems* 23.6].

The whole Euxine Sea will be silted up some day, if such influxes continue; for even now the regions on the left side of the Pontos are already getting shallow; for instance, Salmudessos [coast of modern Bulgaria], and the land at the mouth of the Ister [Danube], which sailors call "the Breasts," and the desert of Skuthia [southern Ukraine and Russia]. [Polubios 4.39–42 also explains this silting up]

Perhaps too the temple of Ammon was formerly on the sea, but is now situated in the interior because there has been an outpouring of the sea. It is likely that the oracle of Ammon became so distinguished and so well-known, if it was on the sea, and its present position so very far from the sea gives no reasonable explanation of its present distinction and fame; and in ancient times Egypt was covered by the sea as far as the bogs about Pelousion, Mt. Kasios, and Lake Sirbonis [Pelousion is Tell el-Farama, at the east edge of the Nile delta; Herodotos 2.6 and Diodoros of Sicily 16.46 describe the silting up; Mt. Kasios, now called Mehemdia, is a 13-meter-high sandhill at the west end of Lake Sirbonis, modern Sabhat el-Bardawîl].

(Jones [1917] 183, 185)

5.4 Eratosthenes of Kurênê (*ca.* –285 to –193) grew up while Kurênê enjoyed a brief autonomy from Ptolemaic rule, and studied philosophy in Athens. He became in *ca.* –245 royal tutor to Ptolemy III ("Euergetes") and head of the

Library at Alexandria. He corresponded with Archimedes and wrote on a wide variety of literary, mathematical, philosophical, and scientific topics (see Chapters 2 and 3), refusing to employ myth as data and eschewing paradoxography. His two geographical works *Measurement of the Earth* (on the sphericity of the Earth, see below in Theon) and the *Geography* (in which he modeled the lands of the earth in geometrical sections called *sphragidês*, "seals," like tiles) are preserved only in extracts.

Measurement of the Earth

(fr. in Kleomedes 1.10) [Aristotle *On Heaven* 2.14 (298a16–20) reports that the "mathematicians" had determined the size of the earth as 40 myriad stades in circumference; then around –304 ± 3 Dikaiarchos repeated the measurement, finding 30 myriads by a method similar to Eratosthenes: see Kleomedes 1.8.3; Archimedes, *Sand-Reckoner* 1.8; Collinder [1964]; Keyser [2001].]

Suênê [Aswan] and Alexandria lie under the same meridian circle [Alexandria is at *ca.* 30° east, and Suênê is at *ca.* 33° east]. Since meridian circles are great circles in the *kosmos*, the circles of the earth which lie under them are necessarily also great circles. Thus, of whatever size this method shows the circle on the earth passing through Suênê and Alexandria to be, this will be the size of the great circle of the earth. It is the fact that Suênê lies under the summer tropic circle. Whenever, therefore, the sun, being in the Crab and making the summer solstice, is exactly in the middle of the heaven, the gnomons [pointers] of sundials are necessarily shadowless, the sun lying exactly vertical above them; and it is said that this is true for a diameter of three hundred stades [i.e., the uncertainty is about 50 km]. But in Alexandria, at the same hour, the pointers of the sundials cast shadows, because this city lies further to the north than Suênê (Figure 5.1).

The two cities lying under the same meridian and great circle, if we draw an arc from the extremity of the shadow to the base of the gnomon of the sundial in Alexandria, the arc will be a segment of a great circle in the hemispherical bowl of the sundial, since the bowl of the sundial lies under the great circle of the meridian. If next we conceive straight lines produced from each of the gnomons through the earth, they will intersect at the center of the earth. Since then the sundial at Suênê lies vertically under the sun, if we further conceive a straight line coming from the sun to the top of the gnomon of the sundial, the line reaching from the sun to the center of the earth will be one straight line. If now we conceive another straight line drawn upwards from the extremity of the shadow of the gnomon of the bowl in Alexandria, through the top of the gnomon to the sun, this straight line and the aforesaid straight line will be parallel, since they are straight lines coming through from different parts of the sun to different parts of the earth [rays of sunlight are nearly parallel at the earth].

On these straight lines, therefore, which are parallel, there falls the straight line drawn from the center of the earth to the gnomon at Alexandria, so that the

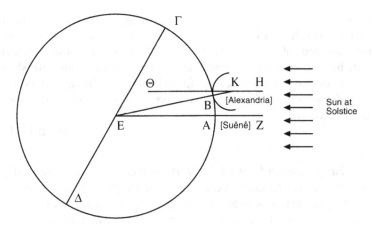

Figure 5.1

alternate angles which it makes are equal [Eukleidês, *Elements* 1.29]. One of these angles is that formed at the center of the earth, at the intersection of the straight lines which were drawn from the sundials to the center of the earth; the other is at the point of intersection of the top of the gnomon at Alexandria and the straight line drawn from the extremity of its shadow to the sun through the point where it touches the gnomon. Now on this latter angle stands the arc carried round from the extremity of the shadow of the gnomon to its base, while on the angle at the center of the earth stands the arc reaching from Suênê to Alexandria. But the arcs are similar, since they stand on equal angles [Eukleidês, *Elements* 3, def. 11]. Whatever ratio, therefore, the arc in the bowl of the sundial has to its proper circle, the arc reaching from Suênê to Alexandria has that ratio to *its* proper circle.

But the arc in the bowl is found to be one-fiftieth of its proper circle. Therefore, the distance from Suênê to Alexandria must necessarily be one-fiftieth part of the great circle of the earth. And the said distance is 5000 stades; therefore the complete great circle measures 25 myriad stades [40,000 km: close to correct; see below Section 5.10 (Poseidonios), Section 5.14 (Heron), and Section 5.17 (Theon)].

(Heath [1932] 109–112)

Geography

Book 1 (in Strabo 1.3.4) [In the first book he rejected Homer as a geographical authority, remarking "you'll find where Odysseus wandered when you find the cobbler who sewed up the bag of winds"; here he wonders whether the division of land and sea is eternal.]

This question in particular has presented a research problem: how does it come about that large quantities of mussel-shells, oyster-shells, scallop-shells, and also salt-marshes are found in many places in the interior at a distance of two thousand or three thousand stades [almost 500 km] from the sea – for instance in the neighborhood of the temple of Ammon and along the road, three thousand stades in length, that leads to it? At that place there is a large deposit of oyster-shells, and many beds of salt are still to be found there, and jets of salt-water rise to a height; besides that, pieces of wreckage from seafaring ships are shown which the natives said had been cast up through a certain chasm, and there are dolphins dedicated on small columns having an inscription of the sacred ambassadors of Kurênê.

(Jones [1917] 181)

Book 3 (paraphrased in Strabo 2.1.22) [His map employed data gathered in the aftermath of Alexander's conquests, and his *sphragidês* served as a general framework to avoid privileging any local perspective; for Eratosthenes, the proper division of humanity was not racial but ethical, as he insisted at the end of book 2.]

Eratosthenes, beginning at the Pillars, divides the inhabited world in half by means of this line, and calls them respectively the Northern Part and the Southern Part, and then attempts to cut each of these parts again into such parts as are possible; and he calls these parts *sphragidês*. And so, after calling India the First *Sphragis* of the Southern Part, and Ariana the Second *Sphragis*, since they had perimeters easy to sketch, he was able to represent not only the length and breadth of both parts but more or less also the shape, as would a geometrician. In the first place, India, he says, is rhomboidal, because the southern and eastern seas, which form shores without very deep gulfs, wash those sides; and because the remaining sides are marked, one by the mountain Tauros and the other by the river Indus, and because on these two sides, also, the rectilinear figure is fairly well preserved. Secondly, Ariana, although he sees that it has at least three sides well-suited to the form of a parallelogram, and although he cannot mark off the western side by mathematical points, on account of the fact that the tribes there mix with one another, yet he represents that side by a rough line that begins at the Caspian Gates [the pass at Firûzkûh, 35° 44′ N, 52° 50′ E] and ends at the edges of Karmania [the region south-east of the Caspian Sea] joined to the Persian Gulf. Accordingly, he calls this side "western" and the side along the Indus "eastern," but he does not call them parallel; neither does he call the other two sides parallel, namely, the one marked by the mountain and the one marked by the sea, but he merely calls them the "northern" and the "southern" sides.

(Jones [1917] 293, 297)

5.5 Agatharchides of Knidos (*ca.* −215 to *ca.* −140) worked in Alexandria under Ptolemy V ("Epiphanes") and Ptolemy VI ("Philometor"), a time of strife and chaos. He wrote several works of history enriched with anthropology and geography, and towards the end of his life, fleeing the chaos in Alexandria, he published *On the Red Sea*, which is lost but paraphrased by Diodoros of Sicily

(3.15–17) the first-century BCE historian and more closely but less completely by Photios the ninth-century bishop of Constantinople (material found only in Diodoros is in italics).

On the Red Sea

Book 5, fr. 30–34 [fisheaters]

30 In the zone south of Egypt there are four great tribes: one living along rivers who cultivate sesame and millet, one dwelling around marshes who feed on reeds and shoots, one freely wandering who build their life on meat and milk, and one preferring the coast and hunting fish.

31 This last people have neither cities nor fields nor any understanding of technical activity, but some claim it is larger than the others. For, from the Autaioi, who inhabit the remotest recess which the Great Sea encloses, as far as India and Gedrosia [south coastal Iran], and even Karmania and Persia and the islands near those peoples, the Fisheaters dwell everywhere, naked, even the women. The begetting of children is in common, *as are their cattle,* and they have a natural understanding of pleasure and pain, adding not the least idea of the shameful and the good. [probably Agatharchides is referring to the Cynic "life according to Nature"]

32 For them all the depths near shore are foreign to life, just like those lying along the great beaches. For such territory provides an abundant catch of neither fish nor other similar creatures. The residences of these people lie along rocky shores, which have deep hollows and irregular ravines and narrow channels and curving inlets, *which are naturally divided by crooked side-channels.* In those fitted for their use, they place un-hewn stones in the hollows and narrow passageways, *by which they catch fish as if in nets.*

Then, when the tide is borne onto the land from the sea (which happens twice each day, at the ninth and at the third hour), the sea covers the whole rocky shore, and it also brings with its force from the main toward the beach many fish *of all kinds* which for awhile stay near shore wandering about the sheltered recesses *and hollows* for food. But as soon as the ebb tide occurs, the water flows back *little by little* through the stones and passageways towards the place that draws and attracts it, but the fish remaining behind in the hollows are ready prey and food for the Fisheaters.

33 At this moment the mass of the natives, children and women too, gather on the rocky shore as if at a single order. They divide into companies and each rush to their own place with a horrific shout, as if surprising prey. The women with the children snatch up the smaller fish and those close to shore, and throw them on the land, while the stronger men lay hands upon the more difficult larger prey. Most fish are easily subdued; but when larger dogfish, seals, "sea-scorpions" [crabs? lobsters? *Scorpaena scrofa?*], eels, and all creatures of this sort fall into the trap, the struggle becomes risky. *These beasts they subdue without artifice of weaponry, but by piercing with sharp goat-horns and slashing with jagged rocks.*

34 But when they have caught however many fish it may be, they carry them to south-facing rocks that are fiery hot and throw them on these. After leaving them for a short time, they turn them all over. Then when they grasp the fish by the tail and shake the entire body, the whole meaty part, softened by the heat, falls off, but the skeletons they pile up, and huge heaps of these are visible from a great distance. After collecting the meat of the fish on a smooth stone, they tread it vigorously for a long time and mix it with the fruit of the *paliouros* [*Zizyphus spina-Christi*, the red-berried Christ's thorn]. When this is mixed in, the whole becomes much more sticky, and it seems to serve as flavoring and binder. When they have worked it with their feet for a sufficient time, they form it into oblong cakes and set them out again in the sun. When the cakes have dried, they all sit down and feast, not according to measure or weight, but each according to his own desire and pleasure.

*(Henry [1974] 156–158)

5.6 Seleukos of Seleukeia on the Red Sea, a practitioner of Babylonian astrology (*ca.* –150) adopted the heliocentric theory of Aristarchos and connected that model with his lunar theory of tides (Neugebauer [1975] 2.610–611, 697); on tides see also Poseidonios (below Section 5.10) and *Voyage on the Red Sea* 45–46.

(title unknown)

(fr. in Strabo 3.5.9) [lunar theory of tides]

Seleukos speaks of a certain irregularity in these phenomena, or regularity, according to the differences of the signs of the zodiac; for if the moon is in the equinoctial signs, the behavior of the tides is regular, but, in the solstitial signs, irregular both in amount and speed, while, in each of the other signs, the relation is in proportion to the nearness of the moon's approach. But although he himself [Poseidonios] spent several days in the Herakleion at Gadeira at the summer solstice, about the time of the full moon, as he says, he was unable to observe those annual differences in the tides. About the time of the conjunction, however, during that month, he observed at Ilipa [Alcalá del Río] a great variation in the back-water of the Baetis [the Guadalquivir river], that is, as compared with the previous variations, in the course of which the water did not wet the banks so much as halfway up, but now the water overflowed to such an extent that the soldiers got their supply of water on the spot (and Ilipa is about seven hundred stades [*ca.* 100 km] distant from the sea). And, although the plains near the sea were covered as far as thirty stades [*ca.* 5 km] inland, to such a depth that islands were enclosed by the flood-tide, still the altitude of the foundations, both the foundation of the temple in the Herakleion and that of the breakwater which lies in front of the port of Gadeira, was, by his own measurement, not covered as high up as ten cubits [4 m]; and further, if one should add the double of this figure for the additional increases which at times have taken place, one might thus present to the imagination the aspect which is produced in the plains by the magnitude of the flood-tide.

(Jones [1923] 153, 155)

5.7 Polubios of Megalopolis (writing *ca.* −140 ± 20) included many geographical sketches in his multi-book history, as well as an entire book devoted to a geographical survey (34, now preserved only in bits). First a hostage then a friend of the rulers of Rome, he wrote to explain to Greeks how and why the Romans managed in less than two generations to conquer half the known world, and geography had for him value in proportion to utility, incarnated in service to history. He also composed a short (lost) work arguing that the "torrid" zone (at the equator) was habitable, based on his trip down the west African coast around −145.

History

2.14.4–12 [Italy and the Po valley]

Italy as a whole has the shape of a triangle of which the one or eastern side is bounded by the Ionian Sea and then continuously by the Adriatic Gulf, the next side, that turned to the south and West, by the Sicilian and Turrhenian Seas. The apex of the triangle, formed by the meeting of these two sides, is the southernmost cape of Italy known as Kokunthos and separating the Ionian Strait from the Sicilian Sea. The remaining or northern and inland side of the triangle is bounded continuously by the chain of the Alps which, beginning at Marseilles and the northern coasts of the Sardinian Sea, stretches in an unbroken line almost to the head of the whole Adriatic, only failing to join that sea stopping at quite a short distance from it. At the foot of this chain, which we should regard as the base of the triangle, on its southern side, lies the last plain of Italy to the north. It is with this that we are now concerned, a plain surpassing in fertility any other in Europe with which we are acquainted. The general shape of the lines that bound this plain is likewise triangular. The apex of the triangle is formed by the meeting of the Apennines and Alps not far from the Sardinian Sea at a point above Marseilles. Its northern side is, as I have said, formed by the Alps themselves and is about two thousand two hundred stades in length, the southern side by the Apennines which extend for a distance of three thousand six hundred stades. The base of the whole triangle is the coast of the Adriatic, its length from the city of Sena [Ancona], to the head of the gulf being more than two thousand five hundred stades; so that the whole circumference of the plain [8300 stades] is not much less than ten thousand stades.

(Paton [1922] 273, 275, 277)

5.8 Hipparchos of Nikaia worked and wrote in Rhodes, in the early years of its subjection to Rome. His astronomical observations date his work to −146 to −126 (see Chapter 3). His *Geography* revised Eratosthenes (whom he criticized for basing conclusions on false or ambiguous evidence), dividing the world into *klimata* (latitude-zones) using latitude measurements and not Eratosthenes' *sphragidês* ("seals"); he suggested the astronomical method of determining longitude, later used by Heron (below Section 14). (Note: the longest-day lengths are evidently precise only to ± ¼ day, and the latitudes are correspondingly imprecise.)

Geography

Book 3 (fr.46–52 D. from Strabo) [determining latitude]

46 In the regions round Meroë and the Ptolemais in the country of the Troglodutes, the longest day is 13 equinoctial hours [*ca.* 16° 30′ N]; and this inhabited region is approximately midway between the equator and the parallel through Alexandria, although the distance to the equator is greater [than from Meroë to Alexandria, which is 10,000 stades, or *ca.* 1600 km] by 1800 stades [*ca.* 280 km].

47 In Suênê and Berenikê on the Arabian Gulf and in the country of the Troglodutes the sun stands in the zenith at the summer solstice, and the longest day is 13½ equinoctial hours [*ca.* 24° N; in Hipparchos' day, the tropic was at 23° 43′ N]; moreover, almost the whole of the Great Bear is visible on the arctic circle, except the legs, the tip of the tail, and one of the stars in the square.

48 In the regions some 400 stades [*ca.* 60 km] south of the parallel through Alexandria and Kurênê, where the longest day is 14 equinoctial hours [*ca.* 30° 30′ N], Arktouros reaches the zenith, but declines a little towards the south. In Alexandria the gnomon bears to its equinoctial shadow a ratio of 5/3 [*ca.* 31° 10′ N]. These regions are 1300 stades [*ca.* 200 km] south of Carthage, if it be true that in Carthage the gnomon has a ratio of 11/7 [*ca.* 32° 40′ N – but Carthage is at *ca.* 37° N] for its equinoctial shadow.

49 In the regions round Ptolemaïs in Phoenicia and round Sidon and Tyre the longest day is 14¼ equinoctial hours [*ca.* 33° 30′ N]; and these regions are about 1600 stades [*ca.* 250 km] north of Alexandria and about 700 [*ca.* 110 km] north of Carthage.

50 In the Peloponnese and in the regions round the center of Rhodes, round Xanthos and in Lukia or a little further south, and also in the regions 400 stades [*ca.* 60 km] south of Syracuse, here the longest day is 14½ equinoctial hours [*ca.* 36° 10′ N]; and these regions are 3640 stades [*ca.* 570 km] distant from Alexandria (about 2440 stades [*ca.* 380 km] from Carthage).

51 In the district round Alexandria in the Troad, at Amphipolis, at Apollonia in Epeiros, and in the district south of Rome but north of Naples, the longest day is 15 equinoctial hours [ca. 41° N]. This parallel is 7000 stades [ca. 1100 km] north of the one through Alexandria in Egypt, over 28,800 stades [ca. 4500 km] from the equator, 3400 stades [ca. 540 km] from the parallel through Rhodes, and 1500 stades [ca. 240 km] to the south of that through Buzantion, Nikaia and the regions round Massalia.

52   In the regions round Buzantion the longest day is 15¼ equinoctial hours [ca. 43° 10' N], and the ratio of the gnomon to its shadow at the summer solstice is one-hundred-twenty to forty-two minus one fifth [or 43° 11' N]. These regions are about 4900 stades [ca. 770 km] distant from the parallel through the center of Rhodes, and about 30,300 [ca. 4770 km] stades from the equator.

(Dicks [1960] 95–97)

5.9 Skumnos of Chios wrote a geographical survey around –160 ± 15 (lost), and his name got attached to the poetic periplus quoted here, which dates to ca. –85 ± 10 (dedicated to King Nikomedes III of Bithunia). Even this is damaged, and only the section on Europe survives intact (it continued around the Pontos, and presumably Asia Minor, Syria, Palestine, Egypt, Libya, and Mauretania, ending where it started at the Pillars). In his lengthy prologue (lines 1–138) he says he will rely on Eratosthenes, Ephoros, and others.

Periplous

139–166 [south coast of Spain]

Th' Atlantic mouth is said to be
one hundred twenty stades in width;
the regions lying round about
are capes of Libya, and of Europe.
Islands lie on either side,
being separated almost thirty
stades apart, and called "Herculean
Pillars"[a]. Near to one, a Massalian
city stands, Maimákê named,
which has the furthest place of all
the Greek foundations throughout Europe[b].
[150] Heading westward round the cape
that lies opposed you sail a day;
and then you come upon the isle
named "Reddish," miniscule in size,
but having herds of cows and cattle,
very like Egyptian bulls,
and Epirote Thesprótian breeds;
they say the West Aithíops settled

127

here, and dwell as colonists.
Nearby, an ancient city took in
[160] Tyrian merchant colonists,
it's Gáddir[c], boasting monstrous sea-beasts.
After this, complete a two day's
sail, and famous greatly blessed
port of Tartessos you find,
which offers placer tin from Keltic
lands and lots of gold and bronze.

[[a] the Pillars are called islands only here; [b] Strabo 3.4.2 describes Maimákê – modern Almuñécar – in a similar way; [c] Gaddir is modern Cadiz]

167–195 [south coast of Gaul]

Then comes the country called the Keltic,
running to the sea of Sardo[a]:
Kelts, the greatest western people.
[170] Almost all the eastern Earth
Indoí do master, southward whence the
warm wind blows, Aithíops dwell,
Keltoí possess from setting sun
to Zéphyr; northward Skúthai lie.
Indoí inhabit lands from summer
up to winter sunrise; Kelts are
under spring and summer setting.
Peoples four, alike in force
and quantity of citizens:
[180] Aithíop land is larger; greater
desert holds the Skúthai: one
all burnt, the desert over-watered.
Kelts now follow Grecian manners,
many visitors from Greece
affording easy adaptation;
Kelts assemble making music,
which they study seeking culture[b].
Legend tells a Northern Pillar's
set among them: vastly tall, its
[190] furthest end is stretching seaward[c].
Final edge of Earth Keltoí are,
dwelling round about this Pillar,
Énetoí and furthest Ístroi
properly Adrían, where
the Danube's spring-fed stream arises[d].

[[a] similarly described by Eratosthenes, see Pliny 3.10; [b] the four peoples and their lands are similarly described by Ephoros, *On Europe* (fr. in Strabo 1.2.28); [c] the "Northern Pillar" is probably Ephoros' term for the northern extremity of a flat Earth – see the prolegomenon; [d] the Danube reaching the Adriatic is from Ephoros]

*(Muller [1855] 199–203)

5.10 Poseidonios of Apamea (in Syria) became a citizen of the "free city" of Rhodes (a Roman protectorate), where he wrote and taught between about – 100 and –50. He was widely traveled and well connected. Most of what we know about his book *On the Ocean* comes from Strabo, who wrote in the generation after him (see below). In one of his works (reported by Strabo 2.2.2, and accepted by Ptolemy, *Geography* 7.5), he estimated the circumference of the earth as 18 myriad stades (28,000 km, about 2/3 of its actual value), or 500 stades per degree, which became the standard value; he thus argued that the sea-route from Spain to India was less than 7 myriad stades (at *ca.* 35° N).

On the Ocean

(fr. 49c E–K in Strabo 2.3.4) [circumnavigation of Africa]

[Eudoxos of Kuzikos (*ca.* –115 to –110) came to Egypt in the reign of Ptolemy VIII, ("Euergetes II"), and went to India guided by a shipwrecked sailor, but Euergetes abstracted his whole cargo on return.]

The king died and his wife Kleopatra took over the sovereignty [Kleopatra II, in – 115]; and so Eudoxos was sent off to India by her too with greater equipment. On the way back, he was carried off course south of Ethiopia by wind; and on making landings he won over the people with gifts of bread, wine and dried fruit cakes, delicacies they had no access to, in exchange for water and guides; and he wrote down some of their words. He found a wooden prow from a shipwreck with a horse carved on it, and was told that this piece of wreckage came from some people sailing from the west. He took it with him when he turned back on his homeward voyage. When he returned safely to Egypt, Kleopatra was no longer in charge, but her son was [Ptolemy IX, ("Soter II, Lathuros" – the "chick-pea," same as Latin "cicer" (as in Cicero), and possibly a euphemism for "testicle")]. Again Eudoxos had all his cargo abstracted; for he was caught having appropriated a lot for himself. As for the prow, he took it to the market and showed it to the ship masters, and they identified it as coming from Gadeira [Cadiz]. There (they said) merchants fit out the large vessels, while poor men fit out small boats, which they call "horses" from the device on their prow; these are used for fishing trips round the Mauretanian coast as far as the river Lixos. But some of the captains recognized the prow as belonging to one particular boat that had sailed too far beyond the river Lixos and not returned safely.

Eudoxos concluded from this that the circumnavigation of Africa was possible, went home, put all his property on a ship and put to sea. [He sailed down the west African coast, was shipwrecked, built a smaller ship from the wreckage,] and sailed on until he encountered a folk speaking the same phrases as he had written down earlier [from east Africa]: and at the same time he recognized that the people there were ethnically similar to his former Ethiopians, while bordering on the kingdom of Bogus [i.e., Mauretania]. Abandoning the Indian voyage, Eudoxos turned back. In coasting along he saw and marked a well-watered, treed and uninhabited island [Madeira? Fuerteventura? (see Keyser [1993a] 156–157)].

[He returned via Mauretania and Spain, refitted and set out again – no-one knows what happened to him (see Thiel [1966]).]

(fr. 217 E–K in Strabo 3.5.7–8) [theory of tides: the diagram is reconstructed]

One day and night is a single revolution of the sun, part of the time below the earth, part shining above it. The movement of ocean undergoes a cycle of a type like a heavenly body, exhibiting diurnal, monthly and annual movement in joint affinity with the moon.

[Diurnal Cycle] When the moon's elevation reaches one sign of the zodiac from the horizon, the sea begins to swell and encroach on the land perceptibly until the moon is in the meridian. When the moon turns, the sea retreats again gradually, until the moon is one sign's elevation from sinking. Then the water level remains stationary the whole period the moon takes to reach its sinking, and still more all the period it moves below the earth to a distance one sign from the horizon. Then the sea encroaches again on the land until the moon reaches the meridian below the earth. It then retreats until the moon turning around in the direction of its rising reaches one sign's distance from the horizon. Then there is slack water until the moon passes in elevation to one sign above the earth; and then the tide again encroaches.

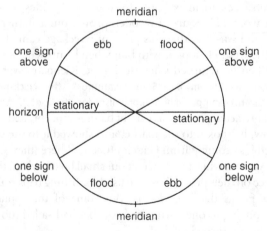

Figure 5.2

130

[Monthly Cycle] For the monthly cycle, the greatest tides occur at the conjunction of the moon, then they lessen until the third quarter; they increase again until full moon, and decrease again until first quarter; then increase until conjunction. Between the third quarter and conjunction, the time intervals between the highs increase and so also the velocity of the tidal waters.

[Annual Cycle] The people of Gadeira told Poseidonios that ebb and flow tides reached their maxima at the summer solstice. Poseidonios himself conjectured from this that the tides then decreased from summer solstice to equinox, but increased from equinox to winter solstice, then decreased until spring equinox, to increase again to summer solstice.

(Kidd [1999] 116–118, 283–284)

5.11 Anonymous The author of this work (traditionally ascribed to Aristotle) seems to have written it around –50 or a little later; he gives a textbook-like summary of geography influenced by the view of Plato (*Phaidôn* 108–111) and Aristotle (*Meteorologika* 1.13[350a15–b20]) that the Mediterranean and Black Sea are the bottom of a deep dimple in the Earth, whose encircling mountains are many miles high.

Kosmos

3 (392b14–393b23) [the Earth's oceans]

The ocean that is outside the inhabited world is called the Atlantic, or Ocean, and surrounds us. To the west of the inhabited world, this ocean makes a passage through a narrow strait called the "Pillars of Herakles," and so makes its entry into the interior sea, as if into a harbor; gradually it broadens and spreads out, embracing large bays adjacent to each other, here contracting into narrow necks of water, there broadening out again. They say that the first of these bays that the sea forms, to starboard, if you sail in through the Pillars of Herakles, are two, called the Surtes, of which one is called the Major, the other the Minor [these are the gulf on the coast of Libya]; on the other side it does not form gulfs at first in the same way, but makes three seas, the Sardinian, Galatian [the sea between Spain and Gaul, or else perhaps named for one of the two islands called Galata, between Sardinia and Africa: Ptolemy, *Geography* 4.3.12, 44?], and Adriatic; next to these, and across the line of them, is the Sicilian sea; after this, the Cretan; and continuing this on one side are the Egyptian and Pamphulian and Syrian seas, on the other the Aegean and Murtoan [south of Asia Minor]. Lying opposite to these is the Pontos, and this has very many parts: the innermost part is called Maiotis, and the outermost part, towards the Hellespont, is joined by a strait to the sea called Propontis.

In the East, the Ocean again penetrates the inhabited world; it opens out the gulf of India and Persia and without a break reveals the Red Sea, embracing these as parts of itself. Towards the other promontory of Asia, passing through a long narrow strait and then broadening out again, it makes the Hurkanian or Caspian sea [Patrokles, a satrap of Seleukos I in *ca.* –285 was read as claiming a Caspian connection to the Arctic Ocean, believed until Ptolemy]; beyond this, it occupies a deep hollow beyond Lake Maiotis. Then little by little, beyond the land of the Skuthians and Kelts, it confines the inhabited world as it passes towards the Galatian Gulf and the Pillars of Herakles, already described, on the farther side of which the Ocean flows round the earth. There are two very large islands in it, called the British Isles, Albion and Iernê; they are larger than those already mentioned, and lie beyond the land of the Kelts. No smaller than these are Taprobanê [Sri Lanka] beyond the Indians, which lies obliquely to the inhabited world, and the island known as Phebol [mentioned only here: Socotra? Madagascar? a reference to Euhemeros' mythical Panchaia?], by the Arabian Gulf. There is quite a number of other small islands round the British Isles and Spain, set in a ring round this inhabited world, which as we have said is itself an island; its breadth, at the deepest point of the continent, is a little short of 4 myriad stades [*ca.* 6300 km], in the opinion of good geographers [his figures are similar to those of Eratosthenes], and its length is approximately 7 myriad stades [*ca.* 11000 km]. It is divided into Europe, Asia and Libya.

(Furley [1955] 355, 357, 359, 361)

5.12 Strabo of Amaseia (–63 to *ca.* 21) believed in the authority of Homer and the moral and political utility of history and geography (1.1.14); he studied under Xenarchos (see Chapter 3), and wished his work to promote mutual Greek and Roman understanding (1.1.16), as well as peace of mind through acceptance of the natural order governed by divine providence. He eschewed marvels and myths, useful only to impress children and the ignorant (1.2.8), and saw foreign wise men as equivalent to Greek philosophers, reliable for information and explanation (Druids 4.4.4; Magoi 11.9.3; Brahmins 15.1.39, 68; Chaldeans 16.1.6; Egyptian priests 17.1.3–5), as had Poseidonios and C. Iulius Caesar. He also rejected the typical Greek geographical determinism, remarking that culture is more due to custom and training (2.3.7). After a two-book introduction (with a long yet shrinking nod to mathematical geography), he devotes 8 books to describing Europe, 6 to Asia, and 1 to Africa (his name ought to be transliterated Strabon).

Geography

[1.1.2: Homer is the ultimate authority]

[1.1.8: the inhabited world is an island; one could sail around the globe: compare Aristotle *Meteorologika* 2.5(362b18–20), *On Heaven* 2.14(298a9–15), Strabo 1.4.6 (from Eratosthenes) and Strabo 2.3.6 (from Poseidonios) expressing the same idea; it is constructed for human and divine benefit: 17.1.36]

[1.2.1b: we know much more thanks to the conquests of the Romans in the west and the Parthians in the east]

[2.2.1–2.3.1: zones according to Poseidonios and Polubios]
[2.5.26: a well-considered description of Europe as a whole]
[4.1.4: an explanation of how the river-system of Gaul greatly enables internal and external communication]

4.5.4–5 [Ireland and "Thoulê," islands at the western edge of the world]

4 There are small islands around Britain, and there is also a large island, Iernê [Ireland], which extends parallel to Britain on the north, of breadth [north to south] greater than length [east to west]. Concerning this island I can tell nothing certain, except that those dwelling there are more savage than the Britons, since they are man-eaters as well as heavy [or "herb-"] eaters, and they also count it honorable to devour their dead fathers, and openly to have sex with women and even with their mothers and sisters; but I say this on the understanding that I have no trustworthy witnesses for it (and yet, as for man-eating, that is said to be a custom of the Skuthians also, and, in cases of necessity forced by sieges, the Keltoi, the Iberians, and several other peoples are said to have practiced it) [cannibalism, public sex, and incest were standard slanders against the uncivilized].

5 Concerning Thoulê our historical information is still more uncertain, on account of its outside position; for Thoulê, of all the countries that are named, they set farthest north. But that the things which Putheas has told about Thoulê, as well as the other places in that part of the world, have indeed been fabricated by him, is clear from the known districts, for he has falsified most of them, as I have said before, and hence he is obviously more false about those outside the inhabited world. And yet, with respect to celestial phenomena and mathematical theory, he might possibly seem to have made adequate use of the facts, since for the people who live near the frozen zone, there is an utter dearth of some domesticated fruits and animals, and a scarcity of others, and that the people live on millet [some local plant that resembled millet] and other herbs and fruits and roots; and where there are grain and honey, the people also get their drink from them. As for the grain, since they have no pure sunshine, they pound it out in large storehouses, after first gathering in the ears thither; for the threshing floors [unroofed in the Mediterranean area] become useless because of this lack of sunshine and the rains.

5.2.7 [Corsica and Sardinia; Theophrastos, *Plant Researches* 5.8.1–2, describes Kurnos as thickly wooded]

But Kurnos is by the Romans called Corsica. It affords such a poor livelihood, being rough and in most parts absolutely impassable, that those occupying the mountains and living from brigandage are wilder than beasts. At any rate, whenever Roman generals have made an assault, and, falling suddenly upon their forts, have taken a large number of slaves, you can at Rome see and marvel at how much bestial and bovine character is manifested in them; for either they cannot endure to live in captivity, or, if they live, they so irritate their purchasers by their apathy and insensibility, that, no matter how much they paid for them, nevertheless they regret the purchase. But still there are some habitable parts, and what might be called towns, namely, Blesinon palisade, Eniconiae and Vapanes ["Venicium" and "Opinum" in Ptolemy, Geography 3.2 (below Section 19)]. The length of the island, says the Chorographer [either Agrippa or perhaps Cornelius Nepos, contemporaries of Strabo], is one hundred and sixty miles, and the breadth seventy; but the length of Sardo is two hundred and twenty, and the breadth ninety-eight. According to others, however, the perimeter of Kurnos is about three thousand two hundred stades [ca. 500 km], and of Sardo as much as four thousand [ca. 630 km].

The greater part of Sardo is rugged and not peaceful, though much of it has also soil that is blessed with all products, especially with grain. There are many cities, and Caralis [Cagliari] and Sulchi [near Caralis] are noteworthy. But the excellence of the sites is offset by a serious defect, for in summer the island is unhealthful [probably malaria: see Chapter 11], particularly in the fruitful districts; and these districts are continually ravaged by those mountaineers who are now called Diagesbes; in earlier times, however, their name was Iolaës; for Iolaos, it is said, came hither, bringing with him some of the children of Herakles, and took up his abode with the foreigners who held the island (the latter were Etruscans). Later on, the Phoenicians from Carthage conquered them, and with them waged war against the Romans; but when they were defeated, everything became subject to the Romans [in –237]. There are four tribes of the mountaineers, the Parati, the Sossinati, the Balari, and the Aconites, and they live in caverns [actually, stone towers called "nuraghi"]; but if they do hold some arable land, they do not sow even this diligently; instead, they pillage the lands of the farmers: not only of the farmers on the island, but they actually sail against the people on the opposite coast, the Pisatae in particular.

[6.4.1: how Italy is marvelously situated for world domination]

(Jones [1923] 259, 261, 263, 359, 361)

5.13 Anonymous The author of the Voyage on the Red Sea seems to have lived in Egypt; he wrote this merchants' guide around 58 ± 12.

Voyage on the Red Sea [(15–18) an accurate description of the coast of
Somalia and Kenya, known as "Azania," down to Dar es Salaam; after
which "the coast bends to the west and joins the western ocean"]

[East coast of Africa]

29 After Kanê [Husn al Ghurab, 13° 59′ N, 48° 19′ E], with the shoreline receding further, there next come another bay, very deep, called Sachalitês, which extends for a considerable distance, and the frankincense-bearing land; this is mountainous, has a difficult terrain, an atmosphere close and misty,

134

and trees that yield frankincense. The frankincense-bearing trees [the shrubby *Boswellia sacra* whose resin is frankincense] are neither very large nor tall; they give off frankincense in congealed form on the bark, just as some of the trees we have in Egypt exude gum. The frankincense is handled by royal slaves and convicts. For the districts are terribly unhealthy, harmful to those sailing by and absolutely fatal to those working there – who, moreover, die off easily because of the lack of nourishment [compare Herodotos' tale of guardian flying snakes: 3.107].

30 On this bay is a mighty headland, facing the east, called Suagros [Ras Fartak, 15° 39′ N, 52° 16′ E], at which there are a fortress to guard the region, a harbor, and a storehouse for the collection of frankincense. In the open sea off it is an island, between it and the Promontory of Spices across the water but nearer to Suagros, called Dioskouridês [Socotra]; though very large, it is barren and also damp, with rivers, crocodiles, a great many vipers, and huge lizards [a now-extinct species of *Varanus*], so huge that people eat the flesh and melt down the fat to use in place of oil. The island bears no farm products, neither vines nor grain. The inhabitants, few in number, live on one side of the island, that to the north, the part facing the mainland; they are settlers, a mixture of Arabs and Indians and even some Greeks, who sail out of there to trade. The island yields tortoise shell [the most frequently mentioned trade-item in the work], the genuine and the land and the light-colored kinds, in great quantity and distinguished by rather large shields, and also the oversize mountain variety with an extremely thick shell, of which the parts over the belly, whichever are useful, do not take cutting [for veneer, the usual Roman use]; besides, they are rather tawny. On the other hand, whatever can be used for small boxes, small plaques, small disks, and similar items gets cut up completely. The so-called Indian cinnabar is found there; it is collected as an exudation from the trees [resin of the *Draco dracaena*, native to Socotra and to the Canaries and Madeira: compare Section 10.6 (Poseidonios)].

[the Indus delta]

38 After this region, with the coast by now curving like a horn because of the deep indentations to the east made by the bays, there next comes the seaboard of Skuthia, which lies directly to the north; it is very flat and through it flows the Sinthos River [the Indus], mightiest of the rivers along the Red Sea [the northern Indian Ocean] and emptying so great an amount of water into the sea that far off, before you reach land, its light-colored water meets you out at sea. An indication to those coming from the sea that they are already approaching land in the river's vicinity are the snakes that emerge from the depths to meet them; there is an indication as well in the places around Persis mentioned above, the snakes called *graai*. The river has seven mouths, narrow and full of shallows; none are navigable except the one in the middle. At it, on the coast, stands the port of trade of Barbarikon. There is a small islet in front of it; and behind it, inland, is the metropolis of Skuthia itself, Minnagar [neither this nor Barbarikon can be located owing to extensive shifts of the Indus]. The throne is in the hands of Parthians, who are constantly chasing each other off it [the first Parthian king, Gondophares, ruled 20 to 46; after that records are scant].

[57: The discovery of the open-ocean route following the monsoons is attributed to captain Hippalos; he may be connected with Eudoxos of Kuzikos: see Poseidonios above.]

[61: Taprobanê is very large and reaches almost to Africa.]

[China and the ends of the earth]

64   Beyond this region [island of Chrusê, possibly Sumatra], by now at the northernmost point, where the sea ends somewhere on the outer fringe, there is a very great inland city called Thina [China] from which silk floss, yarn, and cloth are shipped by land via Baktria to Barugaza and via the Ganges River back to Limurikê [southwest coast of India]. It is not easy to get to this Thina; for rarely do people come from it, and only a few. The area lies right under Ursa Minor and, it is said, is contiguous with the parts of the Pontos and the Caspian Sea where these parts turn off, near where Lake Maiotis, which lies parallel, along with it [sc. the Caspian] empties into the ocean.

[65: where malabathron, cinnamon-tree leaf, comes from]

66   What lies beyond this area, because of extremes of storm, bitter cold, and difficult terrain and also because of some divine power of the gods, has not been explored. [end of book]

(Casson [1989] 67, 69, 73, 75, 91, 93)

5.14 Heron of Alexandria wrote a series of books on science and engineering (*ca.* 55 to 68; see also Chapters 2, 6, 8); the book here excerpted (from 62) explains various surveying techniques, and concludes by explaining how to determine longitude, using Hipparchos' method of lunar-eclipse times. The method was known to the Latin astrologer Manilius, 1.221–9; and to Theon, *Mathematics* 3.2 (below Section 5.17), and Ptolemy, *Geography* 1.4. The few longitude-difference data found elsewhere are grossly excessive and probably map-based: 3 hours between Armenia and Italy (Pliny 2.180) and 4 hours between Persia and Spain (Kleomedes 1.8.3).

Dioptra

35 [determining longitude intervals]

For all places that can be walked to, their extents are found either by the dioptra we've just made, or else by the hodometer [Chapter 34]; but since it's also useful to know how long is the path between two regions when islands and seas and perhaps inaccessible places intervene, it's necessary to have a method besides this, so that our treatise may be complete.

For example, let it be required to measure the path between Alexandria and Rome along a straight line (rather: along a great circle route on the Earth), assuming that the circumference of the Earth is 25 myriad plus 2,000 stades (as Eratosthenes, who treated his much more accurately than others, shows in his book entitled *Measurement of the Earth*) [see also below Section 5.17, Theon]. So let there be observed in Alexandria and Rome the same eclipse of the moon (if one is found in the lists, we'll use that; otherwise we can make the observation and give the result ourselves, because lunar eclipses occur every five or six months). Let this eclipse be found in the given regions like this: in Alexandria at the fifth hour of the night and in Rome the same eclipse at the third hour of the night (of course, the same night). And let this night, that is, the daily circle on which the sun is carried, on the given night [the daily circle, or circle on which the sun is moving, is conceived as perpendicular to and sliding up and down along the axis of the ecliptic circle], be ten days from the spring equinox in the direction of the winter solstice [this is the eclipse of 13 March 62 CE, which is so ill-suited to his method that any later eclipse would have been better, so Heron must have used it because it was very recent: Drachmann [1950]; Neugebauer [1975] 2.844–848].

[Now we transform the local or seasonal hours, which are 1/12 of the nighttime, from one location to another.] Let a hemisphere be drawn through the tropics (if we are in Alexandria, to the *klima* of Alexandria; if we are in Rome, to that at Rome). First, let us be in Alexandria, and let a hollow hemisphere be drawn through the tropics to the *klima* at Alexandria, and let the circle around its lip be ABΓΔ [the horizon], its meridian BEZHΔ, its equator be AHΓ, the pole of the parallel circles [the largest of which is the equator] be E, and the pole of the hemisphere at the lip be Z [local nadir] (Figure 5.3). And let there be inserted on the same surface the circle on which the sun is moving on the given night at the fifth hour (it was 10 days before the spring equinox), and let it be ΘΚΛ [which will almost coincide with the equator at such a time]; and let the arc ΘΚΛ be divided into 12 parts and let five of these be ΘM (since the eclipse was observed at the fifth hour in Alexandria) – then the point M will coincide with the point where the sun is at the eclipse.

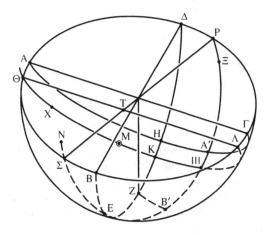

Figure 5.3

And let the *analemma* through Rome be drawn, on which is also drawn the daily circle corresponding to ΘΚΛ (Figure 5.4); the diameter of the horizon is ΝΞ, the gnomon is ΟΠ, the diameter of the daily circle is ΡΣ, and the terminator is ΤΥ. Now with the arc ΥΦΣ representing six seasonal hours, the arc ΥΦ represents 3 such hours (since the observation in Rome was made at the third hour), and [on ΘΚΛ] let ΧΜ be similar to the arc ΥΦ. Then the point Χ is on the horizon at Rome [since the sun is at Μ, and ΥΦ is a three-hour arc, and it is the third hour of night at Rome]. Let the arc ΧΚΙΙΙ be similar to the arc ΥΦΣ [both on the sun's daily circle]; then ΙΙΙ will be on the meridian at Rome [since the arc ΧΚΙΙΙ is a six-hour arc, and Χ is on the horizon]; but Ε is the pole of the parallel circles, so let the great circle ΕΙΙΙΑ′ be drawn through Ε and ΙΙΙ, then this will be the meridian at Rome. Let an axis of the *analemma* be ΨΩ, and let Α′Β′ be laid out similar to the arc ΞΩ [whose angular length is the co-latitude of Rome], [on arc ΕΑ′] from Α′ draw an arc Α′Β′ similar to ΞΩ (and forming a [spherical] quadrilateral Α′Β′ΖΗ); then point Β′ will be the pole of the horizon at Rome (and Ζ is the pole of the horizon at Alexandria). Let the arc Β′Ζ of a great circle be drawn through points Β′ and Ζ, and find out how many times that arc goes into the circle ΑΒΓΔ. [Note: the daily circle contains the points Θ western horizon at Alexandria, Χ western horizon at Rome, Μ sun, Κ meridian at Alexandria, ΙΙΙ meridian at Rome, and Λ; the points Χ and Υ correspond, and the points Μ and Φ correspond, ΙΙΙ is used for the letter "san".]

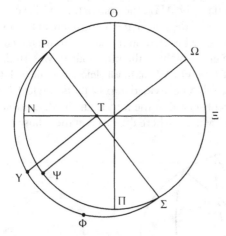

Figure 5.4

For example, let Β′Ζ be found to be 20 degrees, then the chosen path on the Earth between Rome and Alexandria will have 20 degrees (where the great circle has 360) [actually about 17° 24′; "two hours' time" would be 30°; and Ptolemy's data give 21° 02′]. And one degree on the Earth is 700 stades (where the whole perimeter is 25 myriad and 2000), so the 20 degrees becomes 1 myriad and 4000 stades. Thus we have found the number of stades and the length of the given path.

*(Schoene [1903/1976] 302, 304, 306; Rome [1922/1923] 239–240)

5.15 Plutarch of Chaironeia (*ca.* 50 to *ca.* 120), writing under the Flavian emperors, Trajan, and in the early years of Hadrian, was a Platonist teacher of philosophy; much of his copious output is in the form of moralizing essays or dialogues. In this extract he invents a geographical myth about western lands across the Atlantic Ocean, in the tradition of Plato, Euhemeros of Messênê, Hekataios of Abdera, Antonius Diogenes' *Wonders Beyond Thoulê*, and Loukian of Samosata's *True History*.

The Face in the Moon

26 (941a–c) ["Atlantis"]

"An isle, Ogugia, lies far out at sea," [*Odyssey* 7.244] a run of five days off from Britain as you sail westward; and three other islands equally distant from it and from one another lie out from it in the general direction of the summer sunset [west-north-west]. The foreigners tell the tale that in one of these, Kronos is confined by Zeus, and the "ogugian" Briareus ["ogugian" means "primeval": from Oguges, a mythical king of Athens; Briareus is in Hesiod, *Theogony* 729–735], holding guard over those islands and the sea that they call the Kronian main, has been settled close beside him. The great continent, by which the great ocean is encircled, while not so far from the other islands, is about five thousand stades [*ca.* 800 km] from Ogugia, the voyage being made by oar, for the main is slow to traverse and muddy as a result of the multitude of streams. The streams are discharged by the great land-mass and produce alluvial deposits, thus giving density and earthiness to the sea, which has been thought actually to be congealed [compare Putheas, above Section 5.2].

On the coast of the mainland Greeks dwell about a gulf which is not smaller than the Maiotis [Herodotos 4.86 reported it as huge, while Strabo 2.5.23 made it 9000 stades around; it is *ca.* 250 km across] and the mouth of which was roughly on the same parallel as the mouth of the Caspian sea [i.e., the north end, at 47° N: Plutarch, like most geographers between Alexander the Great and Ptolemy, is assuming that the Caspian connects to the northern ocean]. These people consider and call themselves continentals, and the inhabitants of this land islanders because the sea flows around it on all sides. And they believe that with the peoples of Kronos, there later mingled people arriving with Herakles and left behind, and that these latter so to speak rekindled again to a strong, high flame the Hellenic spark which was already being quenched and mastered by the tongue and laws and manners of the foreigners. Therefore Herakles has the highest honors and Kronos the second.

(Cherniss and Helmbold [1957] 181, 183, 185)

5.16 Marinos of Tyre is known to us almost entirely from Ptolemy's *Geography*; around 110 ± 5 he sought to map correctly the known world (using a cylindrical projection), believing it to encompass some 180 to 225 degrees of longitude (from the Canaries to China), and relying on recent Roman exploration of remote lands (Gulf of Tonkin, Lake Chad, Zanzibar, and even Ireland). Since he adopted Poseidonios' estimate of the size of the earth, he concluded that the ocean between China and western Europe was small. Ptolemy describes his work as on the whole excellent and careful.

Geography

Book 3 *(quoted in Ptolemy,* Geography *1.7.4, 1.7.6)* [navigation by the stars]

The Zodiac is considered to lie entirely above the torrid zone and therefore in that zone the shadows alternate [northward and southward], and all the fixed stars rise and set, except Ursa Minor. Ursa Minor begins to be entirely above the horizon 500 stades [ca. 80 km] north of Okelis [Gureira, 12° 42′ N], since the latitude circle through Okelis is at 11⅖° above the Equator. But Hipparchos reports that the southernmost and outermost star of the tail of Ursa Minor [Polaris] is 12⅖° from the Pole. Now if one proceeds from the Equator to the summer tropic, the North Pole continuously rises above the horizon and the South Pole sinks below it; and if one travels from the Equator to the winter tropic, the South Pole continuously rises above the horizon and the North Pole sinks below it.

[Ptolemy notes that Marinos gave other, irrelevant, astronomical data]

Those who travel to Limurikê in India have, as Diodoros of Samos tells in his third book [otherwise unknown, unless the man for whom the islands in *Voyage on the Red Sea* 4 and 25 were named], the Bull in midheaven and the Pleiades above the middle of the yardarm; but sailors from Arabia to Azania direct their course straight toward the South and the star Kanopos [α Car], which is there called the Horse, and is directly south. Stars are visible to them which are not even known by name to us, and the Dog [Sirius] there rises earlier than do Procyon and Orion north of the summer tropic.

Book ? *(in Ptolemy,* Geography *1.11–14)* [the route to China]

[Marinos gives distance intervals from the Fortunate Isles to the Euphrates at Hieropolis (also known as Bambukê; modern Manbij at 37° 31′ N, 37° 57′ E), which add up to 28,800 stades, and thence in order to the Caspian Gates, Hekatompulon (Shahr-i Qumis near Dâmghân 36° 09′ N, 54° 22′ E), the city Hurkania (also known as Zadra-karta or Karta; modern Sârî, 36° 34′ N, 53° 04′ E), the city Antioch Margiana (Gyaur-Kale, 30 km east of Mary (Merv), 37° 42′ N, 61° 54′ E), through Baktria, and to the "Stone Tower" (Tachkurgan south of the Amu-Darya River, at 36° 44′ N, 67° 41′ E) he gives distances which total 26,280 stades. From there eastward to China is 36,200 stades.]

Marinos says that a certain Maês aka Titianus, a Macedonian and a merchant like his father, composed this measurement, although he himself did not travel to China but sent people.

[Ptolemy then paraphrases Marinus' periplous from Tabrobanê to Kattigara; Marinos is recording "loxodromic" triangles, in which from the three known quantities, hypotenuse distance, angle with the meridian, and change in latitude, the change in longitude can be computed – we omit Ptolemy's commentary.] In sailing from India to the region of Sinaros and Kattigara: after the promontory called Koru [opposite Taprobanê: Kalimîr, 10° 18′ N, 79° 52′ E], which closes the bay of Kolchis [gulf of Manar between India and Sri Lanka], the bay of Agarikos [Palk Bay] is next, and it extends to the city of Kurula [Kârikal, 10° 55′ N, 79° 50′ E], a distance of 3400 stades. The city is situated north of the promontory of Koru. From the city of Kurula the course of navigation is toward the winter rising of the sun [the direction is actually not south-east, but north-east] as far as Palura [Ganjâm, 19° 23′ N, 85° 05′ E] and measures 9450 stades; the shore of the Gangetic bay is placed at a further distance of 19,000 stades [much exaggerated]. From Palura to the city of Sada is 13,000 stades by navigating the aforesaid bay toward the equatorial rising of the sun [4/5 the way from Ganjam to Tamala along the coast would indeed be almost due east of Ganjam, placing Sada near Kyaukpyu, 19° 24′ N, 93° 34′ E]. The voyage from Sada to the city of Tamala [near Cape Negrais, itself 16° 02′ N, 94° 13′ E] measures 3500 stades in the direction toward the winter rising of the sun. The next distance from Tamala across the Golden Chersonesos [Malaysia, near Rangoon, 16° 47′ N, 96° 13′ E] is 1600 stades toward the winter rising of the sun. Alexander [otherwise unknown] wrote that the shore-line extends to the south from the Golden Chersonesos to Kattigara [Oc-éo at the Mekong delta is one candidate, another is mediaeval Kiau-tschi, modern Hanoi], and that those sailing along the shore came, after twenty days, to Zaba [east of Taprobanê and possibly the Nicobar Islands]. From Zaba carried southward and toward the left, they came after some days to Kattigara.

[in 1.17 Ptolemy writes that more recent voyagers to China and Kattigara describe the trip differently, placing the capital of China north-east not north of Kattigara]

*(Müller [1883] 17–19, 27–39)

5.17 Theon of Smurna, a Platonist of the era of Hadrian (*ca.* 130 ± 10), wrote a textbook whose full title well describes it: *Aspects of Mathematics Useful for the Reading of Plato* (book 3 is on astronomy); much of it may be from his contemporary Adrastos of Aphrodisias the follower of Aristotle (whom he does often explicitly quote, e.g., 3.1).

Mathematics

3.2 [sphericity of the Earth]

[Compare Aristotle, *On Heaven* 2.14 (297a8–298a20), who offers four "proofs": a thought experiment of allowing the Earth to form from its parts, which are centripetal; that convergent vertical falls are towards a sphere; that lunar eclipses always show convex shadows; and that the rising and setting of the ever-visible stars alters going north or south. Theon retails these in the order 4, 1, and 2, while omitting 3 – as incomprehensible or as inconclusive?]

The risings and settings of stars make the sphericity of the Earth from east to west manifest, the same stars making risings and settings sooner in the east and later in the west; and a given lunar eclipse, completed in a brief time, and seen by everyone at once for whom it is visible, appears at different hours and always for those further east at a later stage [at a given local time], since the sun does not illumine all parts at once (due to the roundness of the Earth), and the Earth's shadow being analogously opposite (since this happens at night).

The Earth is rounded from the arctic and northerly parts to the southerly and midday parts. For those proceeding in this direction in their progress toward the south see many of the ever-visible stars around our upper pole make risings and settings, and likewise some of those always invisible stars around the part hidden from us appear rising and setting. For example, the star called Kanopos, invisible in places north of Knidos, becomes visible in places south of this, and more visible in places further south. Contrarily, for those coming from south to north, many of the stars down there (formerly making risings and settings) become entirely invisible, and some of those rising and setting nearer to the Bears become always visible to the travelers, and these stars always move as the travelers make more progress.

Since she is rounded every way, she is also spherical [omitting the possibility of oblateness or prolateness]. Since things naturally having weight are moved toward the center of All, if we conceive some parts of the Earth at a greater distance from the middle because of their size, it is necessary that the smaller bits surrounded by these be expelled, and weighted down they are pressed back and shoved away from the center, until all are equally distant and under equal forces and in equilibrium and established at rest (just like crossing rafters or wrestlers encompassing one another with equal force). And so if all parts of the Earth on all sides are equidistant from the center, the form of the Earth would be spherical.

Then, since the fall of weights is everywhere toward the center, all of them converging towards one point, and each of them is carried vertically, that is each makes equal angles to the surface of the Earth along its line of fall, this also indicates that the surface of the Earth is spherical.

3.3 [sphericity of ocean]

But the spherical figure also occurs at the surface of the sea and all water at rest. And this is plain to observe there: for if one sits on any shore and contemplates anything across the water, such as a mountain or tree or tower or ship or the Earth itself, then bending down and casting one's gaze along the surface of the water, one will then see either nothing at all or less of the thing formerly more visible, since the convexity of the sea's surface intercepts the sight [compare Strabo 1.1.20]. And often in sailing when from the ship the land or another vessel is no longer seen, the sailors climb the mast and see it, since they are elevated and looking over the top of the convexity of the sea (which intercepted their sight). [Theon sketches the proof in Archimedes, *Floating Bodies* prop. 1]

3.3 [heights of mountains]

Now let no-one consider the height of the mountains or the lowness of the plains as sufficient deviation from this account of the shape of the whole earth. For Eratosthenes shows that the size of the Earth measured around a great circle is 25 myriad stades plus two thousand stades [the extra 2,000 were added to make one degree a round number of stades, 700, rather than 694.4, from the unaugmented 25 myriads], and Archimedes has shown that extending the circumference of the circle in a straight line, it is three times and about one-seventh greater than the diameter [see Chapter 2.2]; so that the whole diameter of the Earth is just about 8 myriads and 182 stades (thrice and one seventh more of which is a perimeter of 25 myriad and two thousand stades).

But the vertical height of the highest mountains from the lowest part of the Earth is 10 stades, as Eratosthenes and Dikaiarchos say they find (they so conclude by means of instruments, measuring such heights from separations, with dioptras) [Keyser [2001]: Dikaiarchos' measurements of mountains in and around Greece gave heights of 8 to 15 stades; he may in fact have been intending to argue for a "dimpled-earth" model like that of Plato or Aristotle – see the *Kosmos* passage above]. So then the height of the greatest mountain is just about one eight-thousandth of the diameter of the whole Earth. [Theon – or Adrastos – goes on to calculate that on a one-foot-diameter sphere, mountains to scale would be smaller than 1/40 of a grain of millet.]

*(Dupuis [1892/1966] 200, 202, 204, 206)

5.18 Arrian of Nikomedia (*ca.* 86 to *ca.* 160) studied philosophy and fancied himself a latter-day Xenophon of Athens; in his mid-twenties he made friends with the future Emperor Hadrian, who appointed him to various offices. His geography includes an extant description of India, plus this work, which was written about 131 (as a letter to Hadrian), while he served as legate of Kappadokia; a work preserved only in quotations, *Physics*, recorded that some mountains were measured to be 20 stades high (compare Theon 3.3, above Chapter 5.17). See also Chapter 10.14.

Voyage on the Black Sea

8 [Phasis River]

Then we sailed into the Phasis [Rioni] (distant 90 stades [*ca.* 15 km] from the Mogros [river]), which, of the rivers I know, has the lightest water and the most altered color. One might prove that it is lightest in the balance and by this, that it floats upon the sea and is not mixed in, just as Homer says the Titaresios floats upon the Peneios "as if atop oil" [*Iliad* 2.754]. And it is possible to dip a pitcher in the surface and draw up sweet water, but if one lets it down into the depths one can draw up salt water. And indeed the whole Euxine has water much sweeter than the sea outside: the rivers are the cause of this, measurable neither in size nor number [compare Straton, above Chapter 5.3]. Proof of the sweetness, if one needs proofs in addition to perceptible phenomena, is that the shore dwellers lead all their cattle down to the sea and water them from it; it is observed that they drink gladly and the account is that this drink is more beneficial to them than sweet water. The color of the Phasis is that of water tinged with lead or tin, but on standing it becomes most pure. Arriving sailors are not allowed to mix water into the Phasis, but before they enter the stream, they must pour out all water on board; if they don't, the account is that those who've neglected this will not fare well at sea. The water of the Phasis does not go foul, but remains pure for more than ten years, except that it becomes sweeter. [Herodotos 2.104 and Hippokrates *Airs, Waters, Places* 18 consider the Phasis' water stagnant and unpotable.]

21 [island of Achilles]

Just opposite the mouth [Psilon, of the Ister, i.e., Danube] there lies an island on the sea, in the path of anyone sailing on the north wind, which some name the Isle of Achilles, some the Course of Achilles [often a distinct, coastal, region], and some White (from its color). It is said that Thetis dedicated it to her son and that Achilles dwells there [the epic poet Arktinos of Miletos *ca.* –600 first told the story; compare Pindar, *Nemeans* 4.49; Euripides, *Iphigeneia in Tauris* 435–438; and Strabo 7.3.16]. There's a temple of Achilles on it, and a wooden idol of ancient style. The isle is empty of people, but a few goats live there, which people coming here are said to offer to Achilles. Many other offerings are hung up in the temple, bowls, rings, and precious stones, all of which are hung up as offerings to Achilles, and inscribed, some Latin some Greek, praising Achilles in various meters; there are also some praising Patroklos, for whoever wants to please Achilles honors Patroklos with Achilles.

Many birds lodge on the Isle, seagulls, shear-waters, and sea-crows, in number immeasurable; these birds care for the temple of Achilles. Every day at dawn they fly down to the sea; then having bathed their wings in the sea, they hasten to fly back to the temple and sprinkle it; when they have done this well, they sweep the pavement with their wings.

*(Silberman [1995] 6–7, 18–19)

5.19 Ptolemy (Claudius Ptolemaeus) of Alexandria (*ca.* 100 to *ca.* 175) wrote on a wide variety of "mathematical" topics (see Chapters 2–8). His *Geography* attempts to achieve Hipparchos' suggestion, with longitude data still based on calculation from travel (not astronomical observation), skewing his map (sites in books 2–7; book 8 describes his 26 maps).

Geography

1.1 [introduction; scope of geography]

Geography is a representation in picture of the whole known world together with the phenomena which are contained therein. It differs from chorography, which selects certain places from the whole to treat by themselves more fully, even dealing with the smallest details, such as harbors, villages, districts, tributaries, and such like. It is the essence of geography to show the known habitable earth as a continuous unit, how it is situated and what is its nature; and it deals with those features relevant to be mentioned in a more comprehensive and general description of the earth, such as gulfs, the larger towns and nations, and the principal rivers.

[goal of chorography]

In geography one must contemplate the shape and extent of the whole earth, and also its position under the heavens, in order rightly to state what are the size and nature of the known part, and under what parallels of the celestial sphere the individual places are located, for so one will be able to discuss the length of its days and nights, the fixed stars which are overhead, the stars which always move above the horizon, and those which never rise above the horizon; in short all information included in an account of our habitations [latitudinal astronomical phenomena: compare Section 3.9 (Theodosios)].

These are part of the highest and most exquisite contemplation (*theôria*), mathematics, which shows to the human intelligence the heaven itself in its own character (which we can see moving around us), yet represents the Earth by a model, for the real earth being very large, and not surrounding us (as does the heaven), can neither wholly nor in part be traversed by one person.

[1.2–5: astronomical measurements are far preferable, but there are very few of them, so we have to rely, as carefully as possible, on distances measured on the surface]

[1.21–24: his mapping projection]

3.2 [Corsica]

[Ptolemy is probably relying mostly on Latin sources here; only Aleria and Mariana have been excavated.]

Kurnos island, which is also called Corsica (Figure 5.5), is surrounded on the west and the north by the Ligusticum sea, on the east by the Turrhenian sea, and on the south by that sea which lies between it and Sardo [Sardinia] (Figure 5.6). The coast of this island, if we begin in the middle on the north side, is described in this order:

mouth of the Volerius River [Ostriconi] 30° 40' E, 41° N
Tilox prom. [Pta. Vallitone or Revellata] 30° 30' E, 41° 10' N
Caesia coast [Golfo Crovani] 30° E, 41° N

Table 5.1 Ptolemy's description of Corsica, west coast

Feature	Longitude	Latitude
Cattius prom. [Pta. Palazzo]	30° E	41° 10' N
Casalus bay [Golfo Pero or Porto]	30° 15' E	40° 25' N
Viriballum prom. [Pta. Puntiglione]	30° 10' E	40° 30' N
Circidius R. mouth [Liamone R.]	30° 10' E	40° 25' N
Rhoitios Mts. [Feno]	30° E	40° 20' N
Rhion prom. [Pta. Parata]	30° E	40° 15' N
Urcinum [Ajaccio or Orchino[a]]	30° 10' E	40° 10' N
Sandy coast	30° 15' E	40° N
Locra R. mouth [Taravo R.]	30° 10' E	39° 55' N
Pauca [Propriano]	30° 15' E	39° 45' N
Ticarius R. mouth [Rizzanese R.]	30° 15' E	39° 40' N
Titianus harbor [Tizzano]	30° 10' E	39° 35' N

[a] 7 km inland.

Description of the south coast:

Ficaria [Figari] 30° 30' E, 39° 30' N
Pitanus River mouth [Ventilegne River or Ortolo River] 30° 45' E, 39° 20' N
Marianum (town and prom.) [Bonifacio] 31° E, 39° 10' N

Table 5.2 Ptolemy's description of Corsica, east coast

Feature	Longitude	Latitude
Palla[a]	31° 20' E	39° 20' N
Syracusan Port [Porto Vecchio]	31° 20' E	39° 25' N
Rubra [on the Oso R.?]	31° 20' E	39° 30' N
Granianum prom. [Guardia, near Pto. Favone]	31° 30' E	39° 40' N
Alista [on the Solenzara?]	31° 20' E	39° 45' N
Philônios port [Palo]	31° 30' E	39° 55' N
Sacer R. [Fiumorbo R.]	31° 30' E	40° N
Aleria colonia [on the Tavignano]	31° 30' E	40° 5' N
Rotanus R. mouth [Tavignano R.]	31° 30' E	40° 10' N
Artemis' harbor [Diane lagoon]	31° 20' E	40° 20' N
Tutêla (altars) [Ordetella]	31° 30' E	40° 30' N
Guola R. mouth [Golo R.]	31° 30' E	40° 35' N
Mariana [La Canonica[b]]	31° 20' E	40° 40' N
Vagum prom. [Pta. d'Arco]	31° 30' E	40° 45' N
Mantinôn [Brando?]	31° 20' E	41° N
Clunium [Pietra Corbara?]	31° 20' E	41° 10' N

[a] 24 km from Bonifacio, on the Golfo di Sta. Giulia; [b] 20 km S. of Bastia.

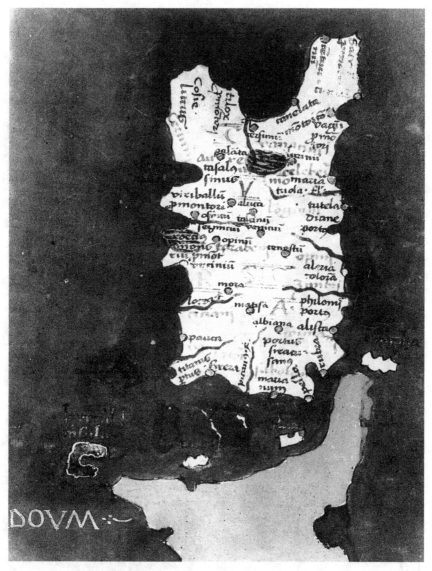

Figure 5.5 Ptolemy's Map of Europe, VII (Corsica)

Description of the northeast coast:

> Sacrum prom. [Cap Corse] 31° 30′ E, 41° 35′ N
> Centurinum [Centuri] 31° 15′ E, 41° 30′ N
> Canelatê [Pta. le Canelle] 31° E, 41° 20′ N

The native races inhabiting the island are the following: the Merveni, occupying

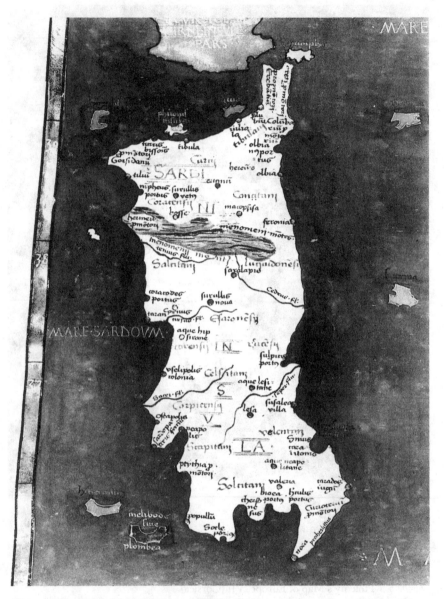

Figure 5.6 Ptolemy's Map of Europe, VIII (Sardinia)

the western part above the Golden mountains [Cinto or d'Oro], and the location is 30° 45′ E, 40° 45′ N. Below these are the Tarabeni [Taravo River], then the Titiani, next the Balatôni, more to the north are the Vanacêni, below whom are the Scilêbênses, then the Licnini and the Macrini, below whom are the Opini; then the Surbi and the Côümaseni; below these but more toward the south are the Subasani [or Tebusani].

148

The towns in the interior are:

Table 5.3 Ptolemy's description of Corsica, interior towns

Town	Latitude	Longitude
Ropicum [Calenzana?]	30° 15′ E	41° N
Cersunum [Belgodere?]	30° 30′ E	41° N
Palania [Evisa?]	30° 20′ E	40° 45′ N
Lurinum [?]	31° E	40° 45′ N
Aluca [near Vico?]	30° 20′ E	40° 30′ N
Asincum [Asingo *or* near Soccia?]	30° 30′ E	40° 30′ N
Sermigium (-tium?) [Serraggio *or* Sermano]	30° 20′ E	40° 20′ N
Talcinum [Corte?]	30° 45′ E	40° 30′ N
Venicium[a] [Venaco]	30° 50′ E	40° 20′ N
Cenestum [San Giovanni]	31° E	40° 15′ N
Opinum[a] [Opino NW of Aleria]	31° 20′ E	40° 25′ N
Moursa [near Bastelica?]	30° 30′ E	40° N
Matisa [?]	30° 45′ E	39° 35′ N
Albiana [?]	31° E	39° 30′ N

[a] compare Strabo 5.2.7, above Chapter 5.12.

(Stevenson [1932] 25–26, 76)

6

MECHANICS

Things move and thus appear animate, possessing autonomous power. Hence, humanity and animals are separated from the rest of the natural world, although moving celestial lights seemed divine (see Chapter 4). Yet fire rises and weights fall without animate guidance, subjects of some pan-terrestrial cause. Divine advice was sought from spontaneous fire-induced cracks in, and scatters of, clay (Vandiver *et al.* [1989]), or of tortoise shells (Shang-dynasty China). Meanwhile powerful devices creating or augmenting motion were devised: spear-throwers, bows, boats, the wheel (in Mesopotamia), and the sail (in Egypt).

Greeks built myths from the magic of autonomous motion, telling of self-moving tripods and living golden girls and dogs wrought by Hephaistos (*Iliad* 18.368–379, 417–420; *Odyssey* 7.91–94). The legendary engineer Daidalos of Crete built a theatre and labyrinth at Knossos, then flew by ornithopter to Sicily to create further marvels; later tales credited him with a puppet of Aphrodite moved by flowing quicksilver (according to Demokritos as recounted by Aristotle, *On the Soul* 1.3[406b16–23]), and manifold other devices (Diodoros of Sicily 4.76–78).

Having asked themselves what everything was made from, Greek thinkers began also to ask how change happens, and so, if there was a beginning, what first set Everything in motion. Parmenides (–470 ± 20) and his student Zenon (–455 ± 10), both from Elea, denied that change and motion were real, as did Melissos of Samos (–455 ± 20). In response, Empedokles of Akragas (also –455 ± 20) interpreted change and motion as manifestations of four eternal "roots" cyclically separated and rejoined by "Love" and "Strife." On the Athenian stage myth and fantasy came to life, aided by machines: from *ca.* –460 the *ekkuklêma* (wheeled platform) exposed "interior" tableaux, while the *mêchanê* (crane) elevated gods (and Sokrates). And in the Athenian state, politics employed the *kleroterion* machine to select individuals at random for official service (Aristophanes, *Women Legislators* 681–686; Euboulos in Athenaios, *Deipnosophists* 10 [450b]; Aristotle, *Athenian Constitution* 63.2; and Boegehold and Crosby [1995] 230–234).

Leukippos (–450 ± 20) in response to Zenon and followed by Demokritos of

Abdera (–410 ± 30) supposed motion was explained as an intrinsic property of the ultimate constituents of matter, "atoms," which eternally speed through the void (objects are aggregates). The works *Fractures* and *Joints* in the medical library ascribed to Hippokrates (–400 ± 20) contain simple devices for setting fractures, the reductions being forced through levers and pulleys or winches (*Fractures* 8, 13, 30–31; *Joints* 8, 43, 47, 72–73). Here the skeleton is clearly seen as a mechanical armature, an insight exploited by puppeteers (Herodotos 2.48).

The Pythagorean Archutas of Taras (–380 ± 20) is credited with a mechanical bird (A. Gellius 10.12.9–10); other Pythagoreans had offered mechanical explanations of sound and music. The invention of artillery is reliably attributed to engineers sponsored by Dionusios I of Syracuse, in –398: various devices greatly augmented the power of the bow (Diodoros of Sicily, 14.41.3–14.42.1); Zopuros of Taras credited with a crossbow by Biton may have preceded that by a few years. Motion appears in the myths of several of Plato's dialogues (–365 ± 20), as the consequence of the heavenly Prime Mover (*Statesman* 269–270), or as the essence of mundane nature (*Theaitetos* 156–157), or as perfected in simple rotation (*Timaios* 34a, *Laws* 10 [893cd]). Ordinary motion probably requires no void, for everything simply rearranges (*Timaios* 79bc, 80c). Meanwhile warfare, partly exacerbated by artillery, encouraged Aineias to compose manuals – one of which, the *Tactics* (–356 ± 2), survives – full of mechanical devices to defend and assault cities.

The system of Aristotle of Stageira (–335 ± 10) reduced all locomotion to "forced" motion and two simple "natural" motions, linear and circular (*On Heaven* 1.2–3 [268b11–270b31]; *Physics* 5.6 [230a19–b23], 8.4 [254b13–6a4], and 8.8 [261b29–2a13]). The natural motion of mundane matter is linear and has speed proportional to weight (*On Heaven* 1.6, 1.8, 4.2: compare below Section 6.6, the spring-catapult), while forced motion has speed inversely proportional to weight (*Physics* 7.5 [249b27–250b8]). Motion through a medium (air or water) has speed proportional to the medium's "thinness," so void cannot exist (*Physics* 4.8 [215a25–6a12]), and the medium is intimately entangled in projectile motion (*Physics* 8.10: compare below Section 6.12). He treats "puppets" (i.e., mechanical automata) as well-known examples (*Movement of Animals* 7 [701b2–13], *Generation of Animals* 2.1 [734b11–19], 2.5 [741b9], *Dreams* 3 [461b15–18]), and devotes two shorter works to animate motion (*Movement of Animals* and *Progression of Animals*).

The *Mechanics* ascribed to Aristotle exhibits skill (*technê*) as struggling to master nature and interprets many devices as levers: on ships (oars, rudder-oars, masts), the sling, the wedge, the pulley, and even tooth-extraction. But the true cause of the lever itself is the marvelous circle which synthesizes opposition. Ktesibios of Alexandria (–270 ± 20) wrote primarily on pneumatics, skillfully deploying levers and wheels to transmit motion to and from water and compressed air (see Chapter 8), but he also designed a scaling ladder (described by Athenaios).

6.1 Eukleidês, who worked in Alexandria under Ptolemy I (*ca.* –300), wrote a textbook on music (as well as on mathematics and astronomy and optics), in which he explains the mechanical origin of sound.

Division of the Scale

Praeface [mechanical production of sound]

If there were stillness and no movement, there would be silence: and if there were silence and if nothing moved, nothing would be heard. Then if anything is going to be heard, impact and movement must first occur [compare Archutas of Taras fr. 1; pseudo-Aristotle, *Audibles* 1 (800a1–17) discusses the mechanical production of sound]. Thus since all sounds occur when some impact occurs, and since it is impossible for an impact to occur unless movement has occurred beforehand – and since of movements some are closer packed, others more widely spaced, those which are closer packed producing higher notes and those which are more widely spaced lower ones – it follows that some notes must be higher, since they are composed of closer packed and more numerous movements, and others lower, since they are composed of movements more widely spaced and less numerous [compare pseudo-Aristotle, *Problems* 19.39; pseudo-Aristotle, *Audibles* 17 (803b26–804a9)]. Hence notes that are higher than what is required are slackened by the subtraction of movement and so reach what is required, while those which are too low are tightened by the addition of movement, and so reach what is required.

(Barker [1989] 191–192)

6.2 Epikouros of Samos (–340 to –269, and a citizen of Athens) adapted the theories of Demokritos, although his own goal was not explanation but peace of mind. Thus, his method was to allow for as many models as possible, so long as all were natural. He rarely offers novel explanations, merely rendering existing ideas into atomist terms.

Letter to Herodotos

38–61 [atomic theory]

38 Having grasped these points, we must now observe, concerning the non-evident, first of all that nothing comes into being out of what is not. For in that case everything would come into being out of everything, with no need for seeds. Also, if that which disappears were destroyed into what is not, all things would have perished, for lack of that into which they dissolved. Moreover, the totality of things was always such as it is now, and always will be, since there is nothing into which it changes, and since beside the totality there is nothing which could pass into it and produce the change.

39   Moreover, the totality of things is bodies and void. That bodies exist is universally witnessed by sensation itself, in accordance with which it is necessary to judge by reason that which is non-evident, as I said above; and if place, which we call "void," "room," and "intangible substance," did not exist, bodies would not have anywhere to be or to move through in the way they are observed to move. Beyond body and void nothing can even be thought of, either by imagination or by analogy with what is imagined, as completely substantial things and not as the things we call accidents and properties of these.

40   Moreover, of bodies some are compounds, others the constituents of those compounds. The latter must be atomic [literally "uncuttable"] and unalterable – if all things are not going to be destroyed into the non-existent but be strong enough to survive the dissolution of the compounds – full in nature, and incapable of dissolution at any point or in any way. The primary entities, then, must be atomic kinds of bodies.

. . .

43   The atoms move continuously for ever, some separating a great distance from each other, others keeping up their vibration on the spot whenever they happen to get trapped by their inter-linking or imprisoned by atoms which link up. For the nature of the void brings this about by separating each atom off by itself, since it is unable to lend them any support; and their own solidity causes them as a result of their knocking together to vibrate back, to whatever distance their inter-linking allows them to recoil from the knock. There is no beginning to this, because atoms and void are eternal.

. . .

46   Moreover, the lack of obstruction from colliding bodies makes motion through the void achieve any imaginable distance in an unimaginable time. For it is collision and non-collision that take on the resemblance of slow and fast. Nor, on the other hand, does the moving body itself reach a plurality of places simultaneously in the periods of time seen by reason. That is unthinkable. And when in perceptible time this body arrives in company with others from some point or other in the infinite, the distance which it covers will not be one from any place from which we may imagine its travel. For that will resemble collision – even if we do admit such a degree of speed of motion as a result of non-collision. This too is a useful principle to grasp.

. . .

61   Moreover the atoms must be of equal velocity whenever they travel through the void and nothing collides with them. For neither will the heavy ones move faster than the small light ones, provided nothing runs into them; nor will the small ones move faster than the large ones, through having all their trajectories commensurate with them, at any rate when the large ones are suffering no collision either. Nor will either their upwards motion or sideways motion caused by knocks be quicker, or those downwards because of their individual weights. For however far along either kind of trajectory it gets, for that distance it will move as fast as thought, until it is in collision, either through some external cause or through its own weight in relation to the force of the impacting body. Now it will also be said in the case of compounds that one atom is faster than another, where they are in fact of equal velocity, because the atoms in the complexes move in a single direction even in the shortest continuous time, although it is not single in the periods of time seen by reason; but they frequently collide, until the continuity of their motion presents itself to the senses [Lucretius, *Nature of Things* 2.308–322 provides the metaphor of a sheep-herd: individuals move while the collection stands still].

(Long and Sedley [1987] 25, 27–28, 37–38, 46, 48)

6.3 Straton of Lampsakos was the third head of Aristotle's school (from –286 to –269), and wrote on a wide variety of scientific topics (so that he was called "the naturalist"): see also Chapters 5.3 and 12.3.

Motion

(quoted by Simplicius, Commentary on Aristotle's "Physics" p. 916 Diels)

["Everyone agrees there is acceleration, but not why; Straton asserts that a falling body completes the last stage of its trajectory in the shortest time."]
In the case of bodies moving through the air by their weight this is clearly what happens. For if one observes water pouring down from a roof and falling from a height, the stream at the top is seen to be continuous, but lower down the water falls to the ground in discontinuous parts. This would not happen unless the water traversed each successive space more swiftly.

. . .

If one drops a stone or any other weight from a height of about an inch, the impact made on the ground will barely be perceptible, but if one drops the object from a height of a *plethron* [about 30 m] or more, the impact on the ground will be a powerful one. Now there is no other cause for this powerful impact. For the weight of the object is not greater, the object itself has not become greater, it does not strike a bigger place, nor is it impelled by a greater force, but it is moving faster. And it is because of this acceleration that this phenomenon and many otherwise take place [Aristotle, *Physics* 5.6 (230b23–28) hints at this view of acceleration; compare Gottschalk [1965] 139].

(Cohen and Drabkin [1958] 211–212)

6.4 Chrusippos of Soloi (*ca.* –280 to –206) wrote so comprehensively and authoritatively on the Stoic teachings that his interpretation became standard;

however, all his writings are lost. He studied and wrote in Athens for about a quarter of a century, and then was head of the Stoic school from −231.

(title unknown)

(paraphrase in Ioannes of Stobi, Selections 1.161)

["place" is incorporeal extension within the kosmos, while "void" is extra-kosmic incorporeal extension]

Chrusippos declared place to be what is occupied through and through by an existent, or what can be occupied by an existent and is through and through occupied whether by one thing or by several things.

. . .

The void is said to be infinite. For what is outside the kosmos is like this, but place is finite since no body is infinite. Just as anything corporeal is finite, so the incorporeal is infinite, for time and void are infinite. For as nothing is no limit, so there is no limit of nothing, as is the case with the void. In respect of its own subsistence it is infinite; it is made finite by being filled, but once that which fills it has been removed, a limit to it cannot be thought of.

(title unknown)

(paraphrase in Ioannes of Stobi, Selections 1.166) [structure of the kosmos]

Everything in the kosmos which is constituted by its own tenor has parts which move towards the center of the universe, and the same holds for the parts of the kosmos itself. It is therefore correct to say that all the parts of the kosmos [including air and fire] move towards its center, and particularly those with weight [earth and water]. There is an identical explanation both for the kosmŏs stable position in infinite void and similarly for the earth's being settled in the kosmos with equipollence at its center. However, body does not have weight absolutely; air and fire are weightless. But they too extend in a way to the center of the whole sphere of the kosmos, and they create the coherence with its periphery. For they are naturally upward-moving owing to their having no share in weight. Similarly they say that the kosmos itself does not have weight since it is entirely composed of elements which do have weight and ones which do not. The whole earth in their view does have weight intrinsically. From its position, by virtue of its central location (and the fact that such bodies move towards the center), it remains in this place.

(Long and Sedley [1987] 294, 296)

6.5 Archimedes of Syracuse (Surakusê) lived from *ca.* −285 to −211, was close to the ruling family of Syracuse, visited Alexandria, and wrote on mathematics, mechanics, optics, and pneumatics, addressing his works to Konon of Samos, Eratosthenes of Kurênê, and others.

Plane Equilibrium

Postulates

1. I postulate that equal weights at equal distances balance, and equal weights at unequal distances do not balance, but incline towards the weight which is at the greater distance.
2. If weights at certain distances balance, and something is added to one of the weights, they will not remain in equilibrium, but will incline towards that weight to which the addition was made.
3. Similarly, if anything be subtracted from one of the weights, they will not remain in equilibrium, but will incline towards the weight from which nothing was subtracted.
4. When equal and similar plane figures are made to coincide with one another, their centers of gravity also coincide.
5. In unequal but similar figures, the centers of gravity will be similarly situated. By points similarly situated in relation to similar figures, I mean points such that, if straight lines be drawn from them to the equal angles, they make equal angles with the corresponding sides.
6. If magnitudes at certain distances balance, magnitudes equal to them will also balance at the same distances.
7. In any figure whose perimeter is concave in the same direction ["convex"], the center of gravity must be within the figure.

With these postulates:

. . .

Proposition 6 [how the balance works]

Commensurable magnitudes balance at distances reciprocally proportional to their weights. Let A, B be commensurable magnitudes with centers A, B, and let EΔ be any distance, and let A:B = ΔΓ:ΓE; then it is required to prove that the center of weight of the magnitude composed of both A and B is Γ (Figure 6.1).

Since A:B = ΔΓ:ΓE, and A is commensurate with B, therefore ΓΔ is commensurate with ΓE (that is, a straight line with a straight line [Eukleidês, *Elements* 10.11]); so that EΓ, ΓΔ have a common measure. Let it be N, and let ΔH, ΔK be each equal to EΓ, and let EΛ be equal to ΔΓ. Then since ΔH = ΓE, it follows that ΔΓ = EH, so that ΛE = EH. Therefore ΛH = 2 · ΔΓ and HK = 2 · ΓE; so that N measures both ΛH and HK, since it measures their halves [Eukleidês, *Elements* 10.12]. And since A:B = ΔΓ:ΓE, while ΔΓ:ΓE = ΛH:HK – for each is the double of the other – therefore A:B = ΛH:HK. Now let Z be the same part of A as N is of ΛH; then ΛH:N = A:Z and KH:ΛH = B:A. Therefore, by equality [of terms taken cross-wise], we have KH:N = B:Z [Eukleidês, *Elements* 5. Def.5; 5.7; 5.22], and therefore Z is the same part of B as N is of KH.

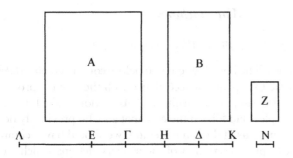

Figure 6.1

Now A was proved to be a multiple of Z; therefore Z is a common measure of A and B. Therefore, if ΛH is divided into segments equal to N and A into segments equal to Z, the segments in ΛH equal in magnitude to N will be equal in number to the segments of A equal to Z. It follows that, if there be placed on each of the segments in ΛH a magnitude equal to Z, having its center of weight at the middle of the segment, the sum of the magnitudes will be equal to A, and the center of weight of the figure compounded of them all will be E; for they are even in number, and the numbers on either side of E will be equal because ΛE = HE [Prop. 5, corollary 2].

Similarly it may be proved that, if a magnitude equal to Z be placed on each of the segments [equal to N] in KH, having its center of weight at the middle of the segment, the sum of the magnitudes will be equal to B, and the center of weight of the figure compounded of them all will be Δ. Therefore A may be regarded as placed at E, and B at Δ. But they will be a set of magnitudes lying on a straight line, equal to one another, with their centers of weight at equal intervals, and even in number; it is therefore clear that the center of gravity of the magnitude compounded of them all is the point of bisection of the line containing the centers of the middle magnitudes. And since ΛE = ΓΔ and EΓ = ΔK, therefore ΛΓ = ΓK; so that the center of weight of the magnitude compounded of them all is the point Γ. Therefore if A is placed at E and B at Δ, they will balance about Γ.

(Thomas [1941] 207, 209, 211, 213, 215)

6.6 Philon of Buzantion (*ca.* −200) wrote in Alexandria a compendium of mechanical science (*mêchanikê suntaxis*): book 1 introduction; book 2 on levers; book 3 on harbor-building; book 4 on war-machines; book 5 on pneumatics (survives only in Arabic and Latin); book 6 on automata; book 7 on defensive siege-works; and book 8 on offensive siege-works (books 1, 7, and 8 are fragmentary; books 2, 3, and 6 are lost).

War-machines

Praeface (pp. 49–50 Wescher) [principles]

Philon to Ariston, greetings. The book we sent you before comprised our *Making of Harbors*. Now is the time to explain (in accordance with the original program we laid out for you) the subject of artillery construction, called engine-construction by some people. Had it been the case that all who previously dealt with this section of mechanics used the same method, we should have required nothing else, perhaps, except a description of the artillery designs which were standard. But, since we see that previous writers differ not only in the proportions of interrelated parts, but also in the prime, guiding factor, I mean the hole that is to receive the spring, it is only right to ignore old authors and to explain those methods of later exponents that can achieve the requisite effect in practice.

I understand you are fully aware that the [artillery-technician's] trade contains something unintelligible and baffling to many people; at any rate, many who have undertaken the building of engines of the same size, using the same construction, similar wood, and identical metal, without even changing its weight, have made some with long range and powerful impact and others which fall short of these. Asked why this happened, they had no reason to give; thus, the remark made by Polukleitos, the sculptor [–435 ± 25, wrote *Kanon*, on his sculpture "Spear-Bearer"], is appropriate for what I am going to say. He maintained that "perfection is achieved gradually in the course of many calculations." Likewise, in this craft, since the processes are completed by means of many calculations, those who make a small discrepancy in particular parts produce a large total error at the end. Therefore, I maintain that we must pay strict attention, when adapting the design of successful engines to our private construction, especially when one wishes to do this after either increasing or diminishing the scale. In this connection, we are sure that those who take the advice quoted above will not go wrong. Now we must speak about dimensional arrangements from the beginnings.

In the old days, some engineers were on the way to discovering that the fundamental basis and unit of measure for the construction of engines was the diameter of the hole. This had to be obtained not by chance or at random, but by a standard method which could produce the correct proportions at all sizes. It was impossible to obtain it except by experimentally increasing and diminishing the perimeter of the hole. The old engineers, of course, did not reach a conclusion, as I say, nor did they determine the size, since their experience was not based on a sound practical foundation; but they did decide what to look for. Later engineers drew conclusions from former mistakes, looked exclusively for a standard factor with subsequent experiments as a guide [Polukleitos, *Kanon* contained the same principle], and introduced the basic principle of construction, namely the diameter of the circle that holds the spring. Alexandrian craftsmen achieved this first, being heavily subsidized because they had ambitious kings who fostered craftsmanship [as had Dionusios I of Syracuse]. The fact that everything cannot be accomplished by the theoretical methods of pure mechanics, but that much is to be found by experiment, is proved especially by what I am going to say.

(pp. 69–70 Wescher) [spring catapult: see Marsden [1971/1999] 175–176]

If one takes two weights, alike in substance and shape, the one of one mina, the other of two minae, and lets them drop simultaneously from a height, I maintain that the two-mina weight will drop far more quickly. In the case of other weights, the same theory applies: the larger always falls proportionately more quickly than the smaller, whether because the heavier weight, as some scientists say, can displace and disrupt the air more easily or because superior downward momentum is a natural attribute of the greater weight, and the superior downward momentum will considerably increase its rate of vertical descent. Again, because the principle mentioned exists, if one takes two weights of one mina, and, having connected and fastened them together as well as possible, lets them drop, I affirm that the weight of two minae will once more drop more quickly than the two weights of one mina joined together. Even if three or more are connected together, they will likewise drop more slowly. It becomes clear from this that, when several forces equal to each other are connected together, their combined velocity will not exceed the standard natural velocity belonging to a single weight by itself [compare Aristotle, *On Heaven* 1.6 (273b27–274a3), 1.8 (277b1–9), 4.2 (309b8–29)].

Since this is so, we have a real proof that the one half-spring does not assist the velocity of the arm at all, because its speed is equal to that of the other. It was considered best, therefore, to remove and scrap the one that cannot benefit the power of the engine. If the springs are made of sinew-cord and have the same tightening arrangement as the old ones, it is impossible to remove one of the half-springs. How could the arm be still held in position by one alone? Another method was required to enable the power of the one half-spring pressing against the heel of the arm to be increased and to remove the other, which could not contribute to the projection of the missile, which is in opposition during pull-back, and which gives much trouble. But they found another method which is now to be described [from Ktesibios].

Bronze plates were constructed for a three-span catapult. At least, they had this name; but they were forged bronze plates, 12 *daktuls* long, 2 *daktuls* wide, and 1/12 *daktul* thick. They were cast of the best possible prepared red copper, properly purified and several times smelted; further 3 *drachmas* by weight of tin – this, too, purified and fully smelted – had been mixed into each *mina* [of copper; the ratio is 3%]. Next, when the plates had been molded, forged, and given the measurements described, we imparted to them a gentle curvature against a wooden shaping-block. After that, we beat them, while cold, continuously for a long period, keeping them uniformly thick, straight along the edge, level on the broad surface, and fitting everywhere against the shaping-block. After that, we placed them together in pairs, putting the concave sides opposite each other, filing the ends down to a nicety, and fastening them to each other with rivets.

Thus, the plates obtained strength through the selection of the bronze: for whitest and purest cast bronze, as far as possible incorporating no foreign matter, is strong, resilient, and powerful. They were beaten cold continuously over a long period with a view to their compressed surface providing resilience. Resting with their concavities against each other, as described, they were placed along-side the heel of the arm, while the arm had its heel pressing against the plates (Figure 6.2). The arm turned about an iron pin, lying outside the face of the arm, and secured in the frame at its ends by iron brackets which also encircled the plates at the same time to eliminate damage to the frame. A bronze finger was placed against the face of the arm and fastened to it; through which arm the iron pin ran, extending its own particular socket up to the ivy-leaf bracket.

When the described arrangement was ready and the bowstring was pulled back, naturally the arm revolved about the iron pin and pressed one of the plates with its heel. Being pressed on its curved section and having been fitted at each end against the other plate, it was straightened out and it straightened out the other plate, for the middle of the other plate was placed against the frame and the iron brackets embracing it. So, in the pulling back, as we have shown, the plates were compelled to straighten themselves out because they were pressed together; but, on being released, they once more resumed their original, designed shape. Thus, in jumping apart with great force, they naturally levered away the heel of the arm.

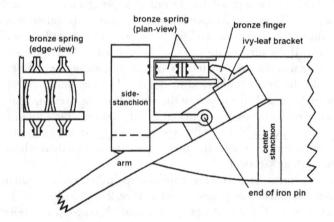

Figure 6.2 Plan of one bronze spring and arm.

(pp. 73–74, 76–77 Wescher) [repeating catapult]

A certain Dionusios of Alexandria constructed in Rhodes what is called a repeating catapult, which has a unique and very intricate arrangement. We shall describe it to you and elaborate the details with appropriate precision. The little scorpion, introduced by him, was not much larger than a one-cubit engine, and not much smaller than a three-span, and it fired a missile one cubit one *daktul* long [total *ca.* 35 cm]. The missiles were unnotched and feathered with three flights. Quite a lot of missiles were loaded all at the same time into the case, and, whenever the block was pulled forward, the claw jumped over the bowstring of its own accord, gripped it, and was automatically secured by the trigger. But, when the claw was pulled back, holding the bowstring, one of the missiles fell on to the groove and, after a little further withdrawal, was fired automatically. It continued to do this until all the missiles were fired. Again, other missiles were loaded all together, so that the gunner had no other job to do after loading the missiles except that of pulling back the block, turning the windlass backwards and forwards with the handspikes, so that the shooting was very rapid [Foley *et al.* [1982]].

...

It fired, at the longest range, slightly more than one stade.

Such a neat arrangement is the design of the repeating catapult, embodying an ingenious and complex plan. Yet it has not found a noteworthy use. We must direct most of our research, as we have often insisted, to achieving long range and to tracking down the features of engines which lead to power. By the means just mentioned I see no advance made in these respects, but merely that several missiles, loaded simultaneously, are fired singly in rapid succession. This is more conducive to inefficiency than to efficiency and does not require a long criticism. In the first place, the target is not stationary, but liable to move out of the way. Who, then, would want to shoot off several missiles to no purpose? Perhaps the argument that it is useful for firing into a group would persuade many people; but this, too, will be found untenable. The missiles will not have a spread, since the aperture has been laid on a single target and produces a trajectory more or less along one segment of a circle; nor will they have a very elongated dropping zone. With engines shooting single shots, we are to pull the trigger whenever we think we have accurately laid on the target; thus, we shall make effective use of most missiles. If, on the other hand, we waste many missiles at random and to no purpose, we shall be voluntarily providing the enemy with them. Someone will observe, of course, that the enemy will not be able to use them because they are unnotched; notching is certainly a major operation, requiring a deal of work.

(Marsden [1971/1999] 107, 109, 139, 141, 147, 151, 153)

6.7 Biton of Pergamon (−160 ± 20) compiled devices built by five fourth- to second-century BCE engineers in a book dedicated to king Attalos of Pergamon. (See Marsden [1971/1999] 90–97 and Lewis [1999].)

War-machines

5 *(pp. 57–60 Wescher)* [scaling ladder]

Following upon what has been already written, we shall describe the construction of a *sambukē* [scaling ladder (Figure 6.3)]. This instrument, in martial engagements, offers opportunities for great exploits. I shall describe for you the one which Damis of Kolophon designed [a Damios of Kolophon is known from –167]. It had the following dimensional arrangement. There was a joist, and it had in it two parallel axles; the axles were fitted with wheels. The wheels were 3 ft. in diameter, while the breadth of the joist was 3 ft., its height 2 ft., and its length 27 ft.; the height of the trestle on the joist was 14 ft. Let the trestle be N, fitted firmly at right angles to the joist and iron-plated where necessary. Then, through the cross-beam of the trestle were passed half-arm-shaped beams; through these and the head of the trestle, let a roller be inserted, of which the length is 15 ft. and the perimeter 19 *daktuls* [*ca.* 35 cm]. At the ends of the roller let there be a capstan instead of the bearing, so that one can turn the roller one way for raising and back again the other way for lowering. Let the capstan be K. Next, above the head of the trestle, in and above the roller, let there be placed a bracket, ΔE, double, with an aperture set to take all ladders, not having a very accurately machined gap, but, as it were, tailored to take anything inserted in it. As far as possible, let its length be in due proportion, so that, when the ladders are in position, the part covered by the actual apertures is about one twelfth of the whole length. Then, through the bracket beside the roller, let a *sambukē* be inserted, 60 ft. long, and made equal in breadth to the bracket that rises from the hole in the roller. Next, let the *sambukē* be fitted with side-walls, so that the men mounting up may make

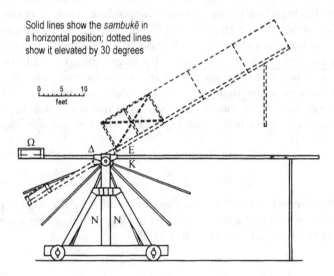

Solid lines show the *sambukē* in a horizontal position; dotted lines show it elevated by 30 degrees

0   5   10
feet

Figure 6.3

their ascent on it confidently. Let it have at Ω a box, 6 ft. all around, in breadth and length. Let the box hold a quantity of lead sufficient to counterpoise the whole balance-beam of the *sambukê* [large enough to hold about 11 times the lead needed, perhaps to allow the use of wood or water instead]. Then again, at the front end, let the *sambukê* be broader, in order that access may be easier for those climbing off on to the wall. Then, at a distance from the front end of about 6 ft., let there be a ladder fitted firmly on to the beams and let it be capable of restrained movement; let it be equal to the trestle in length, so that, when the ladder touches the ground, it is vertical and at right angles to the *sambukê*, and, when it has taken on personnel, then the crew, employing the revolving movement of the roller, can contrive the requisite effect.

(Marsden [1971] 73, 75)

6.8 Hipparchos of Nikaia worked and wrote in Rhodes in the early years of its subjection to Rome. His astronomical observations date his work to −146 to −126, and this is almost all we know of his work on dynamics.

Bodies Carried Down by Their Weight

(paraphrased by Simplicius, Commentary on Aristotle's "On Heaven" *pp. 264–265 Heiberg)* [acceleration]

In the case of earth thrown upward it is the projecting force (so long as it exceeds the downward power) that is the cause of the upward motion, and that to the extent that this dominates, the object moves more swiftly upwards; then, as this force is diminished (1) the upward motion proceeds but no longer at the same speed, (2) the body moves downward under the influence of its proper downward impulse, even though the original projecting power somewhat lingers, and (3) as this force diminishes the object moves downward ever more swiftly, and most swiftly when it is entirely lost [similarly pseudo-Aristotle, *Mechanika* 32 (858a14–16)]. The same cause operates in the case of bodies let fall from above. For the power which held them back remains with them up to a certain point, and this is the restraining factor which accounts for the slower movement at the start of the fall. [Simplicius quotes Alexander of Aphrodisias' arguments against Hipparchos]

Bodies are heavier the further removed they are from their natural places.

[Simplicius comments that some assert bodies move downward more swiftly as they draw nearer their goal because objects higher up are supported by a greater quantity of air, objects lower down by a lesser quantity, and that heavier objects fall more swiftly because they divide the underlying air more easily. That is, heavier objects offset more of air's buoyancy, so that acceleration is due to diminution of the resistant medium. Compare Aristotle, *Physics* 4.8 (216a30–33).]

(Cohen and Drabkin [1958] 209–211;
compare Sambursky [1962/1987] 73–74)

6.9 Athenaios of Kuzikos composed his work sometime in the first century BCE, describing the war-machines of earlier writers; many passages are very similar to the Roman writer Vitruuius, also writing in the late first century BCE.

Mechanics

11 (14.4–15.2 Wescher) [wall-borer]

[Diades traveled with Alexander the Great, then wrote around –320 ± 10; material added from Vitruuius 10.13.7 in *italics*.]

For the "wall-borer" (*truganon*) Diades takes the same tortoise as for the battering ram, with all equipment similar (Figure 6.4). On the foundation he places a case very much like the one in lighter torsion engines [in which the slider ran] *50 cubits long and 1 cubit high* and having an oblique windlass like catapults. At the other end of the case (i.e., at the top) he affixes two pulleys *right and left* though which he runs the yard-arm set upon it *with an iron tip*. Laid in the case under this he places close-set cylinders so it will easily *and more violently* move. And thus he runs the yard-arm, on which there is a battering ram, drawing it with the windlass lying below. It is leather-clad all around with nets over the case, in order for the yard-arm within to be protected.

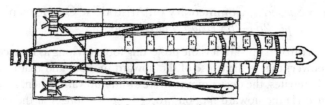

K: 8 cylinders, under the iron-tipped yard-arm

Figure 6.4

18–23 (21.2–26.5 Wescher) ["tortoise" – material added from Vitruuius 10.15.2–7 is in *italics*]

18   A tortoise has been constructed by Hegetor of Buzantion [otherwise unknown] with a platform 42 cubits long and 28 cubits wide [about 19 by 12 m (Figure 6.5)]. Four legs stand fixed in the platform, each formed of two connected beams, having a length of 24 cubits, a thickness of five palms [about 40 cm] and a breadth of one cubit [about 45 cm]. There are eight wheels [*sic*: the MS diagram shows only four] on the platform, on which the whole work moves. The height of the wheels is 4 and ½ cubits [about 2 m], their thickness is 2 cubits. They are constructed *in three layers jointed with tenons* length-wise and breadth-wise alternately, and they are bound together with cold-hammered plates; they turn in axle-sockets.

19   On the platform have been fixed posts 12 cubits long, 3 palms wide, 10

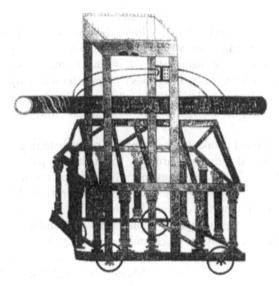

Figure 6.5

daktuls thick [about 20 cm]. Each of the posts stands apart from the next one a distance of 7 palms, and an architrave joins them in a circle at the top, 4 palms wide [about 30 cm], and 3 thick.

20   On the architrave are fixed rafters that rise up 8 cubits high, and on them is fixed an oblique beam onto which all the peaks of the rafters are fixed and forming two reclining sides. And all the rest of the work is covered with planks and is sheltered like armored tortoises. It has a central roof resting on the architrave on which to set a battery of weapons.

21   Behind the frame of the battering ram two compound legs stand upright in the middle of the tortoise and having a length of 20 cubits: their thickness is one cubit, and their width three palms. Fitted on them is a *dovetailed* headpiece, and at the middle another cross-bar through the legs *bound with iron plates.* And between the middle of the headpiece and the cross-bar is fixed an orthogonal beam. And on either side of the affixed beam and the legs are installed turned [i.e., machined] windlasses, from which ropes are fitted to suspend the battering ram. On the headpiece and on the frame of the battering ram is fitted a breastwork, so that *two soldiers* can most safely stand on it looking out for enemy attacks on the battering ram.

22   The total length of the battering ram is 120 cubits [about 53 m]. At the butt-end the thickness is 2 feet [about 60 cm] and the breadth is 5 palms, and towards the front its thickness narrows to one foot, and the breadth is three palms. It has an iron mouth [or "beak"] like an elongated battering ram [or: "like the rams of long-ships"], and a tube-shaped body, and from it four iron helixes extend, riveted into the battering ram, for a distance of 10 cubits. The entire battering ram is under-girded *as ships are from prow to stern* with four ropes, each

eight *daktuls* thick, and they are secured at the middle [of the ram] by three bindings the thickness of chains. The binding that holds the battering ram in the middle is wound on the battering ram with a five-palm pitch to its spiral [Athenaios is unclear; we follow Vitruuius' interpretation]. It is leather-clad all around, when it has been closed up, with raw-hide. The ropes which are tensed up by the winches on the battering ram frame, and hold the battering ram, have their ends formed of four iron chains. And the chains are covered all around with hides so that they cannot be seen.

23   There is a ladder of fixed planks on the front end of the battering ram, and on this there is a net knotted with strong threads, having holes a palm wide *whose coarseness affords slip-free ascent* on which one can easily climb up the wall. The battering ram also has extensions out of each side, since . . . [a gap in the text]. The device performs six motions: forward, backward, those towards the sides, and upwards and downwards. It can elevate up to 70 cubits [about 30 m] and it sweeps sideways for 70 cubits. It is steered by 100 men and has a total weight of four thousand talents [about 160 metric tons].

*(Schneider [1912] 18, 24, 26)

6.10 Apollonios of Kition (–40 ± 20) an Empiricist doctor wrote this commentary on Hippokrates' *Joints* (compare Hipparchos' commentary on Aratos' *Phainomena*, in Chapter 4).

On Hippokrates' "**Joints**", Book 2 [spine]

What Hippokrates conclusively explains about the joints in the spine, I will here set out:

[Apollonios quotes *Joints* 42, "hump-backs caused by falls cannot be cured even by succussion on a ladder"; then adds:]

He does not mean that such a surgical device is completely worthless, but that it can be useful for the setting of a joints when properly applied. Nevertheless, he says that such things are intentionally to be avoided.

[Apollonios then quotes *Joints* 43–44, "succuss downwards for humps low on the spine, and upward for those high on the spine, in order to employ the body's weight; pad the ladder and lay the patient thereupon on his back, carefully tied on; then lower the ladder smoothly and vertically; tie the patient on at the feet and knees for head-down succussions, and tie at the head and chest for foot-down succussions." (Figure 6.6)]

That completes the account of how to perform succussion by the feet. The fact that he is not entirely approving of such stretchings is clear from what he has added [*Joints* 44]:

> *These things must be done in this way, if the patient absolutely must be shaken on a ladder. In every skill and not least of all in the healing art, it is shameful to draw a great crowd, make a great display, express much theory, and then confer no benefit.*

Figure 6.6

Accordingly in what follows, he has written how it is necessary to restore a spine inclining outwards [<> indicate Apollonios' omissions from the standard text of *Joints* 47].

> *One must prepare the apparatus for forcible reduction like this. One can embed in the ground a strong broad plank having a transverse groove, or instead of a plank, one can incise a transverse groove in a wall <either at a cubit> above the ground <or> as high as suitable. Then place athwart something like a rectangular oak support separated from the wall enough for someone to pass between if necessary. Then on the support spread cloaks or something which is soft but does not yield too much. Let the person sweat, when possible, or bathe with much hot water* [to warm and heat the body and thus soften it], *and then recline him face downwards. Bind his arms, extended naturally, to the body, then bind him twice about the middle of the sternum very near the armpits, applying at its middle a strap, broad and strong, but soft, long, and composed of two strands. Then the remainder of the two straps is passed around the shoulders on either side at the armpits. Then let its ends be fastened to a pestle-shaped block, those ends being fitted in length to the underlying board against which the pestle-shaped block is placed, the stretching being supported thereby* [the underlying board is used a fulcrum]. *One must attach another similar bond above the knees and above the heels, and attach the ends of these straps to a similar board; and with another broad strap* . . . [Apollonios' quotation ceases, but Hippokrates continues "bind him at the waist as close as possible to the hips."]

After this has been set out, he adds that:

> *And this method is very mild* [Hippokrates' text has "harmless"]: *it's even harmless for someone to sit on the patient's hump while he is stretched out and perform the succussion by raising himself* (Figure 6.7).

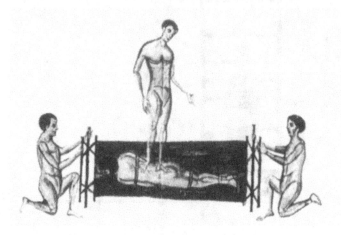

Figure 6.7

In this way it is necessary to restore the projection of the vertebrae to their natural place, using this arrangement or even with a windlass, applying pressure with the "chair" [*Joints* 6; Apollonios, book 1, p. 30 K–K; compare Figure 12 in Phillips [1973/ 1987]] (instead of hand-pressure), or even through succussion. How it must be done has been described.

*(Kollesch and Kudlien [1965] 50, 52, 54, 56, 58)

6.11 Heron of Alexandria wrote a series of books on science and engineering (*ca.* 55 to 68; see also Chapters 2, 5, 7, 8); the book here excerpted survives in an Arabic translation, and a partial Greek paraphrase by Pappos of Alexandria (around 300).

Mechanics

1.20–21 [weights]

There are many who think that weights lying on the ground are only moved by an equal force [contrast Aristotle, *Physics* 7.5 (250a11–19)], wherein they hold wrong opinions. So let us prove that weights placed in the way described are moved by an arbitrarily small force, and let us make clear the reason why this is not evident in fact. Let us imagine a weight lying on the ground, and let it be regular, smooth and with its parts coherent with each other. And let the surface on which the weight lies be flat, smooth and completely joined, and able to be inclined to both sides, i.e., to the right and the left. And let it be inclined first towards the right. It is then evident to us that the given weight must incline towards the right side, because the nature of weights is to move downwards, if nothing holds them and hinders them from movement; and again if the inclined side is lifted to a horizontal position and will be level [i.e., in equilibrium], the weight will come to rest in this position. And if it is inclined to the other side, i.e., to the left side, the weight will again sink towards the inclined side, even if the inclination is very small, and so the weight will need no force to move it, but will need a force to hold it so that it does not move. And if the weight again becomes level without inclination to either side, then it will stay there without a force holding it, and it will not cease being at rest until the surface inclines to one side or another, and then it will incline towards that side. Thus the weight that is ready to incline to whichever side, does it not require only a small force to move it, namely as much force as causes the inclination? And so isn't the weight moved by any small force?

21   Now, water on a surface that is not inclined will not flow, but remains without inclining to either side. But if the slightest inclination occurs, then all of it will flow towards that side, till not the smallest part of the water remains thereon, unless there are hollows in the surface, and small amounts stay in the bottom of the hollows, as happens often in vessels. Now water inclines like this because its parts lack cohesion and are very soluble.

As for the bodies that are coherent, since by their nature they are not smooth on their surfaces and not easily made smooth, it happens through the roughness of the bodies that they strengthen each other, and it happens that they lean upon each other like teeth, and they are strengthened thus, for if the teeth are numerous and closely joined, they require a strong and coherent force [to separate them].

And so from experiment people gained understanding: under tortoises [war machines: see Athenaios above, Section 6.9] they placed pieces of wood whose surfaces were cylindrical and so did not touch more than a small part of the surface, and so only very little rubbing occurred. And they use poles to move the weight on them easily, even though the weight is increased by the weight of the tools.

And some people put on the ground cut boards (because of their smoothness) and smear them with grease, because thus their surface roughness is made smooth, and so they move the weight with smaller force. As for the cylinders, if they are heavy and lie on the ground, so that the ground does not touch more than one line of them, then they are moved easily, and so also balls; and we have already talked about that [1.2–7].

2.1.1 [simple machines (surviving in Greek)]

Since the powers by which a given weight is moved by a given force are five, it is necessary to present their form and their use and their names, because these powers are all derived from one natural principle, though they are very different in form. Their names are as follows: the axle-in-wheel [windlass], the lever [*mochlos*], the pulley [*trochilos*], the wedge [*sphên*], and what is called the "endless" screw [*kochlia*].

[construction of the axle-in-wheel]

[2.2 the lever]

2.3 [pulley (surviving in Greek)]

The third power [pulley] is also called the "multi-lifter" [*poluspaston*].

Whenever we want to move some weight, if we tie a rope to this weight we pull with as much force as is equal to the burden. But if we untie the rope from the weight, and tie one of its ends to a stationary point and pass its other end over a pulley fastened to the burden and draw on the rope, we will more easily move the weight. And again if we fasten on the stationary point another pulley and run the end of the rope through that and pull it, we will still more easily move the weight. And again if we fasten on this weight another pulley and run the end of the cord over it, we will much more easily move the weight. And in this way, each time we add pulleys to the stationary point and to the burden, and run one end of the rope through the pulleys in turn, we will more easily move the weight. And every time the number of pulleys through which the rope runs is increased, it will be easier to lift that weight. The more "limbs" [*kôla*] the rope is bent into, the easier the weight will be moved.

And one end of the rope must be securely tied to the stationary point, and the rope must go from there to the weight (Figure 6.8). As for the pulleys that are on the stationary point they must be fastened to one piece of wood, turning on an axle, and this axle is called *manganon*; and it is tied to the stationary point with another rope. And as for the pulleys that are fastened to the burden, they are on

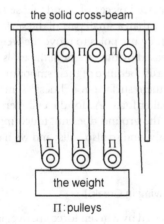

the solid cross-beam

the weight

Π : pulleys

Figure 6.8

another *manganon* like the first, tied to the burden. The pulleys should be so arranged on the axles that the "limbs" do not get entangled and unwieldy. And why the ease of lifting follows from the number of "limbs," and why the end of the rope is tied to the stationary point, we shall explain in the following.

3.2.1–2 [the crane (surviving in Greek): compare
Vitruuius 10.2.8]

1 For the lifting of burdens upwards there are certain machines: some have one mast, and some have two masts, and some have three, and some have four masts. As for the one that has a single mast it is made in this way. We take a long piece of wood, longer than the distance to which we want to raise the burden, and even if this pole is strong in itself, we take a rope and coil it round, winding it equally spaced, and draw it tight. The space between the single windings should not be greater than four palms [*ca.* 30 cm], and the windings of the rope are like steps for the workers and they are useful for anyone wanting to work on the upper section. And if the pole is not elastic, we must estimate the burdens to be lifted, lest the mast be too weak.

2 This mast is erected upright on a piece of wood, and three or four ropes are fastened to its top, stretched and tied to fixed points, so that the beam, however it is forced, will not give way (being held by the ropes). Then they attach to its top pulleys tied to the burden. Then they pull on the rope either by hand or with another engine, until the burden is raised [Figure 6.9].

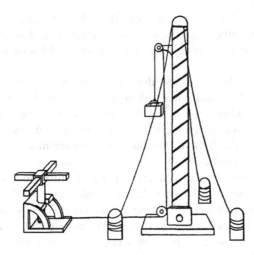

Figure 6.9

[3.3–5: two-, three-, and four-mast cranes]

(Drachmann [1963] 46–47, 50, 53–55, 98–99)

Automatic Theatre [pseudo-Aristotle, *Kosmos* 6 (398b13–20) records
that mechanical marvels were an analogy for *kosmic* motion; here they
serve as an imitation of purposive motion.]

1.1.1–8

1 The study of automaton-making has been considered by our predecessors
worthy of acceptance, both because of the complexity of the craftsmanship
involved and because of the striking nature of the spectacle. For, to speak briefly,
every facet of mechanics is encompassed within automaton-making, in the
completion of its several parts.

2 These are the topics to be discussed: shrines or altars of appropriate size
are constructed, which move forward of themselves and stop at specified
locations; and each of the figures inside them moves independently according to
the argument of the arrangement or story; and then they move back to their
original position. Thus such realizations of automata are called mobile.

[3–7: A "stationary" automaton displays a sequence of dioramas, which is easier.]

8 Therefore, in this book I am going to write about moving automata, and
set out my own complex scenario, which is adaptable to every other scenario, so
that someone who wanted to offer a different presentation would not lack
anything for the implementation of his own scenario. In the following book I talk
about stationary automata.

1.2.1–12

1 First you need a hard, level, smooth surface on which the automaton will
move, so that the wheels will neither sink because they are weighed down by
their load, nor drag on account of the roughness, nor stop and roll backwards
when they hit a bump.

2 If such a surface as has been posited does not exist, you must put
straightened slats on the floor, on which there will be grooves formed by rods
nailed lengthwise to make the wheels run in the grooves. You must make the
mobile automata of light, dry wood, and if there is anything other than wood in
the construction, it too should be as light as possible to prevent the mechanism
from being hindered by its own weight.

3 And anything that turns or moves in a circle must be perfectly round; and
anything that it revolves around must be smooth, not rough, e.g. the wheels
around iron pivots inserted into iron sockets, and the figures mounted on bronze
axles inserted into bronze axle-boxes made perfectly flush and airtight with them.

4 You must oil these fixtures well so that they will revolve freely with no
jamming. Otherwise none of the aforementioned events will happen according to
plan.

The cords which we use on these moving parts must not stretch or shrink, but
remain the length they were at the beginning.

5 We achieve this by stretching the cords tightly between pegs, leaving them for a while, then stretching them again; and after frequent repetitions of this operation, we smear them with a mixture of wax and resin. It is better to put a weight on them and leave them for a good while. When cord is pre-stretched in this way, it will not stretch further, or at most very little. If we find, after setting up the automaton, that one of the cords has stretched, we must trim it again.

6 But you must not use sinew, except perhaps when you need to use a springboard, since this material stretches and shrinks in response to atmospheric conditions. The springboard should be analogous to the arm set in the half-skein of a catapult, as will be made clear at the proper point. All these mobile automata are set in motion either by a springboard or by lead counterweights.

7 Between the source of motion and the part being moved is a cord with one end attached to each. The axle around which the cord is wound is what is moved. Wheels are fixed on the axle so that when the axle is turned and the cord unwound, the wheels, which rest on the floor, are also turned. The base of the moving automaton camouflages the wheels.

8 You must adjust the tension of the springboard or the heaviness of the counterweight so that neither the weight nor the tension is outweighed by the base. Movements in a forward direction result from the action of all the cords, which are looped around the moving parts and attached to the counterweight. The counterweight is fitted into a tube in which it is able to move up and down easily and precisely.

[9–11: motive power is efflux of millet or mustard seeds.]

12 You must avoid old-fashioned scenarios so that your presentation will look modern; for it is possible, as I said before, to create different and varied scenarios while still using the same methods. Your scenario will turn out better if it is well designed. The scenario I am describing is one such.

...

1.4.1–13

1 These preparations made, when the automaton is set down somewhere and we stand back, after a little while it will move to some prearranged spot. When it stops, the altar in front of the Dionusos will blaze up; and either milk or water will squirt from Dionusos' thursos [his wand or sceptre], while wine is poured from his chalice on to the little panther lying beneath it.

2 Each side facing the base's four columns will be decked with garlands. The Maenads will dance in a circle around the shrine, and there will be a clamor of drums and cymbals. Then, when the noise has subsided, the figure of Dionusos will rotate towards the outside and, at the same time, the Nike standing on the rooftop will rotate with him.

3 And when the altar that started behind Dionusos arrives in front of him, it will blaze up again; and again the thursos will squirt milk and the chalice pour wine. The Maenads will again dance around the shrine to the accompaniment of drums and cymbals; and when they stop the automaton will return to the place from which it started.

[6: reversing motions; 7: circular motions; 8: use of a cone – compare Aristotle, *Movement of Animals* 7 (701b3–6); 9: rectangular motion; 10: raising and lowering wheels; 11: general polygonal motion; 12–13: pneumatic effects]

1.14.1–2

1   After Dionusos first pours the libation there should be a banging of drums and cymbals. This is done as follows: in the lower part of the pedestal, where the wheels are, is placed a box containing lead balls, which roll out, one by one, on to the floor of the pedestal. On the floor is a hole the right size for the balls to fall through easily, and this has a closure which is opened as needed by a cord. A little drum is placed at an angle under the hole, and a little cymbal should be fastened to it.

2   Then, as they fall, the balls will strike the drum first and bounce off it on to the cymbal to complete the sound effect. A partition can divide the container into two chambers so that there are balls in each one and those in one section produce the first set of sounds and those in the other, the second, when the stopper there is opened in the same way.

(Murphy [1995] 11–15, 23–25)

War-machines

Praeface (pp. 71–74 Wescher) [why study artillery]

The largest and most essential part of philosophical study deals with tranquillity, about which a great many researches have been made and still are being made by those who concern themselves with learning; and I think the search for tranquillity will never reach a definite conclusion through the argumentative method. But mechanics, by means of one of its smallest branches – I mean, of course, the one dealing with what is called artillery-construction – has surpassed argumentative training on this score and taught mankind how to live a tranquil life. With its aid men will never be disturbed in time of peace by the onslaughts of enemies at home or abroad, nor, when war is upon them, will they ever be disturbed, by reason of the scientific skill which it provides through its engines. Therefore, it is always essential to ensure that complete pre-consideration of this section is undertaken. After peace has continued for a long time, one would expect more to follow, when men concern themselves with the artillery section; they will remain tranquil in their consciousness of security, while potential aggressors, observing their study of the subject, will not attack. But every act of aggression, even the most feeble, will overwhelm the neglectful since artillery preparations in their cities will be non-existent.

Writers before me have composed numerous treatises on artillery dealing with measurements and designs; but not one of them describes the construction of the engines in due order, or their uses; in fact, they apparently wrote exclusively for experts. Thus I consider it expedient to supplement their work, and to describe artillery engines, even perhaps those out of date, in such a way that my account may be easily followed by everyone. I shall speak about the construction of complete engines and the individual parts thereof, about nomenclature, composition, cord-fitting, and, furthermore, their individual use and measurements – after first remarking on the difference between the engines and the original development of each engine.

(Marsden [1971] 19, 21)

Pneumatics

1.43 [organ: here emphasizing mechanics of the device rather than properties of air; compare Chapter 8.5 (Aristokles)]

Construction of an organ to produce the sound of a flute when the wind blows. [Figure 6.10; see 1.42 for the details of the organ itself.] Let A be the flutes, the transverse pipe into which they are bored BΓ, and the "riser" pipe ΔE, and from that let EZ another pipe lead transversely to the cylinder [lit.: "box"] HΘ having an inner surface squared off for a piston. Let the piston KΛ be fitted into this so that it can easily slide inside; let a rod MN be joined to this piston, itself fastened to another rod NΞ as a swing-beam around an axis PΠ, and let the pin at N be

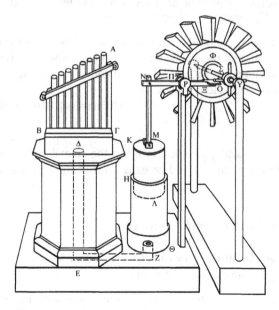

Figure 6.10

175

easily movable. And at Ξ let a little plate ΞO be joined, and on ΞO let an axis Σ be set and let it be moved on iron pivots and in a movable frame. Let there be joined to the axis Σ two drums Y and Φ, on which let Y have little rods hitting against the plate ΞO, and let Φ have plates like what are called "wind-watchers" [weathervanes].

So when they are all hit and pressed by the wind, and turn the drum Φ, the axis will also turn and thus also the drum Y, and the little rods on it intermittently striking the little plate ΞO raise the piston KΛ, and when a little rod is withdrawn from the plate, the piston falls back down and expels the air in the cylinder HΘ into the tubes and the flutes and it will produce the sound. It's always possible to turn the frame holding the axis into the blowing wind so that the turning [of the axis] is stronger and more continuous. [also in Humphrey, Oleson and Sherwood [1998] 26–27]

*(Schmidt [1899/1976] 202, 204, 206)

6.12 Plutarch of Chaironeia (*ca.* 50 to *ca.* 120) was a Platonist teacher of philosophy writing under the Flavian emperors, Trajan, and in the early years of Hadrian; much of his copious output is in the form of moralizing essays or dialogues.

Platonic Puzzles

7.5 (1005) [reporting Aristotle's doctrine of projectile motion: *Physics* 8.10 (266b27–267a22)]

Weights that are thrown cleave the air and separate it because of the impact with which they have fallen upon it; and the air (because of its nature always to seek out and fill up the space left empty) flows around behind and follows along with the object discharged, helping to accelerate its motion.

(Cherniss [1976] 67, 69)

6.13 Ailian the Platonist is possibly identical to Ailian the animal-writer (see Section 10.15), but is anyway of the second or early third century CE (Barker [1989] 230).

Commentary on the "Timaios"

Book 2 (from Porphyry, Commentary on Ptolemy's *"Harmonics")*
[mechanical nature of sound]

33.19–31 Sounds [*phonai*] differ from one another in height and depth of pitch. Let us see, then, what are the principal causes of the difference between notes [*phthongoi*]. The principal cause of all sound is movement. For if sound is air that has been struck, the impact is a movement, and if it is the organ of hearing [being struck], as the followers of Epikouros argue (when a quasi-sound [*para-phonê*] comes from the sounds to the organ of hearing through the agency of certain fluxes), in this case too movement is the cause of the experience. We shall therefore consider what difference there is in respect of movement, and what sort of movement is the cause of one sort of sound, and what sort is the cause of another. Our predecessors paid attention, first of all, to the phenomena, and taking their starting point from them they provided us with the answer we are seeking. For they found that swift movement is the cause of high-pitched sounds and slowness that of low-pitched: and this can be confirmed from the phenomena by anybody, by the use of the senses.

36.9–12 Since we have demonstrated that a swift movement produces a high note and a slow movement a low one, it is clear that the movement, or the speed of the movement from which the high note arises is in four-thirds ratio with the movement, or the speed of the movement from which the low note arises [Archutas, fr. A19–DK6, said a weak and slow blow generates a low note while a hard and fast blow generates a high one].

(Barker [1989] 231, 234)

6.14 Ptolemy (Claudius Ptolemaeus) of Alexandria (*ca.* 100 to *ca.* 175) wrote on a wide variety of "mathematical" topics (he is quoted in each of Chapters 3–8).

Harmonics

1.3.3 How the height and depth that relates to sounds is constituted

Among sounds, as among all other things, there are differences in respect both of quality and quantity, and to which of the two classes mentioned (difference relating to heights and depths) a sound belongs is not something that can be shown off-hand, before the causes of this kind of occurrence have been investigated, causes which seem to me to be shared in some way with variations in other sorts of impact. For the modifications [*pathê*] arising from them become different in accordance with the force of the striker, and with the bodily constitutions of both the thing struck and the thing with which the impact is made, and again in accordance with the distance of the thing struck from the origin of the movement. For it is clear that if the other factors involved remain the same, each of the things mentioned, when it is varied in one way or another, has its own specific effect on the modification [*pathos*]. In the case of sounds the difference related to the constitution of the thing struck either does not occur at all, or is not perceptible, because variations in the air are also imperceptible to the senses, while difference related to the force of the striker is the cause only of loudness, and not of height or depth. For if other factors are the same, we find no alteration of this sort arising in sounds when, for instance, people make utterances more gently or more loudly, or again when they blow in or pluck more mildly or more vigorously or strongly, but only that the greater follows upon the more forceful, the lesser upon the weaker.

...

Through diffuseness or density and thickness or fineness it makes qualities in accordance with which we again call sounds by the same words, dense or flabby, thick or thin; and from here it also makes heavinesses and sharpnesses, since each of these, being a quality of the kinds of composition mentioned, comes about in correspondence with the quantity of substance. For a denser thing is one that has more substance in an equal bulk, and a thing that is thicker than things of the same constitution is one having more substance in an equal length. The denser and the finer are creative of the sharper, the more diffuse and the thicker of the heavier. In all other things too, the sharper is described as such because it is finer, just as is the blunter because it is thicker. For finer things strike more compactly because they can penetrate more quickly, and denser things because they penetrate further. It is for this reason that bronze makes a sharper sound than wood, or gut-string than flax (for they are denser): of pieces of bronze with the same density and equal length it is the finer that is sharper in tone, of strings with the same density and equal length it is the more slender, while hollow things have a sharper tone than solids, and of windpipes, again, it is the denser and finer that do so.

In each case this happens not through the density or fineness as such, but through the high tension, since it is an attribute of things like this that they are tenser, while what is tenser is more vigorous in its impacts: the more vigorous is more compacted, and the more compacted is sharper. Hence if a thing is tenser in some other way, for instance by being harder to a greater degree than it is larger overall, it makes a sharper sound; and where there exists in both of two things something that has the same effect, victory goes to the excess of the one ratio over the other – as when bronze makes a sharper sound than lead, since it is harder than lead in a greater degree than lead is denser than it. And again, any larger and thicker piece of bronze makes a sharper sound than the smaller and finer, whenever the ratio in respect of magnitude is greater than that in respect of thickness. For sound is a sort of continuous tensing of the air, penetrating to the outer air from the air that immediately surrounds the things making the impacts, and for this reason, to whatever degree each of the things making the impacts is tenser, the sound is smaller and sharper to the same degree.

(Barker [1989] 279–281)

6.15 Galen of Pergamon (129 to 210 ± 5), viewing himself as philosopher and physician, considered Hippokrates and Plato as the most worthy authorities; he wrote voluminously on medicine and much else. (This work survives only in extracts in a work preserved only in Arabic.)

Motion

(from Alexander of Aphrodisias' Refutation 62b21–63a1, 63a5–7, 63a9–17)

In order that the discussion with which the argument is resumed be clear, let us say that things whose source of motion is present within them move primarily, and for things in which something similar is not present, their source of movement is accidental and not primary. It is clear that when we say that a thing moves *essentially*, we have indicated no more than that it moves primarily. This is because both those expressions merely refer to the things whose source of motion exists in them and whose motion is not basically due to anything external. [the italicized word is restored by Rescher and Marmura]

...

The thing that moves primarily and essentially according to Aristotle is one whose source of motion is in it and is not basically in some other external thing.

...

When some magnitude, which moves all together, moves essentially and primarily, it must then be one of the simple first bodies, i.e., the bodies whose parts are similar, since these alone are the things that move essentially and primarily. Since the things whose natural principle of motion is in them are the first simple bodies [fire, air, water, earth], and since these consist of similar parts, the part in these things is no other than the whole. Hence Aristotle was definitely not right with respect to continuous things in holding that when one part of them stops, the whole stops. For the part in these things is no other than the whole.

(from the same, 63b23–64a1)

When Aristotle, however, set down in the statements we have presented and described that one of the things that move essentially is AB, he made the exception that some part of it, i.e., ΓB, does not move. And there is no difference between this and our setting down that the whole moves and simultaneously does not move. This is so because in as much as all of it is of one substance, it being one of the things that move primarily, at no time is there any part in it which is not moving basically [Alexander records that this argument is from Chrusippos of Soloi].

(Rescher and Marmura [1965] 33, 34, 36)

7

OPTICS

Our earliest lights were sun, moon, and stars; soon we made our own lights from fire. All gave us sight, and thereby knowledge, and so came to symbolize perception and enlightenment. Reflecting in the light of those fires, animal eyes seemed to hold inner fire which sparked a belief that light dwelt in living eyes. People often felt eyes were emissive, leading to the legend of the "evil eye." Animals ignore or attack their mirror-images, while to us they seemed marvelous or even magical, a window on a wonderland. For Egyptians, the sun (Ra^c) was chief among the gods, while for Greeks he was the all-seeing eye of heaven (*Iliad* 3.276–277; Aischulos, *Prometheus Bound* 91). Light was granted for a symbol to initiates in the Eleusinian mysteries (and in many religions), and one celestial light, Iris the rainbow, seemed a divine message. Some early Mediterranean peoples perhaps knew the magnifying property of lenses (rounded quartz or glass: Sines and Sakellarakis [1987]).

For Greeks (as for many people), seeing was always close to believing, and Greek thinkers developed three main models of vision. Demokritos and the atomists supposed *eidôla* (images) streamed from objects into our eyes, Aristotle believed that light was the actualization of the potential called transparency, and Plato and others saw rays emitted from eyes mingling with external fire and light to produce sight. Those theories arose in the context of the traditions sketched above, and the speculations we now describe.

Earlier writers rarely discuss light or sight, but Alkmaion of Kroton (*ca.* –500) may have dissected the eye, and in any case conceived sight as transmitted from eye to brain along the optic-nerve "tube." Xenophanes of Kolophon (–510 ± 30) saw the rainbow not as messenger but as colored cloud. Empedokles of Akragas (–455 ± 20) depicted light as traveling from the sun throughout the *kosmos* (reflecting from sky and moon) and sight as occurring by interaction of likes in the eyes (seeing fire by fire or water by water), color being an effluence of shapes. Playwrights or their assistants in Athens at this time are said to have written on perspective (in scene-painting), presumably using elementary geometrical optics (Vitruuius 7.pr.11); while the comedian Aristophanes mentions the use of lenses as burning glasses (*Clouds* 766–768). Demokritos of Abdera (–410 ± 30) supposed that atoms continuously emanate from all objects as images, and that this effluence impinges upon eyes to cause sight (possibly by mixing with similar effluences from the eye).

Plato offers an account, which he considers no more than probable, that seeing occurs through the emission of ocular fire mixing in sympathy with external light, weakened by night (*Timaios* 45). Mirrors reverse right-to-left through reflection of this visual stream, but curved ones retain handedness (*Timaios* 46), and light has four primary colors, white, black, red, and *lampron* ("bright," *Timaios* 67–68). Aristotle of Stageira (writing –335 ± 10) tried to subsume sight and light under his theory of potential and actual being (*On the Soul* 2.7 [418b3–419a15]). Using the Platonic vision-ray model, he explained the circular form of the rainbow as a geometrical consequence of proportional reflection from particles of water (*Meteorologika* 3.2–5 [371b26–377a28]), producing the primary colors red, green, and purple through sequential weakening of white to black (Sorabji [1972]). The pseudo-Aristotelian *Colors* (1 [791a1–792a4]) attributes primary color to the four elements, which are white or yellow (Fire), black being the absence of color in matter undergoing elemental transformation, or the simple absence of light; other colors are produced as proportional mixtures of those (black plus fire gives red, for example).

Color and sunlight continued to seem almost magical – although normally the sun bleaches colors out, the famous purple dye only develops fully under the sun (*Colors* 5 [795a29–b22]), just like the green of plants (*Colors* 5 [794b11–795a29]). Other material transformations (compare Chapter 9) were effected by the sun: the poison used by Deianeira (Sophokles, *Trachiniai* 685–704) or the decomposition of cinnabar (Vitruuius 7.9.2, doubtless long known).

7.1 Eukleidês, who worked in Alexandria under Ptolemy I (i.e., ca. –300), wrote a textbook on optics (and others on mathematics and astronomy and music). He expounds what had become the standard model, vision produced by "vision rays" (*opseis* or *aktînës*) which emanated (lit. "fell forward") from the eye, explaining the often-paradoxical alterations in appearances (compare Berryman [1998]).

Optics

Definitions

1. Let it be assumed that lines extended directly from the eye pass through a space of great extent;
2. and that the figure included within our vision-rays is a cone, with its apex in the eye and its base at the limits of our vision;
3. and that those things upon which the vision-rays fall are seen, and that those things upon which the vision-rays do not fall are not seen;
4. and that those things seen within a larger angle appear larger, and those seen

within a smaller angle appear smaller, and those seen within equal angles appear to be of the equal size;

5.   and that things seen within the higher vision-rays appear higher, while those within the lower vision-rays appear lower;

6.   and, similarly, that those seen within the vision-rays on the right appear on the right, while those within those on the left appear on the left;

7.   but that things seen within multiple angles [i.e., subtending a larger angle] appear to be more clear.

1 Nothing that is seen is seen all at once

For let the thing seen be AΔ, and let the eye be B, from which let the vision-rays emanate, BA, BΓ, BK, and BΔ (Figure 7.1). So, since the vision-rays diverge, they could not emanate continuously upon AΔ [because there are not infinitely many of them]; so that there would also be spaces along AΔ upon which the vision-rays would not emanate. So AΔ will not be seen in its entirety, at the same time. But it seems to be seen all at once because the vision-rays traverse rapidly [a Greek argument for which is to gaze at stars and blink: they reappear instantly].

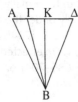

Figure 7.1

2 Nearby objects are seen more clearly than distant objects of equal size

Let B represent the eye and let ΓΔ and KΛ represent the objects seen; and we must understand that they are equal and parallel, and let ΓΔ be nearer to the eye; and let the vision-rays emanate, BΓ, BΔ, BK, and BΛ (Figure 7.2). For we would not say that the vision-rays emanating from the eye upon KΛ will pass through the points Γ and Δ. For [if they did,] in the [hypothetical] triangle BΔΛKΓB, the line KΛ would be longer than the line ΓΔ; but they are supposed to be equal. So ΓΔ is

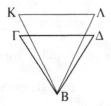

Figure 7.2

seen by more vision-rays than KΛ. So ΓΔ will appear more clearly than KΛ; for objects seen within more angles appear more clearly.

12 Of magnitudes extending forward, those on the right seem inclined toward the left, and those on the left toward the right [perspective of roads and buildings]

Let two magnitudes be seen, AB and ΓΔ, and let the eye be indicated by E, from which let the vision-rays emanate EΘ, EK, EA, EZ, EH, and EΓ (Figure 7.3). I say that EZ, EH, and EΓ seem to be inclined toward the left, and EΘ, EK, and EA toward the right. For since EZ is more to the right than EH, and EH more than EΓ, hence EΓ seems to be inclined to the left of EH, and EH to the left of EZ. Similarly, also EA, EK, and EΘ seem to be inclined to the right.

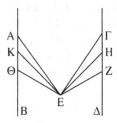

Figure 7.3

16 As the eye approaches objects of unequal size which rise one upon another above the eye, the shorter one appears to gain height, but as the eye recedes the taller one appears to gain [e.g., mountains]

Let there be the lines AB and ΓΔ, of unequal length, and of these let AB be the taller. Let E represent the eye, and from this let the vision-ray EZ emanate through Γ (Figure 7.4). Now, since the lines ZB and ΓΔ are included under the vision-ray EZ, thus BZ and ΓΔ appear equal to each other. So, AB appears taller than ΓΔ by the length of AZ. Now let the eye be moved nearer, and let it be H, from which let the ray HΘ emanate through Γ. Now, since BΘ and ΓΔ are included under the vision-ray HΘ, and ZB and ΓΔ under the vision-ray EZ, and ZA is longer than AΘ, as the eye approaches the objects the shorter one appears to gain height, but, as the eye recedes, the taller one appears to gain.

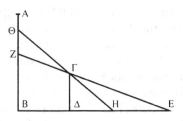

Figure 7.4

19 To know how great is a given height when the sun is not shining [18 gives
Thalês' method; 19 gives Dikaiarchos' method: see Chapter 5.4]

Let there be a certain height, AB, and let the eye be Γ, and let it be necessary to
know how high is AB when the sun is not shining (Figure 7.5). Let a mirror be
placed, ΔZ, and let ΔB be produced in a straight line from EΔ, to the point where
it meets B, the limit of AB, and let a vision-ray emanate, ΓH, from the eye Γ, and
let it be reflected back to the point where it meets A, the limit of AB, and let EΘ be
produced from ΔE, and from Γ let the perpendicular ΓΘ be drawn upon EΘ.
Now, since the vision-ray ΓH has emanated and the vision-ray HA has been
reflected back, they have been reflected at equal angles, as is said in the *Mirrors*
[survives in a probably fourth-century-CE edition; compare pseudo-Aristotle, *Problems*
16.4, 16.13, where the rebound of solid objects is at equal-angles; and Heron, *Mirrors* 3–
4 (below Section 7.7)]; thus, the angle ΓHΘ is equal to the angle AHB. But also the
angle ABH is equal to the angle ΓΘH; and the remaining angle HΓΘ is equal to
the remaining angle HAB. So, the triangle AHB is equiangular with the triangle
ΓHΘ. But the sides of equiangular triangles are proportionate. Thus, as ΓΘ is to
ΘH, so is AB to BH. But the ratio of ΓΘ to ΘH is known; and the ratio of BA to
BH is known. But HB is known, and so AB is known.

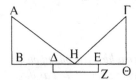

Figure 7.5

*23 If a sphere is however seen by one eye, always less than a hemisphere is seen,
and the part of the sphere that is seen appears as a circumference* [compare
Aristarchos, *Sizes and Distances* 2, Chapter 3.3]

Let there be a sphere whose center is A, and let B be the eye. And let AB be
joined, and let the plane through line BA be produced. So it will make a circular
section [in the sphere]. Let it make the circle [passing through] ΓΔΘH, and around
the diameter AB let the circle ΓBΔ be drawn, and let the straight lines be joined
ΓB, BΔ, AΔ, and AΓ (Figure 7.6). Now, since AΓB is a semicircle, the angle AΓB is
a right angle; similarly, also the angle BΔA. So, the lines ΓB and BΔ are tangent.
Now, let ΓΔ be joined, and let HΘ be drawn through the point A parallel to ΓΔ.
So the lines at K are orthogonal. Now, with AB remaining in its place, if the
triangle BΓK is revolved about the right angle K, and is restored to its same start,
the line BΓ will touch the sphere at one point and the line KΓ will make a circular
section. So, a circumference of a circle will be seen in the sphere. And I say that it
is less than a hemisphere. For, since HΘ is a semicircle, ΓΔ is less than a semicircle.
And the same part of the sphere is seen by the vision-rays BΓ and BΔ. So ΓΔ is less

than a hemisphere; and it is viewed by the vision-rays BΓ and BΔ. [24–33 are similar, covering sphere, cylinder, and cone.]

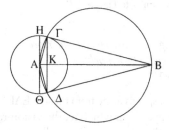

Figure 7.6

51 If, when several objects move at unequal speed, the eye also moves in the same direction, the objects moving with the same speed as the eye will seem to stand still, those moving slower will seem to move in reverse, and those moving faster will seem to move ahead [racers passing]

For let B, Γ, and Δ move at unequal speed, and let B move most slowly, and Γ at the same speed as the eye K, and Δ more quickly than Γ. And from the eye K let the vision-rays emanate KB, KΓ, and KΔ (Figure 7.7). So Γ moving with the eye, will seem to stand still, and B, left behind, will seem to move in reverse, and Δ supposed faster will seem to move forward; for it will separate from these.

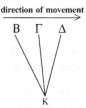

Figure 7.7

52 When some objects are moved, and one is obviously not moved, the unmoved object will seem to move backward [a boat anchored in a river: Ptolemy, *Optics* 2.131–3]

For, let B and Δ move, and let Γ remain unmoved, and from the eye let the vision-rays emanate ZB, ZΓ, and ZΔ. So B, as it moves, will be nearer to Γ, and Δ, receding, will be farther away; therefore, Γ will seem to move in reverse.

...

54 When objects move at equal speed, the more remote seem to move slower [Ptolemy, *Optics* 2.100]

For let B and K move at equal speed, and from the eye A let vision-rays be drawn

AΓ, AΔ, and AZ. So, B has longer vision-rays than K drawn from the eye. Therefore it will cover a greater distance, and passing the line of vision later AZ will seem to move more slowly. [MS figure very similar to Figure 7.7]

55 If the eye remains at rest, while things seen are moved, the more
remote of the things seen will seem to be left behind
[moving ships]

Let A and Γ be the things seen [moving at equal speeds away from the line EΔB] on the straight lines, AB and ΓΔ, and let E be the eye, from which let the vision-rays emanate, EΓ, EΔ, EA, and EB (Figure 7.8). I say that the object at A will seem to be left behind. Let EΔ be extended until it meets AB, and let it be EB. Now, since the angle ΓEB is greater than the angle AEB, the distance ΓΔ appears greater than AB [the manuscript diagram exaggerates the difference]. So that if the eye remains at E, the vision-rays, moving in the direction of A and Γ, will pass A more quickly than Γ. So AB will seem to be left behind.

Figure 7.8

57 When things lie at the same distance and the edges are not in line with the
middle, it makes the whole figure sometimes concave, sometimes convex

For let ΓBΔ be seen by the eye located at K, and let the vision-rays emanate KΓ, KB, and KΔ. So the whole figure will seem to be concave (Figure 7.9). Now let the thing seen in the middle be moved back, and let it be nearer the eye. So ΔBΓ will seem to be convex.

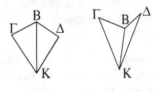

Figure 7.9

(Burton [1945] 357–362, 367, 371–2)

7.2 Epikouros of Samos (–340 to –269, and a citizen of Athens) adapted the theories of Demokritos, although his own goal was not explanation but peace of

186

mind. Thus, his method was to allow for as many models as possible, so long as all were natural, and he rarely offers novel explanations, merely rendering existing ideas into atomist terms. An opinion could be held, he believed, so long as nothing was known to contradict it.

Letter to Herodotos

46–48 [objects give off emanations]

Moreover, there are delineations which represent the shapes of solid bodies and which in their fineness of texture are far different from things evident. For it is not impossible that such emanations should arise in the space around us, or appropriate conditions for the production of their concavity and fineness of texture, or effluences preserving the same sequential arrangement and the same pattern of motion as they had in the solid bodies. These delineations we call "images" [*eidôla*]. Next, that the images are of unsurpassed fineness is uncontested by anything evident. Hence they also have unsurpassed speed, having every passage commensurate with themselves [argument by analogy], in addition to the fact that infinitely many of them suffer no collision or few collisions, whereas many, indeed infinitely many, atoms suffer immediate collision. Also that the creation of the images happens as fast as thought. For there is continuous flow from the surface of bodies – not revealed by diminution in their size, thanks to reciprocal replenishment – which preserves for a long time the positioning and arrangement which the atoms had in the solid body, even if it is also sometimes distorted. [for the immediately following section see Section 12.2]

(Long and Sedley [1987] 72–73)

Letter to Puthokles

91 ["The sun is peculiar in always appearing the same size from any distance"; compare below Archimedes, Agatharchides, and Poseidonios on the sun's apparent size]

The size of the sun and the other stars as far as appearances go is the same as it appears. For no other distance [than our distance from them] is a more appropriate distance for viewing them [to estimate their size]. And in reality the sun is either bigger or a little smaller than how it happens to appear. And earthly fires [are no counter-argument, but] observed from a distance appear just the same [as closer]. And every counter-argument to this point will be easily dissolved, if we hold fast to what is clear, as I show in my books *On Physics* [book 11, preserved in scraps, stated (a) "we can walk to the place at which the sun seemed to rise," and (b) "if distance changed the sun's size it should change its color even more, but does not"].

*(Sedley [1976] 48)

7.3 Archimedes of Syracuse (Surakusê) lived from *ca.* –285 to –211, was close to the ruling family of Syracuse, visited Alexandria, and wrote on mathematics, mechanics, optics, and pneumatics, addressing his works to Konon of Samos, Eratosthenes of Kurênê, and others.

Sand-Reckoner

1.10–17 [the apparent size of the sun: see Dijksterhuis [1956/1987] 360–373]

I have attempted to measure by means of instruments the angle subtended by the Sun and having its vertex at the eye in the following way. It is, however, not easy to measure this angle with precision, for neither the eye, nor the hands, nor the instruments for measuring it are reliable enough for determining it exactly; but this is scarcely the occasion to discuss this subject further, especially as this has often been pointed out. In order to demonstrate my proposition, it is sufficient for me to find an angle which is not greater than that which is subtended by the sun and has its vertex at the eye, and then to find another angle which is not smaller than that which is subtended by the sun and has its vertex at the eye.

Having first placed a long, straight rod on a vertical stand, the stand was put in a position where the rising sun could be seen. A small cylinder, turned on a lathe, was placed vertically on the rod immediately after the rising of the sun. Then, when the sun was still near the horizon and could be looked at directly, the rod was directed towards the sun and the eye was placed at one end, while the cylinder, placed between the eye and the sun, concealed the sun. The cylinder was then shifted away from the eye, and when the sun showed slightly on each side of the cylinder it was set in place. If it were so that the eye sees from a single point, then the tangents drawn to the cylinder from the end of the rod, where the eye was placed, would form an angle smaller than the angle subtended by the sun, and having its vertex at the eye, because a part of the sun was seen on each side of the cylinder. But as eyes do not see from a single point, but from a certain area, a cylindrical magnitude was chosen that is not less than that of the pupil. This magnitude was placed at the end of the rod where the eye was located, and straight lines were drawn tangent to both this magnitude and the cylinder. The angle subtended by these lines was therefore less than the angle subtended by the sun, and having its vertex at the eye.

A magnitude not less than that of the pupil is found in the following way. One takes two thin cylinders of the same thickness, one white, and the other not, and places them in front of the eye in such a way that the white one is removed from the eye, and the non-white one is brought as close as possible to the eye so that it touches the face. Then, if the cylinders chosen are smaller than the pupil, the closer cylinder is encompassed by the pupil and the white is seen behind it. If the cylinders are much smaller, the white is seen in full. If they are not much smaller, a part of the white is seen on each side of the cylinder nearer the eye. But, if cylinders of a suitable magnitude are chosen, then one covers the other without covering a larger area. It is therefore certain that a magnitude equal in size to the thickness of the cylinders which are determined in this way is not less than that of the pupil.

The angle which is not smaller than the angle subtended by the sun, and having its vertex at the eye, was found in the following way. The cylinder is placed on the rod, away from the eye, so that it entirely conceals the sun. Lines are drawn tangent to the cylinder from the end of the rod where the eye is located. The angle subtended by the lines is not smaller than the angle subtended by the sun and having its vertex at the eye.

When it was determined how many times the angles thus found are contained in a right-angle, it turned out that the angle which terminates at a point is found to be smaller than the 1/164th part of a right-angle, and the smaller is found to be larger than the 1/200th part of a right-angle. The result evidently is that the angle subtended by the sun, and having its vertex at the eye, is smaller than the 1/164th part of a right-angle, and greater than the 1/200th part of a right-angle [Hipparchos found 1/650 of a circle (Ptolemy, *Syntaxis* 4.9), and the modern value is close to 1/180 of a right-angle].

(Shapiro [1975] 82–83)

7.4 Diokles, active around −185 ± 5, wrote this work which survives only in Arabic. This is the earliest evidence of knowledge of the focal property of the parabola, discovered by Dositheos of Pelousion (to whom Archimedes addressed his later works).

Burning-Mirrors

1 Praeface

Puthion the Thasian geometer wrote a letter to Konon [of Samos] in which he asked him how to find a mirror surface such that when it is placed facing the sun the rays reflected from it meet the circumference of a circle. And when Zenodoros [Toomer [1972]] the astronomer looked towards Arkadia and arrived there, he asked us how to find a mirror surface such that when it is placed facing the sun the rays reflected from it meet a point and thus cause burning. So we want to explain the answer to the problem posed by Puthion and to that posed by Zenodoros; in the course of this we shall make use of the premises established by our predecessors. Dositheos was the man who constructed one of those two problems, namely the one requiring construction of a mirror which makes all the rays meet in one point. The other problem, since it was only theoretical and there was no application which deserved to make it well-known, was not constructed. We have set out a compilation of the proofs of both these problems and eluci-dated them [Prop. 1].

The burning-mirror surface submitted to you is the surface bounding the figure produced by a section of a right-angled cone [parabola] being revolved about the line bisecting it. It is a property of that surface that all the rays are reflected to a single point, namely the point [on the axis] whose distance from the surface is equal to a quarter of the line which is the parameter of the squares on the perpendiculars drawn to the axis. [compare Archimedes, *Conoids and Spheroids* 3; Toomer [1976] 141] Whenever one increases that surface by a given amount, there will be an increase in the above-mentioned conic section. So the rays reflected from that addition will also be reflected to exactly the same point, and thus they will increase the intensity of the heat around that point. The intensity of the burning in this case is greater than that generated from a spherical surface, for from a spherical surface the rays are reflected to a straight line, not to a point, although people used to guess that they are reflected to the center; the rays from this surface which meet at one place on that surface when they are reflected from the surface consisting of a spherical segment less than half the sphere, and if the mirror consists of half the sphere or more than half, only those rays reflected from less than half the sphere are reflected to that place.

[It is reasonable to treat the Earth as a point, and thus we can treat the solar rays as parallel]

Perhaps you would like to make two examples of a burning-mirror, each having a diameter of two cubits, one constructed on the circumference of a circle, the other on a section of a right-angled cone [parabolic], so that it may be possible for you to measure the burning-power of each of them by the degree of its efficiency. So one knows the amount by which the stronger of the two in burning exceeds [the other], and then measuring the burning of one and that of the other is a matter requiring observation: that is to say, if the mirror-surface with a diameter of the amount of one foot burns the whole of what is near the place of burning in some wood, then it is more likely to burn easily when its diameter is nine times that amount. For when the burning-power is multiplied by nine, the difference between it (such a mirror) and the original mirror must be very great and strong. [Arabic "seven" and "nine" look identical in this manuscript, but "nine" is more likely for Greek.]

Prop. 1 [construction of a parabolic mirror]

Let there be a parabola KBM, with a line cutting it in two AZ, and let half the parameter of the squares on the ordinates be line BH [a given length, which is the distance from the parabola's vertex to the apex of the right-angled cone generating it; this is Archimedes' way of generating cones: compare Apollonios, *Konika* 1.11, Dijksterhuis [1956/1987] 59–63; the parabolic invariant is $2 \cdot BH \cdot B\Gamma = \Theta\Gamma \cdot \Theta\Gamma$]. Let BE on the axis be equal to BH, and let BE be bisected at point Δ. Let us draw a line tangent to the section at an arbitrary point, namely line ΘA, and draw line $\Theta\Gamma$ as ordinate to AZ [i.e., $\Theta\Gamma$ is perpendicular to AZ (Figure 7.10)]. Then we know that AB = BΓ [in Euclid's *Conics*, see Archimedes, *Method* 1 (Chapter 2.2), and Apollonios, *Conics* 1.35, 2.49] and that the line drawn from Θ perpendicular to ΘA meets AZ beyond

190

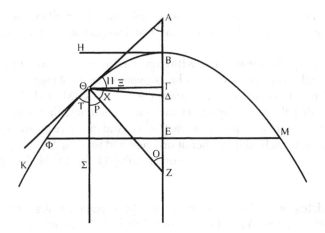

Figure 7.10

E [BE = BH; so ΓZ = BE as below; the perpendicular from the tangent falls at Γ below B; so Z must be below E]. So let us draw ZΘ perpendicular to ΘA, and join ΘA.

Then ΓZ = BH [since ΘΓ is a perpendicular from the right angle AΘZ to the opposite side, it forms two similar right triangles, AΓΘ and ΘΓZ, in which side AΓ of AΓΘ corresponds to ΘΓ of ΘΓZ, and side ΓΘ of AΓΘ corresponds to ΓZ of ΘΓZ, so AΓ/ΘΓ = ΘΓ/ΓZ; using that and AΓ = 2 · BΓ in the parabolic invariant yields the stated equality], and HB = BE, so ΓZ = BE.

We subtract ΓE, common [to ΓZ and BE], then the remainder ΓB = EZ.

But ΓB = BA, so AB = EZ.
And BΔ = ΔE, because BE is bisected at Δ, so the sum AΔ = ΔZ.
And because triangle AΘZ is right-angled and its base AZ is bisected at Δ, AΔ = ΔΘ = ΔZ [Eukleidês, *Elements* 3.31].
So angle-O = angle-X and angle-A = angle-ΠΞ [Diokles employs an angle-notation similar to Aristotle, *Prior Analytics* 1.24 (41b5–22)].
So let a line parallel to AZ pass through Θ, namely line ΘΣ.
Then angle-O = angle-P, which is alternate to it [ΘZ cuts parallel lines], and angle-O = angle-X, so angle-X = angle-P also.
And angle-ΠΞX = angle-PT, right angles, so angle-T = angle-ΠΞ, the remainders.

So when line ΣΘ meets line AΘ it is reflected to point Δ [the focal point], forming equal angles, angle-ΠΞ and angle-T, between itself and the tangent AΘ. Hence it has been shown that if one draws from any point on KBM a line tangent to the section, and draws the line connecting the point of tangency with point Δ, e.g. line ΘΔ, and draws line ΣΘ parallel to AZ, then in that case line ΣΘ is reflected to point Δ, i.e., the line passing through point Θ is reflected at equal angles from the tangent to the section. And all parallel lines from all points on KBM have the same property, so, since they make equal angles with the tangents, they go to point Δ.

Hence, if AZ is kept stationary, and KBM revolved until it returns to its original position, and a concave surface of brass [probably arsenical copper, as in Plato, *Kritias* 116c] is constructed on the surface created by KBM, and placed facing the sun, so that the sun's rays meet the concave surface, they will be reflected to point Δ, since they are parallel to each other. And the more the surface is increased, the greater will be the number of rays reflected to point Δ.

(Toomer [1976] 34, 36, 42, 44, 46, 48)

7.5 Agatharchides of Knidos (*ca.* −215 to *ca.* −140) worked in Alexandria under Ptolemy V ("Epiphanes") and Ptolemy VI ("Philometor") (a time of strife and chaos). He wrote several works of history enriched with anthropology and geography, and towards the end of his life, fleeing the chaos in Alexandria, he published *On the Red Sea*, which is lost but paraphrased by Diodoros of Sicily, the first-century-BCE historian, and more closely but more briefly by Photios the ninth-century bishop of Constantinople (material found only in Diodoros is in italics).

On the Red Sea

Book 5, fr.107 [appearance of the sun; see Diodoros 3.48.2–4]

They say that the appearance of the sun is peculiar and different in the regions beyond the Ptolemaïs [on the Red Sea]. First, there is not, as here, that twilight we see for a long time just before dawn and then sunrise. Instead, while night is still dark, the sun suddenly shines forth, and it is not day there before one sees the sun. Second, the sun appears to rise from the middle of the sea. Third, when it does this, it is like a coal of the most fiery kind, scattering great sparks, some into the circle of illumination and some away. Fourth, they also say that the sun does not have a disc-shape but at first closely resembles a thick column which appears a little weightier at the top like a head. Fifth, no ray or beam shines at all, neither on the earth nor the sea, until the first hour, but the sun is a lightless fire in the dark. At the beginning of the second hour, the whole star rises taking a shield-shape, and it casts an image and light of this shape onto the land and sea which is so strange and fiery that both are thought to be extremely enormous. Sixth, they say that the opposite events concerning the sun are seen at evening, reporting that after the sun has sunk below the earth, it illuminates for not less than three hours after sunset, which among those people they consider the pleasantest time of day *the heat being lowered because of the setting of the sun.*

*(Henry [1974] 188)

7.6 Poseidonios of Apamea (in Syria) became a citizen of the "free city" of Rhodes (a Roman protectorate), where he wrote and taught between about –100 and –50. He was widely traveled and well-connected. Most of what we know about his book *On the Ocean* comes from Strabo, who wrote in the generation after him (see Chapter 5.12).

On the Ocean (?)

fr. 119 E–K (in Strabo 3.1.5) [appearance of sun during sunsets]

["Don't believe Artemidoros of Ephesos" – geographer around –100 – "who clung to common tales".]

It is actually the common story that the sun when it sets along the coast of the ocean is larger and makes a noise as if the sea was sizzling in extinguishing the sun because it was falling into its depths. That is untrue, and so too is their view that night follows instantly on sunset; it is not instant, but a short time after sunset, as is the case too with all other great seas. You see, where the sun sets behind mountains, a longer period of daylight after the setting occurs arising from diffused light, but after oceanic sunsets that longer period does not follow on; not, however, that darkness conjoins instantly either, just as that also doesn't happen in the great plains.

And the impression of increased size of the sun alike at sunset and sunrise at sea, derives from the increase of exhalations rising from the water; the light refracted through these as through glass vessels filled with water gives a broader, flatter image, just as occurs too when a setting or rising sun or moon is seen though a dry, thin cloud; at such times the star also appears reddish.

(Kidd [1999] 174–175)

7.7 Heron of Alexandria wrote a series of books on science and engineering (*ca.* 55 to 68; see also Chapters 2, 5, 6, 8); this work survives only in sparse quotations, and an incomplete thirteenth-century Latin translation.

Mirrors

Praeface [nature of vision, and what Heron covers]

There are two senses through which one may reach understanding, according to Plato, I mean hearing and vision, and a science (*theôría*) of both. As to hearing, music involves the science of concords and harmonies, and in sum, the theory of the nature of melody and harmonious composition. For reason proposes much about the *kosmos* being arranged according to musical harmony: the entire heaven is divided into eight spheres (seven spheres of the planets and one containing all those and bearing the fixed stars), and so it happens that the onward motion of the stars in these spheres occurs melodiously and harmoniously because of the coherent power of their mutual motions, just as also chords harmonize in the lyre [see Chapter 3.11]. We ought to understand that there are certain sounds arising from the onward motion of the stars through the *aither*, some of them being deeper, some sharper, as some of the spheres move slowly, some faster. Just as we perceive, when a chord is struck, the air vibrating, so also we ought to see, as the stars are carried through the zodiac, that the continuously altered and changed *aither* provides us a well-tempered sound.

Our undertaking, vision, is divided into optics (i.e., sight), dioptrics (i.e., perspective), and katoptrics (i.e., reflection). Optics has been amply described by our predecessors, and especially by Aristotle. Dioptrics has been treated by us elsewhere as fully as seemed good [the *Dioptra*]. Now we have composed this work, seeing that katoptrics is also an undertaking worthy of study and has an admirable theory. Thus mirrors are constructed showing the right on the right, and likewise the left on the left, while ordinary mirrors are contrary to nature and show us the opposite [*Mirrors* 11–12]. It's moreover possible through mirrors to see people behind us, and ourselves inverted, and having three eyes and two noses, and with facial expression deranged as in mourning [also *Mirrors* 11–12]. Katoptrics is useful not only for theory but also for ordinary needs. For how would someone not think it right useful to see people staying in the neighboring house, e.g., and how many people are in the street and doing what? [*Mirrors* 16] Or how will someone not think it equally marvelous to see the current time, both night and day, via images? (For however many hours of the day or night have gone by, that many images appear. Likewise if some part of a day has passed, an image appears.) [a clock whose motions projected changing images; lost, but see *Mirrors* 17] Or how will someone not think it marvelous to see neither one's-self nor another, but whatever someone might have chosen? [*Mirrors* 18] Given such an undertaking, I suppose it is necessary to dignify with description what is accepted by our predecessors, so our undertaking will lack nothing. [*Mirrors* 2–10]

3–4 [reflection at equal angles]

3   So therefore it is sufficiently shown that we see along straight lines [*Mirrors* 2]. Moreover, that these vision-rays, falling on mirrors as well as on water and any plane surface, are reflected, we will now make clear. It is a property of polished bodies that their surfaces are compact. Mirrors therefore before polishing have some pores, from which incident vision rays cannot be thrown back. But when polished by rubbing, each porous place is filled with a fine substance, so the incident vision-rays are thrown back by the dense body. In the same way a forcefully-projected stone striking against a compact body leaps back, for example from some wood or a wall, but against a soft body, such as wool or the like, it remains at rest; this is because the force of the projection follows through and impinges in the one case upon a hard thing, and the projectile is unable to

proceed and move, but when falling on a soft thing, the force remains there and leaves the projectile – in the same way also the vision-rays, carried away from us with great speed, as has been shown [*Mirrors* 2], and striking against a compact body are reflected. However, those striking against water and glass are not all reflected because both substances have pores and are composed of solid and miniscule bodies. For through glass and water, we see ourselves and what lies beyond; and in shallow waters we see what lies at the bottom. All vision-rays incident on solid bodies are thrown back and reflected, but they penetrate porous bodies and see what lies beyond. So therefore what is represented in such bodies is imperfectly seen because not all vision-rays are reflected to the same places, but some as we said are terminated by pores.

   4   [preserved in Greek by Olumpiodoros, sixth-century ce; compare Thomas [1941] 497–503] It is agreed by everyone that Nature does nothing in vain, nor exerts herself needlessly; but if we don't grant that reflection (*anaklasis*) occurs according to equal-angles, then Nature exerts herself needlessly for unequal angles, and instead of the vision-ray attaining the seen object on the shortest path, it will arrive to reveal the object on a long path. For the straight lines comprising the unequal angles, running from the eye to the mirror and thence to the seen object, are greater than the straight lines comprising the equal angles [a principle of least action]. Now we show that this is true.

   Let the mirror be supposed to be at the straight line AB, and let the viewer be Γ, and the seen object be Δ, and E the point on the mirror where the incident vision-ray is reflected to the object. And let the points ΓE and EΔ be joined (Figure 7.11). I say that the angle AEΓ is equal to angle ΔEB. For if it is not, let the other point on the mirror, where the incident vision-ray is reflected at unequal angles, be Z, and let ΓZ and ZΔ be joined. It is clear that the angle ΓZA is greater than the angle ΔZE. I say that the straight lines ΓZ and ZΔ, which comprise unequal angles on the straight line AB, are greater than the lines ΓE and EΔ (which comprise equal angles with line AB).

   For let there be drawn the perpendicular from Δ to AB at the point H, and let it be extended as a line to point Θ. Now it is clear that the angles at H are equal, for they are right-angles. And let ΔH be equal to HΘ, and let ΘZ and ΘE be joined.

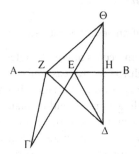

Figure 7.11

Well, since the line ΔH equals the line HΘ, but also the angle ΔHE equals the angle ΘHE, and the side of the two triangles HE is shared, and the baseline ΘE equals the baseline EΔ ["baseline": side of triangle from which its height is being measured], thus also the triangle HΘE equals triangle ΔHE, and the other angles are equal to the other angles which the equal sides subtend. Thus the line ΘE equals the line EΔ.

Again, since the line HΔ equals the line HΘ, and the angle ΔHZ equals the angle ΘHZ, and the side HZ of the two triangles ΔHZ and ΘHZ is shared, therefore the baseline ΘZ equals the baseline ZΔ, and the triangle ZHΔ equals the triangle ΘHZ. Therefore the line ΘZ equals the line ZΔ [i.e., now lines no longer considered as part of a triangle – perhaps a note by Olumpiodoros?]. And since the line ΘE equals the line EΔ, let EΓ be added to both. Therefore the two lines ΓE and EΔ are equal to the two lines ΓE and EΘ. Then the whole line ΓΘ equals the lines ΓE and EΔ. [Heron omits to prove that ΓE and EΔ are collinear: since angle ΔEH equals angle ΘEH, and since we constructed angle AEΓ equal to angle HEΔ, the angles ΘEH and AEΓ are equal; then since AB was constructed as a line, ΓEΘ must also be a line.]

And since the two sides of every triangle are greater than the remaining side, no matter how taken, then the two sides ΘZ and ZΓ of the triangle ΘZΓ are greater than the one side ΓΘ. But the line ΓΘ equals the lines ΓE and EΔ. Thus the lines ΘZ and ZΓ are greater than the lines ΓE and EΔ. But the line ΘZ equals the line ZΔ, thus the lines ZΓ and ZΔ are greater than the lines ΓE and EΔ. And the lines ΓZ and ZΔ comprise the unequal angles [of reflection] – thus the lines comprising unequal angles are greater than the lines comprising equal angles.

*(Nix and Schmidt [1900/1976] 316, 318, 320, 322, 324; 368, 370, 372; compare Cohen and Drabkin [1958] 261–265)

7.8 Plutarch of Chaironeia (*ca.* 50 to *ca.* 120), writing under the Flavian emperors, Trajan, and in the early years of Hadrian, was a Platonist teacher of philosophy; much of his copious output is in the form of moralizing essays or dialogues.

The Face in the Moon

17(930a–d) [apparent exceptions to the equal-angles reflection law]

Yet it must be said that the proposition, "all reflection occurs at equal angles," is neither self-evident nor an admitted fact. It is refuted in the case of convex mirrors when the point of incidence of the visual ray produces images that are magnified in one respect; and it is refuted by folding mirrors, either plane of which, when they have been inclined to each other and have formed an inner angle, exhibits a double image, so that four likenesses of a single object are produced, two reversed on the outer surfaces and two dim ones not reversed in the depth of the mirrors [images formed by single and double reflections]. The reason for the production of these images Plato explains, for he has said that when the mirror is elevated on both sides the visual rays interchange their reflection because they shift from one side to the other [*Timaios* 46ac]. So, if some of the visual rays revert straight to us, while others glance off to the opposite sides of the mirrors and thence return to us again, it is not possible that all reflections occur at equal angles. Consequently they take direct issue and maintain that they confute the equality of the angles of incidence and reflection by the very streams of light that flow from the moon upon the earth, for they deem this fact to be much more credible than that claim. Nevertheless, suppose that "equal-angles" must be conceded as a favor to dear geometry. In the first place, it is likely to occur only in mirrors that have been polished to exact smoothness; but the moon is very uneven and rugged, so that the rays from a large body striking against considerable heights which receive reflections and diffusions of light from one another are variously reflected and intertwined and with the refulgence itself combines with itself, coming to us, as it were, from many mirrors.

(Cherniss and Helmbold [1957] 107, 109, 111)

7.9 Ptolemy (Claudius Ptolemaeus) of Alexandria (*ca.* 100 to *ca.* 175) wrote on a wide variety of "mathematical" topics (he is quoted in each of Chapters 2–8). His *Optics* survives (minus book 1) in a medieval Latin translation of an Arabic version, and includes a number of experiments (e.g., to confirm the equal-angles law and to find a law of refraction).

Optics

2.13–14 [color]

A sole proper sensible ["thing sensed"] can be found that is appropriate to each of the senses; e.g., the quality of "resisting the hand" for touch, savors for taste, sounds for hearing, and odors for smell [Aristotle, *On the Soul* 2.6 (418a11–14)]. But among the things that are common to the senses according to the origin of nervous activity [the *hegemonikon*: see Chapter 12], sight and touch share in all except color, for color is perceived by no sense but sight. Thus, color must be the proper sensible for sight, and that is why color is taken to be what is primarily visible after light.

In view of this, it seems that color is not really a proper sensible, as certain people have supposed [Demokritos and the atomists], claiming that color is something accidental to the visual flux and to light and that it has no real subsistence, because none of the proper sensibles needs anything extrinsic [to make it sensible], whereas colors need light. Hence, it seems that this missing subsistence is provided by the visual flux, not by the visible objects. For objects that are seen and that are affected by the visual flux are not of such a nature as to appear to the visual faculty without light.

2.28–31 [binocular vision]

It seems, moreover, that nature has doubled our eyes so that we may see more clearly and so that our vision may be regular and definite. We are naturally disposed to turn our raised eyes unconsciously in various directions with a remarkable and accurate motion, until both axes converge on the middle of a visible object, and both cones form a single base upon the visible object they touch; and that base is composed of all the correspondingly arranged rays of the cones.

But if we somehow force our sight from its accustomed focus and shift it to an object other than the one we wanted to see, and if the object toward which our sight is directed is somewhat narrower than the distance between our eyes, and if the visual rays that fall together from our eyes on that object are not correspondingly arranged, then that same object will be seen at two places [parallax]. But when we close or cover either of our eyes, then the image in one of the two locations will immediately disappear, while the other image will persist, sometimes the one directly in front of the covered eye and sometimes the one directly in front of the other eye. This point will be easily understood if we try to explain it in the following way:

Let a short ruler be set up, let two long, thin cylindrical pegs be stood vertically upon it, and let the distance between the two pegs themselves and between the pegs and the edges of the ruler be moderate. Let either edge of the ruler be placed between the eyes so that the pegs lie in a straight line at right angles to the line connecting the eyes. [compare Archimedes, above Section 3]

Accordingly, if we focus our eyes on the nearer of the pegs, we will see it as one, whereas we will see the other, which is farther away, doubled. And if we close either eye, the peg that appears directly in front of that same eye and that forms one of the doubled images will disappear. On the other hand, if we focus our eyes on the farther of the pegs, we will see it as one and will see the nearer one doubled. And if we close either eye, the one of the doubled pegs that appears to be opposite to this same eye and that forms one of the two images will disappear.

[2.32–44: experiments using the apparatus]

2.107 [after-images, or residual coloring from mirrors and lenses]

"Anterior coloring" is generated by itself when we have looked for a long time at some very bright color and then look away to something else. For in that case what is last looked at seems to possess something of the color of what was first observed because the impression of bright colors lasts a long time in the visual faculty. And so, after we have looked at bright colors, we see neither clearly nor without some impairment. [pseudo-Aristotle, *Problems* 31.19: we prefer to gaze on easier-to-eyes green] Anterior coloring also happens when we look at something through thin, threadbare cloths of a red or purple hue. The visual flux passes through the weft and warp of the cloths without breaking and, in the process, takes on something of the color of the threads it brushes by. Thus the visible object appears to be tinged with the color of media traversed by the visual flux. [compare Aristotle, *Dreams* 2 (459b8–19), pseudo-Aristotle, *Colors* 3.2 (793b–4a)]

2.133 [eyes of a painted face "following" viewer]

It is also assumed that the image of a face painted on panels follows the gaze of [moving] viewers to some extent, even though there is no motion in the image itself, and the reason is that the true direction of the painted face's gaze is perceived by means only of the stationary disposition of the visual cone that strikes the painted face. The visual faculty does not recognize this, but the gaze remains fixed solely along the visual axis, because the parts themselves of the face are seen by means of corresponding visual rays. Thus, as the observer moves away, he supposes that the image's gaze follows his.

3.59 [illusions in depth-perception]

Generally speaking, in fact, when a visual ray falls upon visible objects in an unnatural or unaccustomed way, it perceives less clearly all the characteristics belonging to them [perception attributed to the vision ray itself]. So too, its perception of the distances it apprehends will be diminished. This seems to be the reason why, among celestial objects that subtend equal visual angles, those that lie near the zenith appear smaller, whereas those that lie near the horizon are seen in another way that accords with custom. Things that are high up seem smaller than usual and are seen with difficulty [the still-debated "moon-illusion": see Sabra [1987]].

4.109–113 [concave mirrors magnify]

Accordingly, in concave mirrors, when the image lies behind the mirror, the distance of the visible object [from the eye along the vision ray] will be smaller than that of its image [along the incident ray] if the visual radiation were to continue behind the mirror [i.e., the image seems closer].

Let ABΓΔ be the arc of a circle inscribed in a concave mirror with center-point E. Let point Z be the eye, and point H the visible object to which ray-couple ZBH is reflected at equal angles. Let EHΓΘ be drawn [perpendicular to the surface], and let BZ and HE be extended to intersect at point Θ beyond the mirror (Figure 7.12). Therefore, the image of H will lie at point Θ. We say, then, that: BZ + BH < ZΘ, whereas ΓH < ΓΘ.

Let EB be drawn, and let line KBΛ be drawn tangent to the circle at point B. Therefore, [tangent] angle ABZ = [tangent] angle ΓBH, and [horn] angle ABK = [horn] angle ΓBΛ [see Chapter 2.1], so that the whole angle KBZ, which is equal to angle ΘBΛ, is equal to angle ΛBH. Now angle BΛH is acute, since angle ΛBE is right. Therefore, angle BΛH < angle BΛΘ. And if we posit angle BΛM = angle

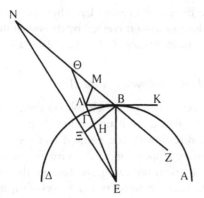

Figure 7.12

BΛH, then, since angles MBΛ and ΛBH in triangles MBΛ and ΛBH are equal, while side BΛ is common, it follows that MB = BH. Consequently, BH < BΘ. And if we take BZ as common, then ZB + BH < ZBΘ.

Moreover, since ΛH:ΛΘ = BH:BΘ, while BH < BΘ, then ΛH < ΛΘ, and ΓH will be much smaller than ΓΘ. [To see that the given proportion is correct, note that the angle ΛBH = angle ΛBΘ (label it "α"), and the angle BΛH = 180 degrees – angle BΛΘ (label angle BΛH as "β"); then Ptolemy uses the "law of chords" – similar to our law of sines: compare Toomer [1984b] 7, n.10, and 462, n.96 – to find that ΛH:chord(2α) = BH:chord(2β), and ΛΘ:chord(2α) = BΘ:chord(2(180−β)); of course chord(360 −2β) = chord(2β), for which compare Ptolemy, *Syntaxis* 1.10, in Section 2.9; and ΛH:BH = chord(2α):chord(2β) = ΛΘ:BΘ; from ΛH:BH = ΛΘ:BΘ to the stated proportion is easy.]

From what we have established, it is evident that, if the distance between the objects and the same viewpoint increases, then the distance between the images and the eye increases. For if we extend BH to Ξ and continue EΞ until it meets the prolongation of ZBΘ at point N, then the image of Ξ will lie at point N, and so BΞ + BN > BH + BΘ.

5.3–6 [experiments on angles of refraction]

At this point we ought to investigate the quantitative relationship between the angles of incidence and refraction according to specific intervals. But we should start by discussing the phenomena that such refractions have in common with reflections. First, in either case, whatever is seen appears along the continuation of the incident ray (i.e., along the continuation of the ray that emanates from the eye to the surface at which it is broken); second, the object appears on the straight line dropped perpendicularly from the visible object to the surface where the breaking occurs. It therefore follows that, just as was the case for mirrors, so in this case, the plane containing the broken ray-couple must be perpendicular to the surface where the breaking occurs [compare 3.5].

We have already shown in the place where we laid out the principles governing mirrors that the above points are in the nature of observable phenomena and that what happens is quantifiable. [compare 3.14–17]

That this is clear and indubitable we can understand on its own terms by means of a coin that is placed in a vessel called a *baptistir* [cylindrical trough]. For, if the eye remains fixed so that the visual ray passing over the lip of the vessel passes above the coin, and if water is then poured slowly into the vessel until the ray that passes over the edge of the vessel is refracted toward the interior to fall on the coin, then objects that were invisible before are seen along a straight line extended from the eye to a point higher than the true point. And it will be supposed not that the ray is refracted toward those lower objects but, rather, that the objects themselves are floating and are raised up to the ray. For this reason, such objects will be seen along the continuation of the incident visual ray, as well as along the normal dropped from the visible object to the water's surface – all according to the principles we have previously established [already in Eukleidês, *Mirrors* def. 6: see Knorr [1985]].

Now, let us suppose that point A is the eye, ZHE the common section of the plane containing the refracted ray-couple and the surface of the water, and ABΔ the ray passing over the vessel's lip at B. Let us also suppose that there is a coin at Γ, which lies toward the bottom of the vessel (Figure 7.13). Then, as long as the vessel remains empty, the coin will not be seen, because the body of the apparatus at B blocks the visual ray that could proceed directly to the coin. Yet, when just enough water is poured into the vessel so that its surface reaches line ZHE, ray ABH is deflected along line ΓH, compared to which AHΔ is higher. In that case, then, the coin will appear to be located along the perpendicular dropped from point Γ to EH – i.e., perpendicular ΛKΓ, which intersects line AHΔ at point K. Moreover, its image-location will lie on the radial line passing from the eye and continuing rectilinearly to point K, that radial line being higher than the actual ray HΓ and nearer the water's surface; so the image will appear at point K.

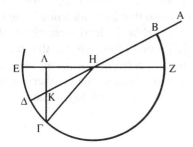

Figure 7.13

[Note that 5.14–18 is the experiment trying to find a law of refraction: Smith [1982] 234–236]

(Smith [1996] 74–75, 83–84, 114, 124, 151, 205–206, 230–231)

7.10 Galen of Pergamon (129 to 210 ± 5), viewing himself as philosopher and physician, considered Hippokrates and Plato as the most worthy authorities; he wrote voluminously on medicine and much else.

Opinions of Hippokrates and Plato

7.5 [optical *pneuma*]

A body that is seen does one of two things: either it sends something from itself to us and thereby gives an indication of its peculiar character, or if it does not itself send something, it waits for some sensory power to come to it from us. Which of these alternatives is the more correct may best be judged in the following way. We see through the perforation at the pupil; if this perforation waited for some portion or power or image or quality of the external bodies underlying (our perception) to come to it, we would not discern the size of the object seen, which might be, for example, a very large mountain. An image the size of the mountain would have come from the mountain and entered our eyes, which is utterly absurd. It is also absurd that at one moment of time the image should reach every viewer, even though they are countless. And the optic *pneuma* cannot extend itself and acquire such a stream as to envelop the whole object being viewed; for this is comparable to the Stoic drop that mixes with the whole sea [on Stoic theory of mixtures, see Chapter 9.15].

We are left, then, with the view that at the time when we look at something, the surrounding air becomes for us the kind of instrument that the nerve in the body is at all times. It seems that the effect produced on the air around us by the emission of the *pneuma* is of the same sort as the effect produced on it by the light of the sun. For sunlight, touching the upper limit of the air, transmits its power to the whole; and the vision that is carried through the optic nerves has a substance of the nature of *pneuma*, and when it strikes the surrounding air it produces by its first impact an alteration that is transmitted to the furthest distance – the surrounding air being, of course, a continuum, so that in a moment of time the alteration spreads to the whole of it. This sort of thing is clearly seen also in the case of the sun's power. If we interpose some solid object, we see that the air beyond immediately loses its brightness, the reason being that the air becomes luminous as it changes but does not remain in the changed state; otherwise the light would remain in it for a long time even if the source of light is removed.

7.7

The organ of sight, for example, since it had to discriminate colors, was made luminous, for only such bodies are by nature capable of being altered by colors. The surrounding air shows this: when it is especially clear, then it is altered by colors. Thus when a person reclines under a tree in such air as that, you can see the color of the tree enveloping him. And often when bright air touches the color of a wall, it receives the color and transmits it to another body, especially when the wall is blue or yellow or some other bright hue. And just as at a mere touch all the air is assimilated all at once by the sunlight, in the same way it is immediately changed by color.

...

The Stoics, then, must not say that we see by means of the surrounding air as with a walking-stick. This latter kind of discernment is of resistant bodies, and it is besides more inferential than perceptive, whereas the perception of our eye is not perceptive of a thing as close packed, or of its hardness or softness, but of its color, size, and position, and none of these can be discerned by a walking-stick. Therefore Epikouros' view – considering that both views are mistaken – is much better than that of the Stoics. For the latter do not bring anything of the visual object up to the visual power, but Epikouros declared that he did so. Aristotle is much superior to him, since he does not bring a corporeal image but a quality from the visual object to the eyes through an alteration of the surrounding air. He avoided making the surrounding air sensitive at all, although he saw clearly that flesh becomes sensitive as a result of the power that comes to it from the *hegemonikon*. But what difficulty is there in supposing that the sunlight is sensitive, much as the *pneuma* in the eyes that is brought in from the brain is clearly seen to be? For it is luminous. And if we must speak of the substance of the soul, we must say one of two things: we must say either that it is this, as it were, bright and *aitherial* body, a view to which the Stoics and Aristotle are carried in spite of themselves, as the logical consequence, or that it is an incorporeal substance and this body is its first vehicle, by means of which it establishes partnership with other bodies. We must say, then, that *pneuma* itself extends through all the brain, and that by partnership with it the optical *pneuma* becomes luminous.

(deLacy [1984] 453, 455, 471, 475)

8

HYDROSTATICS
AND PNEUMATICS

The world mostly endures and children soon learn that objects are permanent, but some things flow for ever: streams and clouds and drafts. Of unseen origins, such events are conceived as beings: thus winds and rivers become gods with dwellings, of whom the greatest is Ocean, with unfathomable currents and fluxes. Some objects float on water, so we (like elephants, otters, and dogs) can swim, and ships can travel. Myths of endless jugs (Elisha in *Kings II* 4.1–7 = Iosephos, *Antiquities* 9.4.2(47–50) and Ovid, *Metamorphoses* 8.679–680) or sacks of winds (*Odyssey* 10.1–27) may have influenced the Boiotian trick vases which decanted marvels (Kilinski [1986]). Fluids almost lived, two even seeming the stuff of life: blood and breath.

Greek thinkers soon set air at the center of the speculations: Anaximenes of Miletos (–545 ± 20) supposing it to be the divine *kosmic* essence, producing water and earth and all by its condensation or rarefaction. Herakleitos of Ephesos obscurely quipped that rivers in a world of flux could not be entered twice (–500 ± 20). A generation or more later, Empedokles of Akragas and Anaxagoras of Klazomenai (both –455 ± 20) asserted air was body, since air in a closed container blocked the influx of water which pressed against it; moreover, air was the source of life (pseudo-Aristotle, *Respiration* 6–7 [473–474], pseudo-Aristotle, *Problems* 16.8). Leukippos (–450 ± 20) and Demokritos of Abdera (writing –410 ± 30) hypothesized that flows and compression were explained by atoms spontaneously moving through void. Diogenes of Apollonia (Demokritos' contemporary) espoused a theory of divine air as the essence of everything, parodied in Aristophanes' *Clouds* (produced –422).

In the medical library ascribed to Hippokrates (–400 ± 20), *Joints* (47, 77) briefly describes the reduction of dislocations by inflated bladders (pseudo-Aristotle, *Problems* 25.1 notes the pain caused). Plato's probable account of breath and blood (similar to Empedokles') depicted animals perfused with life-giving hot blood cooled by air (*Timaios* 77–79). Aristotle strictly denied the earlier view that wind was air in motion (*Meteorologika* 1.13 [349a18–22]), and promoted a dual-"exhalation" (*anathumiasis*) model, sublimed from earth to sky by sun and productive of wind and weather (*Meteorologika* 2.4 [359b27–

361b14]). He was aware of currents in the Mediterranean (*Meteorologika* 2.1 [354a5–35]), and one of his followers remarked upon the differing buoyancy of salt and sweet waters (*Problems* 23.13; similar is Strabo 15.3.22).

Ktesibios of Alexandria who worked under Ptolemy II (–282 to –246) is considered the founder of *pneumatics* as a defined discipline (Drachmann [1948]). His inventions (now preserved in paraphrases in Philon, Heron, and Vitruuius 9.8, 10.7–8) included the *hudraulis* ("water-flute," our "organ"); the two-piston water-pump (for fire-engines and bilge-pumps); the constant-flow valve for a regulated water-clock; and the piston-powered catapult. His contemporary Straton of Lampsakos (writing –286 to –269) modified the atomist theory of void and atoms to propose that there was no "bulk" void, only disseminate "micro-voids," which explained condensation and rarefaction, as well as mixing and transparency (Gottschalk [1965]).

8.1 Theophrastos of Eresos (lived –370 to –286) succeeded Aristotle as the head of the Lukeion "school," and wrote widely, amassing data questioning Aristotle's system (Keyser [1997c]). The work here excerpted was one of a set of works on natural philosophy (see also Chapters 9–12).

Winds

1–12 [see Murray 1987]

1 We have earlier considered the nature of the winds: of what they consist, in what way they come to be, and by what they are caused. We must now try to explain that each wind is systematically accompanied by effects and in general by phenomena whereby the winds are differentiated from one another. These differences refer to matters like force, temperature, storm or calm, and (speaking loosely) rain or fair weather, frequency or rarity, seasonal occurrence or failure, continuous and consistent or discontinuous and irregular. In fact, what happens in the sky, in the air, on earth and on the sea is due to the wind. And to put it briefly, our inquiries deal with matters which also concern the life and well-being of plants and animals.

2 [pseudo-Aristotle, *Problems* 26.15, 35] Since each wind has its own proper place and this fact is, as it were, a part of its nature, from this, to put it simply, are derived the differences and the powers of each, such as force, temperature, amount, and most other qualities. Opposing winds have both opposing qualities and the same qualities, and rightly so. Example: the north wind and the south wind are strong winds and blow the longest time, because most air is forced to the north and to the south, since both districts are athwart the motion of the sun from east to west. The air there is dislodged outwards by the effect of the sun. This accounts for the extreme density and cloudiness of the air. It is collected in quantity on both sides, so that the flow is greater, more continuous, and more frequent. This is the cause of the force, the continuity, the amount and other such features of these winds.

3 [pseudo-Aristotle, *Problems* 26.36: winds are like rivers] It seems very clear that the cold and the heat arise from the geographical source. The lands to the north are cold, those to the south warm. The winds which flow from the respective lands are similar. At the same time the less open the succeeding district, the less diffuse the movement, so that wind passing through a narrow confine with a greater rush is colder, but that which is diffused over a greater distance is gentle. Hence the south wind is colder in the south than in our climate, some say, and more than the north wind. The change adds to the sensation if the place was warm already.

4 Generally this holds for all winds. Whether a wind brings rain or shine, whether it is gusty or steady, whether it is frequent or continuous, regular or irregular, forceful or weak when it starts and when it drops away, depends on the distance of the place from which it blows and that to which it displaces the air. The former enjoys sunshine, the latter has clouds and rain. That is why the north wind and, even more, the etesians ["periodic" winds, from northwest in summer] are rainy in the south and the east, while the south wind and generally the winds from that area bring rain for those who dwell in the north.

5 [pseudo-Aristotle, *Problems* 26.2: the north wind is periodic, but the south wind continuous] It is not unimportant, nay, it is highly important if the district has elevations. For wherever the clouds run into an obstacle and stop, rain is generated. And so in adjoining districts different winds are rainy in different areas. But the rains have been discussed elsewhere at length. For the same reason the north wind is forceful at its inception, while the south wind is so as it ceases; hence the proverb about voyages by sea [pseudo-Aristotle, *Problems* 26.20 "sail when south wind starts and north wind dies"]. The north wind hovers, as it were, directly above those who live in the north, while the south wind is far off. The flow from far off takes more time and happens when a mass has been accumulated. For those who live in Egypt and nearby the situation is reversed; the south wind has force. As a result, the proverb is stated the other way around.

6 Thus, down there, the south wind is more frequent, uniform, continuous, and regular [a symmetry argument, not an observation]. For all winds are like that for those who live nearby, but they are irregular and intermittent for those who live at a distance. These then must be reckoned to be the causes of the above phenomena, causes which are apparent also in other, smaller areas and less distant from one another. This may seem incorrect, however, for the south wind is invariably fair in places proper to it, while the north wind is cloudy in areas near, when there is a great storm, but cloudless in places farther away.

7 The reason is that because of its force it moves a mass of air and freezes this air before it can move it on; when the clouds freeze, they do not move because of their weight. It is the force rather than the coldness which is transmitted to distant places and there does its work. The south wind, having less material and not freezing it but moving it on, creates fair weather for the area nearby, but is always rainier in districts farther away. It blows with force when coming to an end rather than when beginning because at the beginning it moves little air, but moves more as it advances. The air thus collected becomes cloud and thus condensed becomes rainy. It also makes a difference that the wind begins from a smaller source, not a greater. If the source is small, the wind is fair; when the source is great, the wind is cloudy and rainy because it forces more air together.

8 [harmonizing with tradition, as in pseudo-Aristotle, *Problems* 26.44: "there is no south wind on the Egyptian coast (that's how Menelaus' ship got becalmed there on his long way home from Troy)"] The statement of some that the south wind does not blow at the coast of Egypt nor for a day and a night's travel from the shore, roughly speaking, while it is vigorous in the area above Memphis and likewise in places equally distant from the sea, is said to be false. But the fact is it does not blow as strongly but weaker. The reason is that Lower Egypt is sunken so that the wind bypasses it at a height. Upper Egypt is higher. Somehow or other, proximity makes for force. Such phenomena are to be explained chiefly by reference to topography when they are in the order of nature. And these winds remain cloudy or fair in a way similar to that which we have just explained.

9 [pseudo-Aristotle, *Problems* 26.47: the north wind follows the south wind, but not vice versa] The fact that the north wind blows after the south wind, while the south wind does not follow the north wind, must be ascribed to its place. This is what happens in our land generally and in those to the north, but it is the reverse in the south. The reason is the same for both: the north wind is nearer to the former, the south wind to the latter, so that as soon as they get under way, they are perceptible, but they reach distant places more slowly.

10 The north winds and the south winds being the most frequent, as we have said, there is a certain orderliness about their periods. The north winds blow in the winter, in the summer, and in late autumn until the end of the season, while the south winds blow in winter, at the beginning of spring, and at the end of the late autumn [this correlates to nautical wind-lore]. The motion of the sun itself contributes to both winds, and a reciprocating interchange takes place, with the air flowing back and forth, as it were. For what is dislodged in the winter (in general there are more north winds blowing) and earlier in the summer by the etesians and their successors, is restored to those places in the spring, at the end of late autumn, and at the setting of the Pleiades [early November] in proportion.

11 [see *Winds* 5 above] From this arises the puzzlement as to why there are northerly etesians but not southerly etesians, as though this were a fact; but it appears that there are southerly etesians. For the south winds in the spring are etesians, those which are called white south winds; they are fair-weather winds and cloudless on the whole. At the same time, being remote from us, they are not noticed. But the north wind is right at hand. Why they blow at this season and in such strength; why they die down when the day dies and do not blow at night, is explained more or less by the following causes: the breeze occurs because of the melting of the snow; when the sun begins to prevail and to dissolve the frost, there come the "forerunners" and then the etesians.

12 The reason that these winds cease when the sun goes down and do not blow at night is that as the sun is sinking, the snow ceases to melt, and at night, when the sun has set, it no longer melts [Aristotle, *Meteorologika* 2.5 (362a1–11, 17–31)]. Yet sometimes these winds do blow, when the melting is greater than usual. This must be accounted the reason for the irregularity. Sometimes the winds blow strong and continuously, sometimes weakly and intermittently, because of the irregularity of the melting. The strength of the wind is as the amount of material. This irregularity can be ascribed to the lay of the land, the distance near or far, and other such variations.

16–17 [*Meteorologika* 2.5 (361b14–20); pseudo-Aristotle, *Problems* 26.8, 21, 33–35: the sun affects winds]

16 Sometimes the sun, by removing the motion, as it were, which it has imparted, halts the wind when it sets. Plainly, the motion must have some proportion so that the wind is not consumed nor kept moving an overlong time. Nothing prevents some winds from blowing more strongly when the sun goes down, such as those which are repressed by the heat and, as it were, dried up and burned out. For this reason these winds are quiet at midday generally and gain strength when the sun is sinking.

17 [pseudo-Aristotle, *Problems* 26.18: the moon affects winds] The moon has this effect also, but not to the same degree, being a kind of weak sun. Therefore the breezes are more powerful at night and the weather stormier at the full moon. And so, when the sun is rising, the winds now rise, now abate. It is the same with the setting sun; sometimes it halts the winds, sometimes it lets loose. The question whether these things happen in conjunction, as at the risings and settings of the stars, must be looked into.

20 [a standard Greek analogy, as in pseudo-Aristotle, *Problems* 26.48]

A good example of this is the breath we release from our mouths, which is said to be hot and cold, mistakenly, since it is always hot but differs in the way it is projected. When we release it mouth agape and in a body, it is warm, but when it is released through pursed lips vigorously, it impels the nearby air and this air the next, which is cold, and so the breath and the motion are cold. This same thing happens with the winds. When the first motion passes through a constriction, the wind is at first not cold, but the matter set in motion by it is hot or cold according to its condition when encountered. If it was cold, it remains cold; if it was hot, it remains hot. For this reason winds are hot in summer and cold in winter. For the air is in keeping with the season.

22–24 [air is not self-moving: compare Aristotle, *Meteorologika* 1.4 (342a24–27)]

22 That the air is not self-moving nor takes this direction because overcome by heat is clear from the following; if air were self-moving, being cold and vaporous by nature, it would move downwards; if it were moved by heat, it would move upwards. For the motion of fire is naturally upwards. In fact, the motion is in a sense a compound of both because neither prevails.

23 [*Odyssey* 5.469; pseudo-Aristotle, *Problems* 26.30: rivers spawn cold winds] And this general principle that the winds will have the same temperature as the air or the exhalation [Aristotle's *anathumiasis*, in *Winds* only here and in 15] has in any given place is supported by the following: all winds emanating from rivers and lakes are cold because of the moisture of the air. When the sun fails, the vapor is cooled and at the same time condensed, and all the more if in the immediate vicinity, with the result that when such a wind reaches home to us, we shiver.

24 And for this reason some places, which are depressed and protected against winds from outside, are cold because of local winds. For the vapor raised by the sun cannot remain there but moves and creates a breeze. And so breezes from rivers and lakes and in general those off the land blow at dawn, when the vapor is becoming cool from the lack of heat. It is logical that this kind of breeze should arise because of fair weather among other reasons. When drizzles and light rains fall, the breezes blow more. For material is added from all sides, and after that the breezes blow off the land more.

26–27 [mechanical anti-Aristotelian explanation; compare pseudo-Aristotle, *Problems* 26.40: bays are full of variable winds]

26 Now reversals originate from land breezes and their like through a concentration of cold air. For the reversing wind is a kind of back-flow of the wind, like the waters in situations of ebb and reflux. For when there is a concentration and build-up, there is a change in the opposite direction. This occurs mainly in hollow places and wherever land breezes blow. Each of these has a scientific basis. When the air comes into hollow places, it is concentrated, in open areas it is diffused. Land breezes are weak by nature so that they cannot force their way far. The compensatory motion takes place in proportion to the amount and force of the land winds as they blow. So too with the time of day, that is, whether they blow early or late.

27 There also occurs a backlash of winds so that they blow back against themselves when they flow against high places and cannot rise above them. Therefore the clouds sometimes move in the opposite direction to the winds as in the neighborhood of Aigai in Macedonia, when north wind blows against north wind. The reason is that when the winds blow against the high mountains near Olumpos and Ossa and do not surmount them, they lash back in the reverse direction, so that the clouds moving on a lower level move in reverse direction. The same thing happens in other places [windshear].

[29: the movement of air is wind]

35–36 [common features of winds]

35   The following features are common to all winds, such as what signs indicate the coming of each wind. For air, by altering with respect to density and rarity, heat or cold, or any other condition, always reveals the coming breeze. For the winds are in key with the air, and the air precedes the wind in affecting our senses. Similarly, it is possible for some people to take in these same signs on the ocean and on fresh-water bodies. [Pseudo-Aristotle, *Problems* 23.12, 28.] The waves, by rising and falling in advance, signify the coming winds. The waves are pushed forward at intervals, not uninterruptedly; one wave pushes the other, which pushes a third; in turn, as the first wave dies, the next wave is pushed by another breeze. Thus then they come on as they are pushed forward. When that which is being moved appears, it is clear that the moving force will arrive. And it also happens that waves persist after the winds, for they crest and break up later, since they are more difficult to set in motion or to bring to a halt once they are in motion.

36   [Pseudo-Aristotle, *Problems* 26.23: meteors portend wind; 26.12, 32: Sirius affects winds.] The following are also common to many winds, such as the appearance and the fading or breakup of stars, moon, haloes, mock-suns, and any other such phenomenon. For what happens to the upper air foretells what the wind will be like. There is also the fact that winds are very strong at the end, and this is common to many winds. For when they blow themselves out in concentrated mass, there is little left [pseudo-Aristotle, *Problems* 26.25]. Such phenomena then are in some way common to the nature of the winds, as has been said.

44–46 [features of individual winds]

44   The common and general aspect of this wind has been presented. The individual aspects here and there must be studied under the light of local conditions and other attendant circumstances. The differences will prove to lie more or less in such considerations as the following for example, in Lokris in Italy and the neighboring territory the west wind brings fertility because it comes from the sea. Another district is not so fortunate, and some districts even suffer harm. The district of Gortuna in Crete in turn is fertile, for it is open, and the west wind reaches it from the sea. Some other region, visited by this wind from the land and some mountains, is desolate.

45   In the Malic Gulf this wind destroys all growth, both the annuals and the fruit of trees, and so too in Thessalê near the Pierian Gulf. The lay of the land in both cases is the same, so too the environs. Both are open to the east and are walled off by high mountains, the Malic Gulf by Mt. Oítê and the adjoining ranges, the Pierian by Mt. Pieros. The west wind, blowing from equinoctial sundown, takes the heat from the sun, which has fallen on the mountains, and deflecting it earthwards, launches it straight into the plain and scorches it. It is the

same for other regions where this or its like occurs, and conversely with opposite situations ["fallwinds," warm downdrafts like Colorado's "chinook"].

46 [Pseudo-Aristotle, *Problems* 26.48: all winds are somewhat cold.] In general, the frequently made statement holds true, that it makes a great difference through what areas and from what areas a wind passes, both with regard to other matters and with regard to temperature. For this reason the south wind also is cold, no less than the north wind, as is stated in the proverb [pseudo-Aristotle, *Problems* 26.46], because, with the air still cooled and moistened by the winter, the wind must arrive in the same condition as the air. The north wind which comes upon the mud of the south wind, which the proverb again says makes stormy weather, does so for the same reason. For air which is humidified is cold. And so it is with breezes from rivers, as has been already stated.

(Coutant and Eichenlaub [1975] odd pp. 3–13, 17, 21, 23–25, 27, 35–37, 43–47)

8.2 Straton of Lampsakos was the third head of Aristotle's school (from –286 to –269), and wrote on a wide variety of scientific topics (so that he was called "the naturalist"). His theory of matter allowed for "micro-voids," and was employed by writers on pneumatics.

Void

(fr. paraphrased in Heron, Pneumatics *1 praeface [pp. 24.20–26.23 Schmidt])*
[micro-voids: compare pseudo-Aristotle, *Problems* 11.49, 58; see Furley [1985]]

That there are voids one could grasp from the following. If there weren't, neither light nor heat nor any other bodily power could penetrate through water or air or any other body. How could the rays of the sun penetrate through water to the bottom of the vessel? If the water lacked pores, and the beams forcefully parted the water [assuming light is corporeal], full vessels would consequently overflow: which they do not. It's also clear from this: if they forcefully parted the water, some rays would not reflect upward while others penetrated downward. But in fact, whichever beams impinge upon the particles of water are reflected upward as if rejected, but whichever penetrate the voids of the water, encounter few particles, and penetrate to the floor of the vessel. It's also clear that there are voids in water from this, that when wine is poured into water, one sees the influx spreads to every part of the water. This wouldn't happen if there weren't voids in the water. Also, when you light many lamps they all illuminate more, the beams interpenetrating in every direction. And other things penetrate even through bronze and iron and other bodies, like what happens with the marine torpedo-fish [the Mediterranean electric fish, *Torpedo marmorata*: Aristotle, *Animal Researches* 8(9).37 (620b19–29); Thompson [1947] 169–171].

*(Gottschalk [1965] 115)

8.3 Archimedes of Syracuse (Surakusê) lived from *ca.* –285 to –211, was close to the ruling family of Syracuse, visited Alexandria, and wrote on mathematics, mechanics, optics, and pneumatics, addressing his works to Konon of Samos, Eratosthenes of Kurênê, and others. (Material in *italics* survives intact only in a medieval Latin translation.)

Floating Bodies 1

Postulate 1 [essential nature of liquids]

Let it be supposed that a liquid is of such a character that, its parts being in equilibrium and continuous, that part which is thrust the less is driven along by that which is thrust the more; and that each of its parts is thrust by the liquid which is above it in a perpendicular direction, if the liquid is not contained in anything or compressed by anything else.

Proposition 1

If a surface be cut by a plane always through the same given point, and if the section *be always a circumference whose center is that point, the surface is that of a sphere. For let some surface cut by a plane through K always make the circumference of a circle whose center is the same K (Figure 8.1). If then that surface were not the surface of a sphere, the lines from the center to the surface would not all be equal. Let there be points ABΓΔ on the surface and let unequal lines be drawn AK and KB. Now KA and KB determine a plane and make a section on the surface, the line ΔABΓ, which is thus a circle with center K, by hypothesis. So the lines KA and KB are not unequal, and therefore it's necessary that the surface is that of a sphere.*

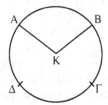

Figure 8.1

Proposition 2

Every liquid's surface, so long as it remains unmoved, has the shape of a sphere whose center is the Earth's center. For consider a liquid at rest and remaining unmoved, and let its surface be cut by a plane through the center of the Earth, and let the center of the Earth be K, and the line which is the section of the surface be ABΓΔ (Figure 8.2). I say the line ABΓΔ is the circumference of a circle whose center is the same K.

For if it weren't, straight lines running from K to ABΓΔ wouldn't be equal. Consider one

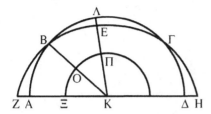

Figure 8.2

straight line of those from K to ABΓΔ which is larger than some and smaller than others of the straight lines, and draw a circle around K whose radius is that of the given straight line. The circumference of this circle will have some part outside ABΓΔ and some part inside, because some of the lines running from the center at K to the line ABΓΔ are greater and some are smaller [than its radius]. So let the circumference of the circle drawn be ZBH, and let a straight line be drawn from B to K, and join ZK and KEΛ making equal angles, and draw a circumference around K at ΞOΠ which lies in the plane and in the liquid. Then the parts of the liquid lying along ΞOΠ are in equilibrium and continuous. And the parts of the liquid along the circumference ΞO are thrust by the parts along ZB, while the parts along OΠ are thrust by those along BE; thus the parts of the liquid along ΞO are thrust differently than those along OΠ. Thus those parts thrust out the less are driven out by those driven the more, and so the liquid will not remain at rest. But we supposed that the state was that it remained unmoved – so it is necessary that the line ABΓΔ be the circumference of a circle with its center at K.

Likewise it can be shown that, if the surface of the liquid be cut by a plane through the center of the Earth in any other way, that the section will be the circumference of a circle and its center will be the same point as the center of the Earth. So it's clear that the surface of a liquid remaining at rest has the shape of a sphere concentric with the Earth, since it's the case that the section cutting through a given point makes the circumference of a circle having for center the point through which the cutting plane passes. [compare Aristotle, *On Heaven* 2.4 (287b5–14)]

Proposition 5

Any solid object [literally "magnitude"] less dense than the liquid, lowered into the liquid, will sink down sufficiently so that a volume of water equal to the volume of the immersed part will have a weight equal to the whole object. [compare Aristotle, *Physics* 4.8 (216a27–30)]

Arrange the same construction [as in propositions 3 and 4], and let the liquid be at rest, and let the object less dense than the liquid be EZHΘ. Since the liquid is at rest, its parts in equilibrium will be thrust equally, so that the liquid beneath the surface along the circumferences ΞO and ΠO will be likewise equally thrust, so that the weight by which they are thrust is equal. And the weight of the liquid in

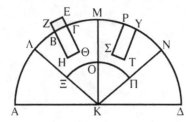

Figure 8.3

the first pyramid [ΛKM] minus the solid BHΘΓ equals the weight of the other pyramid [MKN] minus the liquid PΣTY [since PΣTY is the same volume as BHΘΓ]. So it is clear that the weight of the object EZHΘ equals the weight of the PΣTY, and it is then apparent that a volume of liquid the size of the immersed part of the solid object has a weight equal to the whole object.

Proposition 6

Solids less dense than a liquid, when forced into it, rise up ["are carried up"] with as much force as the weight by which liquid of volume equal to the object is heavier than the object ["excess buoyant force equals weight of displaced liquid minus weight of object"].

Let a given magnitude A be less dense than the liquid, and let B be the weight of the magnitude A, and let BΓ be the weight of the liquid having the same volume as A (Figure 8.4). We are to show that the magnitude A forced into the liquid will rise up with a force as great as the weight Γ. Let a magnitude be assumed which has a weight equal to Γ; so the magnitude formed of both objects A and Δ together is less dense than the liquid, for the weight of the magnitude formed from the two is BΓ, but the weight of the liquid having a volume equal to the compound magnitude is greater than the weight of BΓ because that is the weight of liquid having a volume equal to that of A [completely immersed]. Thus if the magnitude formed of both A and Δ is lowered into the liquid, it will be submerged until [it displaces] a volume of liquid (equal to the submerged part of the magnitude [AΔ]) which will have a weight equal to the whole magnitude (as was proved).

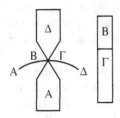

Figure 8.4

Now let the surface of the liquid be the circumference ΑΒΓΔ [an unusual reuse of letter-labels!], since the volume of liquid equal to the magnitude A has a weight equal to the magnitude A + Δ, it is clear that the submerged part of it will be the magnitude A, and the rest, Δ, will be completely above the surface of the water. *For if it were otherwise submerged, it would contradict what has been proven. Thus it's clear that by whatever force the magnitude A is carried upward, by the same force is the upper portion Δ of the magnitude pressed downward*, since neither one is thrust out by the other. But Δ is driven downward by as much weight as Γ has, for it was hypothesized that the weight of Δ is equal to Γ. Q.E.D.

Proposition 7

A solid heavier than a fluid will, if placed in it, descend to the bottom of the fluid, and the solid will, when weighed in the fluid, be lighter than its true weight by the weight of the fluid displaced. [proof omitted]

*(Heiberg [1913/1972] 318–320, 328, 330, 332)

8.4 Philon of Buzantion (*ca.* −200) wrote in Alexandria a compendium of mechanical science (*mêchanikê suntaxis*): book 1 introduction; book 2 on levers; book 3 on harbor-building; book 4 on war-machines; book 5 on pneumatics (surviving only in Arabic and Latin – material here preserved only in Latin is in *italics*); book 6 on automata; book 7 on defensive siege-works; and book 8 on offensive siege-works (books 1, 7, and 8 are fragmentary; books 2, 3, and 6 are lost).

War-machines

(pp. 77–78 Wescher) [air-spring catapult]

The air-spring [*aerotonos*] catapult was invented by Ktesibios and comprised a very ingenious and scientific design. He observed, in the so-called pneumatic demonstrations (which we will also treat later), that air is exceptionally strong, resilient, and mobile – moreover, when it has been enclosed in a sturdy vessel, it can admit compression and, vice versa, swift expansion to fill up its former space in the vessel. With his experience in engineering, he realized quite well that this movement could supply great resilience and high velocity to the arms. Therefore, he constructed cylinders, similar in shape to doctor's pill-boxes without lids. They were of beaten bronze in order to be resilient and strong, but had previously been made pliable and smelted to get thickness. Their interior was bored out by machine and made with surface level, straight (tested by ruler), and smooth. Thus, the piston of bronze, when inserted, could run through and rubbed against the circumference, being itself level and smooth so that the joint between them was so perfect that the air could not force its way through even when under full pressure.

Do not wonder at, or doubt, the possibility of such workmanship. In the pipe played by hand which we call a water-organ, the bellows, which forced the air into the compartment in the water, was bronze and manufactured like the cylinders mentioned above. Ktesibios, it was explained to us, demonstrated the natural property of air namely that it has strong and swift movement, and at the same time, the fabrication entailed by the cylinders which contain the air; he smeared the cylinder with carpenter's glue, set a protective edging over its circular mouth, and with wedge and mallet drove in the piston with very great force. It was possible to see the piston making gradual progress; but, once the air contained inside was compressed, it yielded no more even to the strongest blow on the wedge. When the wedge was forcefully removed, the piston shot out with very violently from the cylinder. It often happened that fire came out, too, since the air rubbed against the cylinder in the speed of its motion.

He constructed two such cylinders, as described, resembling pill-boxes, and he made the hole-carriers of a design conforming to the standard ones. He fastened them securely together, encircling them with wooden framework, iron plates, and rivets; he aimed not only at strength, but also fine appearance so that it should seem up to the full engineering standard. He also enclosed the heels of the arms with sleeves, slightly bent, which he rested against the pistons. The arms were hinged like those mentioned in the bronze-spring engine, working around iron pins and fitted with fingers. When he had constructed the parts mentioned, stretched across the bowstring, and fitted the sling, he drew it back in the manner customary for other engines. As the bowstring was being pulled back, the arms, with their heels rubbing against the pistons, naturally pushed them in and the air, cut off in the cylinders, compressed as I have said and with its density excessively increased, had a desire for its natural state. On the loading of the stone and the releasing of the claw, the arms recoiled with great resilience and flung out the stone, achieving a very gratifying range. [for further discussion, see Marsden [1971/ 1999] 153–155]

Pneumatics

1–2 [demonstration that air is corporeal]

1   Dear Ariston: Your interest in ingenious devices has been known to me, therefore I replied to what you asked me, by writing this book so that you will have what you need for everything you asked me about concerning devices, and I will begin with the making of pneumatic devices, and I will mention every art known to any prior scholar. Natural philosophers know that a vessel, which many people think is empty and void, is not like people think, but is filled with air, since they are only ignorant about that because they aren't certain that air is a body like all others. I hate to have to mention their opinions concerning this and their disagreements about it, but it's clear that air belongs to the elements not only from theoretical reasoning but also from experience, and that it's observable by us, real to our senses. I will mention from their disputes whatever suffices to reach my goal and to establish that air is a body.

2 [from Anaxagoras, and repeated again by Heron] Let us take an empty vessel, similar to the carafes made in Egypt, and plunge it in deep water, inverted on its mouth, while pushing it down. Not the least bit of water enters, until some of the air escapes. Upon the escape of air there follows the entry of water. Which I can demonstrate by this experiment. Let us take an Egyptian carafe, one not with a wide mouth. Let its bottom have a minute hole, and let it be closed with wax. Then let it be turned over, and let it be put in deep water, taking care to hold it straight and not inclined at all (Figure 8.5). Let it be pushed down in the water and then taken out slowly and gently. Let one look inside, and the wall will be found dry: it is wet only on the mouthpiece. Through this experiment it is proven that air is a body, for if it were not a body, and the space inside were void, water would have run into it, as it would have found no resistance. Let's prove that.

Let us again take a vessel and invert it on its mouth like before, and push on it with one's hands until it's deep under the water awhile. Then let the wax, placed on the hole, be removed. There then ensues an escape of air from the hole, which is manifest to the senses. One can see it in form of bubbles above the hole. The vessel then fills with water, as air is moved and leaves through the hole. What forces the air to escape is the motion of the water, which presses the air while we push the vessel down. The description as well as its presentation is proof that air is a body.

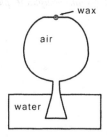

Figure 8.5

4 [demonstrations that water is "attached" to air]

So it is manifest that water sometimes rises because it is drawn up with rising air which is connected with it, just like what happens to the tube used for wine tasting. When someone holds its tip in his mouth and gently sucks out the air that is in it, he also attracts with it the liquid body which is down below, since it adheres to the air as if it were at that time held to it with glue or with some other bond.

This can also be shown with another kind of vessel set up for what is needed. Let us take a bull's horn, well hollowed out inside to make it very wide, thin and polished [to be transparent: prior to the invention of blown glass]. Let it be of medium height; let its form be tubular, with its "lower" end cone-shaped [the "lower" end was once attached to the animal's head]. Around the cone-shaped part let another wooden cap be tightly fitted, to achieve the construction that we need, that is, let them be so conformed that air cannot escape anywhere and the entire horn unit resembles a medicine can.

Then invert this horn and let it stand in some wide-mouthed vessel similar to a beaker. Place another can, made of lead, at a level below and also alongside this one. Let it be well sealed so that, when necessary, air may not exit. Between the tops of these two cans adapt a pipe, very tight, with inverted end pieces, going down to the bottom of either can. Let the lead can have a mouth, a little raised, so that water can be poured in from some vessel. Let this can have a lower pipe, well-attached and small, so that water can be evacuated when you wish. Let the horn, for example, be A, the beaker B, the lead can Γ, the small pipe thereof Δ, the large common pipe E and the small mouth on top of the lead can Z (Figure 8.6).

Arrange all this as we have said. Then close the small pipe below; fill the lead can with water through the mouth on top; close that mouth so that nothing will pass; pour water into the beaker-shaped vessel, in such amount that it can fill the horn standing in it. Then open the closed lower pipe, and water will come from it. As it flows freely from the lead can, air is attracted into the can from the horn. As the air leaves there, it draws with it water from the beaker into the horn. What enters here replaces what leaves through the pipe. You may see it through the horn if you give it due consideration. According to the amount of air leaving the horn, a portion of water is raised, being held to the air, which can lift it. After being lifted it descends, which occurs by nature, as we have demonstrated. The flowing water follows the air tenaciously. According to the amount of air that leaves, a portion of water follows and occupies its place, filling it; and when that water leaves, air enters to refill its place.

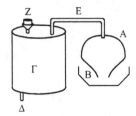

Figure 8.6

6 [demonstrating "attachment" of water to air using a siphon]

Take a bent tube, which is sometimes called Egyptian Compass [i.e., an arched tube]. *It raises water, which stands in a pool, to a high place.* The water then stands where it was at first, and thus is proven that the water does not rise without this device. When the water is raised with this device, it is only for the reason which we want to mention.

When we put one end of this bent tube in a vessel full of water, and from the other side suck the air with our mouth, the air [presumably Philon meant, and maybe wrote, "water"] will rise just as we have described above. When it once begins to ascend, the outflow is continuous until the vessel is empty and it ends up at this tube because the water in it doesn't split up, unless air be introduced. If the siphon breaks and air enters the pipe, this interrupts the water's continuity. That which remains there then stands still, for the reason that we have indicated.

The following experiment which we want to mention proves this. Let there be a long, tight vessel. Let it be put in water and pressed down until it is full. When it has thus been filled, overturn it quickly under the water (so that its bottom is in place of the top), while keeping it full. Then lift it from the water until only its lip stays under water. When this has been done, the vessel still is full. That such is the case is readily seen if it is of glass or horn or other transparent matter. There is no means for water to escape, unless the vessel has a small hole so that air can enter it; the water then returns down to the place where it had been.

Thus it is manifest that the water follows the air and is held with it. They mutually follow one another.

(Prager [1974] 127–134)

7 ["thermoscope," named *libas* or "dripper" by Heron, *Pneumatika* 2.8]

Fire is also naturally mixed with air, and for that reason air is drawn along with it. This will be shown by what follows.

Take a leaden sphere, hollowed out, so that there is room within, and of moderate size. It should not be too thin, lest it be easily broken, nor should it be porous, but for what we want, it should be quite dry. Pierce the sphere on top and insert a bent tube reaching almost to the bottom, so that the water will flow; this tube should also be very tight; and then put the sphere in a sunny place, and set a vessel with water at the mouth of the other end of the tube. Let the sphere be A, the tube B, and the vessel Γ.

Then I say that when the sphere becomes hot, part of the air enclosed in the tube will pass out. This will be evident because the air will descend from the tube into the water, agitating it and producing a succession of many bubbles.

Now if shadow falls over the sphere, the water will rise and pass through the tube from the vessel until it descends into the sphere. If you then put the sphere back in the sun the water will return to the other vessel; but it will flow back to the sphere once more if you place the sphere in the shade. This will always happen as we described it.

In fact, if you heat the sphere with fire, or bring fire close to this sphere in any way that can heat it, the demonstration that we described occurs, and if it's cooled, the water returns as it was, or even if you pour hot water over it, the result will be the same. And if the sphere is then cooled, water passes from the vessel to the sphere. And this is the proof of it.

(Cohen and Drabkin [1958] 255–256)

8.5 Aristokles (−110 ± 10) wrote on the history of the choros in tragedy, including the use of musical instruments; here he mistakenly attributes to Plato a device only possible over a century later.

Choruses

(fr. in Athenaios, Deipnosophists *4[174–5])* [water-organ: compare
Section 6.11 (Heron)]

The question is debated whether the water-organ belongs to the wind or the stringed instruments. Now Aristoxenos [of Taras, writer on music –330 ± 20], to be sure, does not know it; but it is said that Plato imparted a slight hint of its construction in having made a time-piece for use at night which resembled a water-organ, being a very large water-clock. And in fact the water-organ does look like a water-clock. Therefore it cannot be regarded as a stringed instrument or a percussion instrument, but perhaps may be described as a wind instrument, since it is winded [blown] by water. For the pipes are set low in water, and as the water is pumped by a slave, the pipes blow through the organ as air passes through the valves producing a pleasant sound. The organ is shaped like a round altar and they say it was invented by Ktesibios, a barber who lived there in Aspendia [a neighborhood in Alexandria] during the reign of Ptolemy Euergetes II [*sic*: Athenaios meant "Philadelphos II"]; and they say that he became very famous, and even taught his wife Thaïs.

(Gulick [1927] 291, 293)

8.6 Heron of Alexandria wrote a series of books on science and engineering (*ca.* 55 to 68; see also Chapters 2, 5–7); the *Dioptra* was written about 62 and the *Pneumatics* dates to 65 ± 3 (see Keyser [1988]).

Dioptra

31

Given a spring, to determine its flow, that is, the quantity of water which it delivers.

One must, however, note that the flow does not always remain the same. Thus, when there are rains the flow is increased, for the water on the hills being in excess is more violently squeezed out. But in times of dryness the flow subsides because no additional supply of water comes to the spring. In the case of the best springs, however, the amount of flow does not contract very much.

Now it is necessary to block in all the water of the spring so that none of it runs off at any point, and to construct a lead pipe of rectangular cross section. Care should be taken to make the dimensions of the pipe considerably greater than those of the stream of water. The pipe should then be inserted at a place such that the water in the spring will flow out through it. That is, the pipe should be placed at a point below the spring so that it will receive the entire flow of water. Such a place below the spring will it determined by means of the dioptra. Now the water that flows through the pipe will cover a portion of the cross-section of the pipe at its mouth. Let this portion be, for example, 2 digits [in height]. Now suppose that the width of the opening of the pipe is 6 digits; $6 \cdot 2 = 12$. Thus the flow of the spring is 12 [square] digits.

It is to be noted that in order to know how much water the spring supplies it does not suffice to find the area of the cross section of the flow which in this case we say is 12 square digits. It is necessary also to find the speed of flow, for the swifter is the flow, the more water the spring supplies, and the slower it is, the less. One should therefore dig a reservoir under the stream and note with the help of a sundial how much water flows into the reservoir in a given time, and thus calculate how much will flow in a day. It is therefore unnecessary to measure the area of the cross section of the stream. For the amount of water delivered will be clear from the measure of the time.

(Cohen and Drabkin [1958] 241)

Pneumatics

1 Preface (pp. 2–4.13 Schmidt)

The undertaking called *Pneumatics* was thought worthy of zealous attention by the ancient philosophers and mechanicians, the former demonstrating its potentiality by argument, the latter by experiences demonstrating its actuality. Therefore we consider it necessary ourselves to set in order what has been passed down from the ancients and to add thereto what we have discovered in addition. And thereby it will happen that those hereafter who want to devote themselves to the study of mathematical sciences will be benefited. We believe that it follows naturally upon our published work in four books on *Water-clocks* [only a short extract survives] to write next about this, as was said: for by the weaving together of Air and Fire and Water and Earth, and the conjunction of three or four elements, a variety of arrangements are actualized, some providing the most urgent necessities of life and others producing a certain awestruck amazement.

As a prologue to what we're saying, the void must first be treated. Some deny in general that there is any extended void, others that by nature there is no continuous void but that disseminate void in small parts exists in Air and Water and Fire and in other bodies: with these it seems best to agree, for this state of affairs is shown in the following from phenomena and perceptible occurrences. In fact, vessels which seem empty to most people are not as they suppose empty, but full of Air. And as those who undertake natural philosophy agree, Air is composed of fine and minute bodies usually invisible to us. [Anaxagoras' experiment demonstrating the corporeality of air follows]

(pp. 6.23–8.22 Schmidt)

The bodies of Air rub against one another and do not mesh at every part, but have some void spaces between just like sand on the shore. The particles of sand

221

must be supposed like the bodies of air, and the air between the sand-particles like the voids between the air-bodies. Thus it happens that the air is thickened by applied force and as the bodies are pressed against one another unnaturally, they move into the zones of void. When the force releases, the air again returns to its proper arrangement by the resilience of the bodies, just as happens with horn shavings and dry sponges: when they are squeezed and released, they return again to the same place and display the same volume. Similarly, if by some force the air-bodies are pulled apart and a greater void space unnaturally exists, they again run back together: for the motion of the bodies goes quickly through the void, there being no resistance or obstacle, until they rub against one another. [compare pseudo-Aristotle, *Problems* 11.58]

Thus if one takes a very light vessel with a narrow mouth, applies it to one's mouth, and sucks out and exhales the air, the vessel will be suspended from the lips, the void attracting the flesh to fill up the emptied space. From this it is apparent that continuous void exists in the vessel.

1.4 [constant-flux siphon: compare Drachmann [1976]]

From what has been shown [1.1–3], it is apparent that the flow through a stationary siphon is irregular. The same thing happens in a vessel pierced in the base when there's a flow, for then also the flow is irregular, since at the beginning of the flow a greater weight drives the efflux of the water, and as it's emptied the weight is lesser. In the same way, as the outlet of the siphon has a greater excess length, the flow becomes quicker, for again, the efflux through its mouth is driven by a greater weight than when there's a smaller excess (by which the surface of the water in the vessel exceeds the height of the outlet mouth of the siphon). Thus it was said that the flow through the siphon is always irregular; but we want to find an always regular flow through the siphon (Figure 8.7).

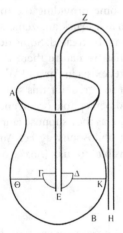

Figure 8.7

Let a vessel AB hold water in which let a little saucer ΓΔ float, covered at its mouth with a cover ΓΔ; and through this lid and the base of the saucer let one leg of the siphon be inserted and let it be sealed to the holes with tin [solder]. Let the other leg of the siphon be outside the vessel AB having its mouth lower than the surface of the water in the vessel AB [Figure 8.7]. Then if we suck the air in the siphon out through the outer mouth of the siphon, the liquid will follow after because of the impossibility of a continuous void occurring in the siphon. Once the siphon starts its flow, it flows until it has emptied all the water in the vessel. And the flow will be regular because the excess of the outlet of the siphon below the surface of the water remains always the same, since the saucer with the siphon descends as the vessel empties. The greater the excess length of the outlet of the siphon, the swifter will be the flow, but regular throughout. (Let the siphon be EZH and the surface of the water be along the straight line ΘK.)

1.20 [the toilet-bowl or chicken-waterer feedback device]

If we want to render this [1.19: ever-full mixing-bowl of wine] useful so that a full mixing-bowl [kratêr] is set up somewhere, and water is drawn from it and it always remains full, this is the arrangement. Let there be a vessel AB in which let there be enough water for the intended use; let there be a spout coming out ΓΔ, let there be a trough HΘ set beneath it; and let a rod EZ be set up as a swing-beam by the spout, from whose end at E let a cork float K be hung in the trough; and on the end at Z bind a small chain holding a lead weight Ξ [to balance the cork (Figure 8.8)]. Let this be arranged so that the cork floats in the water in the trough ΘH to shut off the spout, but when water is taken from the trough, the cork sinks and opens the spout, so that as the water flows back in it raises the cork and reseals the spout (the cork will have to be heavier than the weight at Ξ). Let the mixing-bowl be set in some place ΛM with its lip at the surface of the water in the trough when no water is flowing out of the spout because of the floating cork. Also let a tube ΘN lead from the trough to the base of the mixing-bowl.

Then when the mixing-bowl is full and someone draws water, the trough ΘH will also be emptied of water, and the cork will descend and open the spout, and

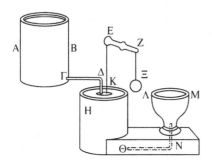

Figure 8.8

the influx to the trough will also enter the mixing-bowl, and will raise the cork, so that again there will be no influx. This will happen however often anyone draws water from the mixing-bowl. [similar feedback devices: 1.21, a "vending machine" and 1.34, a self-trimming lamp, for which see Humphrey, Oleson and Sherwood [1998] 66–68]

1.28 [the two-pistoned water-pump by Ktesibios, used as a fire-engine in Alexandria; Humphrey, Oleson and Sherwood [1998] 320]

1.42 [see also Vitruuius 10.8; the *"hudraulis"* water-organ by Ktesibios in a new version by Heron: Keyser [1988]]

1.43 [the windmill: see Section 6.11]

2.11 [anti-Aristotelian demonstration of rotary motion without friction and through reaction not contact; a cosmological model like 2.6 (sphere floating on stream of air) or 2.7 (sphere suspended by film of water); compare pseudo-Aristotle, *Kosmos* 6 (398b13–20); see Keyser [1992a]]

To spin a ball on pivots by heating a cauldron. Let a cauldron AB holding water be heated, and let its opening be closed by the lid ΓΔ; let there be bored through the lid a bent tube EZH, at the end of which let a hollow sphere ΘK be fitted; directly opposite to its end H let there be a pivot ΛM joined to the lid ΓΔ (Figure 8.8). Let the sphere have two small tubes bored into it, diametrically opposite and bent crosswise. Let the bends be right angles and transverse to the line HΛ. Now it will happen when the cauldron is heated that the steam will rush through the tube EZH into the sphere, and will rush out through the bends and turn the sphere (just like with the dancing figures [2.3: turned by hot air]). [2.11 also in Humphrey, Oleson and Sherwood [1998] 28]

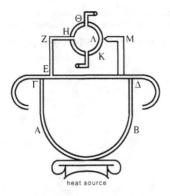

Figure 8.9

*(Schmidt [1899/1976] 2, 4, 6, 8, 42, 44, 46, 106, 108, 110, 228, 230, 232)

8.7 Ptolemy (Claudius Ptolemaeus) of Alexandria (*ca.* 100 to *ca.* 175) wrote on a wide variety of "mathematical" topics (he is quoted in each of Chapters 2–8).

Weights

(reported by Simplicius, Commentary on Aristotle's *"On Heaven" pp. 710–711 Heiberg)*

[Aristotle, *On Heaven* 4.4 (311b6–14) asserted inflated bladders outweigh empty ones; compare also pseudo-Aristotle, *Problems* 25.13; Thomas [1941] 411.] Ptolemy, the mathematician, in his work *Weights* holds a view contrary to Aristotle's and tries to show that neither air nor water has weight in its own place. That water has no weight in its own place he shows from the fact that divers, even those who go down to considerable depth, do not feel the weight of the water above them.

One can say in reply that it is the cohesion of the water being a support not only above the diver but below him and on either side that prevents him from feeling the weight, as with animals in the holes of walls. For though they may be touching the wall on all sides yet they are not weighed down by it because it supports itself on every side. In the case of the water, if a separate mass of it rested upon one, its weight would likely be felt. [compare pseudo-Aristotle, *Problems* 32.5; Heron *Pneumatics* 1. praef. (pp. 22–24 Schmidt)]

Now Ptolemy seeks to prove the proposition that air has no weight in its own medium by the same experiment of the inflated skin. Not only does he contradict Aristotle's view that the skin when inflated is heavier than when uninflated, but he maintains that the inflated skin is actually lighter.

I performed the experiment with the greatest possible precision and found that the weight of the skin when inflated and uninflated was the same. One of my predecessors who tried the experiment wrote that he found the weights to be the same, or rather that the skin was a trifle heavier before inflation, which agrees with Ptolemy.

Now if the result of my experiment is correct, it follows, clearly, that in their respective natural places the elements are without weight, having neither heaviness nor lightness. Ptolemy agrees with this so far as water is concerned. And this is reasonable, for if natural weight is a striving toward the proper place, things which are there should not show any striving or tendency in that direction, since they already are there, just as that which is sated does not reach for food.

But if, as Ptolemy holds, the skin when inflated is lighter than when uninflated, not only does air possess lightness in its own natural place but it would seem to follow by similar reasoning that water possesses heaviness in its own natural place.

(Cohen and Drabkin [1958] 247–248)

9

ALCHEMY

The world is full of marvelous materials, from stones of varying color and frangibility through water that freezes and melts, to brilliantly-colored plants and animals. Our mastery of fire marked a radical break with prior technology and allowed us to transform and improve stuffs – food became cooked and not rotten, flint more knappable, and clay hardened and even exploded (Vandiver *et al.* [1989]). Later, clay-like dough was infected with the invisible power of leavening yeast to produce bread, a second substance with properties entirely dissimilar from the parent stuff. Shiny yellow pyrites was our "fire-stone" from which we expelled inherent sparks; other bright heavy stones were found soft enough to carve like tough wood or bone, then pound out like stiff dough, and at last, in the fire, not harden like wood or clay but melt like water. First copper and then gold were cast for decoration and tools, and soon red copper was smelted from green rocks employed as pottery colorings.

Out of this first artificial production of a natural substance grew the smelting and alloying of yellow bronze (red copper plus varicolored arsenic or tin ores) in Sumer (Mesopotamia). Color-transformations were long and widely familiar from pottery, and dyes applied to woven cloth: now metals too could be colored. The first artificial stone had been fired clay; next came plasters created out of fired white stones ground and mixed with water; still later the Mesopotamian marvel of colored glass, and the beauty of "white gold" (silver) extracted from leaden ore. The first artificial food was bread or wine, produced from ordinary food by mysterious and fickle forces; these were mastered to produce a wine-like potion from bread itself: beer. Somewhere (north of Mesopotamia or west of China) the smelting of silvery iron (which had long been known as a star-stone, i.e., meteorite, dropped from heaven) was achieved from red ocher, long a personal and ceramic pigment. At some point Mesopotamians came to associate their chief metals with their chief gods, who were moving stars: golden sun, silvery moon, leaden Saturn, iron Mars, and coppery Venus (Jupiter was tin or electrum, and Mercury was tin or quicksilver).

All the earliest Greek thinkers wondered what was the ultimate stuff of the *kosmos*, guessing water or air or fire or earth – or all four. Possibly as early as Alkmaion of Kroton (around –500) there was a guess that the human body was a blend of several internal juices (see p. 291). His contemporary, the famously cryptic aristocrat Herakleitos of Ephesos (−500 ± 20), argued the inherent unity and dynamic balance of opposites, manifested in the continual and eternal changes of the *kosmos* and its stuffs: "Fire" (almost "energy") is the ever-living fundamental stuff of the world. The aristocrat Empedokles of Akragas (−455 ± 20) hypothesized that the whole *kosmos* had been cooked up from just four essential divine ingredients: Earth, Water, Air, and Fire. Anaxagoras of Klazomenai (perhaps −455 ± 20), working in Athens, proposed instead that every sensible quality was evidence of infinitesimal "seeds" pervading each part of the *kosmos*. Demokritos of Abdera (writing *ca.* −410 ± 30) opted for indivisible units (atoms) strewn and mixed (in a void) to generate sensible qualities by their mutual arrangements and motions: e.g., lead was softer and heavier than iron because its atoms were more closely and more regularly packed. Plato of Athens (writing −365 ± 20) sought to found a maximally-probable account of colors and stuffs upon a mathematical basis, constructing the four Empedoklean elements from fundamental triangles grouped into four of the five regular polyhedra (the dodecahedron representing the zodiac); he explained elemental transformation and chemical combination by their rearrangements, and sensible and physical qualities by their shapes (*Timaios* 52–64). Plato, like Pindar before him (*Olympians* 1.1, *Isthmians* 5.1–3), saw metals as watery because fusible.

Aristotle of Stageira (writing −335 ± 10) returned to the primitive and unanalytic notion that qualities such as color, density, hardness, and so on, were essential and primary entities which attached to quality-less substrate matter; in his model, every stuff was mixed from some of the four Empedoklean elements, which were themselves explained as products of inseparable primary opposites Hot/Cold and Wet/Dry (Bolzan [1976]; Figure 9.1). Elemental transformation occurs because of a change in a substance's constituent opposites, so Water becomes Earth when Wet departs so that Dry arrives. Most ordinary substances of the earth are formed by the agency of a pair of ill-defined "exhalations" (*anathumiaseis*), metals primarily by the moist one, and stones by the dry (or "smoky") one. All metals except gold are affected by Fire and contain some Earth (from the dry exhalation); apparently the baser metals were "dryer." Color was a secondary quality produced either by the interaction of our eyes with stuff or perhaps from the particular mixture of the four elements in the colored thing (see p. 197).

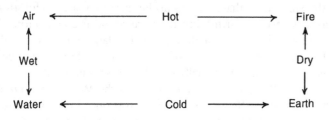

Figure 9.1

It is Bolos of Mendes who is usually credited with combining Greek theories with Egyptian techniques to found "alchemy"; he wrote around –200 apparently often under the name Demokritos, but earlier writers (such as Theophrastos) were already tending in the same direction. (For an expanded version of this prolegomenon, see Keyser [1990].)

(See the prolegomenon to Chapter 7 for materials affected by light.)

9.1 Theophrastos of Eresos (lived –370 to –286) succeeded Aristotle as the head of the Lukeion "school," and wrote widely, amassing data questioning Aristotle's system (Keyser [1997c]). The works here excerpted were part of a set of works on natural philosophy (see also Chapters 8 and 10–12).

Stones [written around –312 ± 2]

1–2 [properties]

1 Of the substances formed in the earth, some are made of water and some of earth. The metals obtained by mining, such as silver, gold, and so on, come from water [hence can melt: see *Stones* 9]; from earth come stones, including the more unusual kinds, and also the types of earth that are peculiar because of their color, smoothness, density, or any other quality. As the metals have been discussed in another place [lost work], let us now speak about the stones.

2 In general we must consider that all of them are formed from some pure and homogeneous matter as a result of a conflux or percolation, or because the matter has been separated in some other way, as has been explained above [lost]. For perhaps some are produced in one of these ways, and some in the other way, and others in a different manner. Hence they gain their smoothness, density, brightness, transparency, and other such qualities, and the more uniform and pure each of them is, the more do these qualities appear. In general, the qualities are produced according to the accuracy with which the stones are composed and solidified.

16–17 [coal]

16 Among the substances that are dug up because they are useful, those known simply as *anthrakês* are earthy, but they are set on fire and burnt like charcoal. They are found in Liguria [north-west coast of Italy], where amber also occurs, and in Elis as one goes by the mountain road to Olumpia [nearby at Brouma lignite is mined]; and they are actually used by workers in metals.

17 In the mines at "Dug-Stuff" [*Skaptê Hulê*: near the Macedonian–Thracian border between the Strumon and Nestos rivers] a stone was once found which was like rotten wood in appearance [possibly palygorskite, which forms matted felted masses that closely resemble woven cloth, and is called "Mountain Leather"]. Whenever oil was poured on it, it burnt, but when the oil had been used up, the stone stopped burning, as if it were itself unaffected. These are roughly the differences in the stones that burn.

23–24 ["emerald"]

23   But the *smaragdos* [emerald and other greenish translucent stones] also has certain powers, for it makes the color of water just like its own, as we have said before; a stone of moderate size affects a small amount of the water in which it is placed, the largest kind the whole of the water, and the worst kind only the part close to it.

24   It is also good for the eyes, and for this reason people carry seals made of it, to be viewed [Ptolemy, *Optics* 2.107].

45–46 [touchstone, to test gold]

45   The nature of the stone which tests gold is remarkable, for it seems to have the same power as fire, which can test gold too. On that account some people are puzzled about this, but without good reason, for the stone does not test in the same way. Fire works by changing and altering the colors [compare *P. Leyden X* 42], and the stone works by friction, for it seems to have the power of picking out the essential nature of each metal [Herodotos 7.10].

46   They say that a much better stone has now been found than the one used before; for this not only detects purified gold, but also gold and silver that are alloyed with copper, and it shows how much is mixed in each stater [coin and weight: about 8.7 g]. And indications are obtained from the smallest possible weight. [also in Humphrey, Oleson and Sherwood [1998] 228]

49 [glass]

And if glass is formed, as some say, from vitreous earth, this too is made by firing [Herodotos 2.69; long known in Egypt and Mesopotamia]. The most peculiar earth is the one mixed with copper, for in addition to melting and mixing, it also has the remarkable power of improving the beauty of the color [known by the early fourth century BCE: Theopompos of Chios, *Philippika* 13 (fr. in Strabo 13.1.56); probably zinc oxide added to make brass]. . . . [49 is also in Humphrey, Oleson and Sherwood [1998] 224]

53–54 [production of red ocher]

53   It is also made by burning yellow ocher, but this is an inferior kind and is a discovery of Kudias [painter around –360 ± 20]; for it is said that he became aware of it when a general-store burnt down, as he noticed that some yellow ocher was half-burnt and had become red in color.

54 New earthen vessels are luted with clay and placed in ovens; for when the vessels become red-hot, they heat the ocher, and as they become hotter in the fire, they make its color darker and more glowing. And its origin is itself a proof of this; for it would seem that all these substances change under the influence of fire, if it is right to consider that the red ocher made in this process is the same as the one made by nature or very similar to it [dehydrating yellow ocher creates red ocher, iron oxide].

56 [making white lead]

Lead about the size of a brick is placed in jars over vinegar, and when this acquires a thick mass, which it generally does in ten days, then the jars are opened and a kind of mold is scraped off the lead, and this is done again until it is all used up. The part that is scraped off is ground in a mortar and filtered frequently, and what is finally precipitated is white lead [lead acetate, used as make-up: Vitruuius 7.11.1, Dioskourides 5.88].

57 [making verdigris]

Verdigris is made in much the same way. Red copper is placed over grape-residues and the matter that collects on it is scraped off; for it is verdigris that appears there [copper acetate, used as paint; Dioskourides 5.79, Pliny 34.110–113, and *P. Stockholm* 74 give longer versions]. [56–57 are also in Humphrey, Oleson and Sherwood [1998] 385]

60 [quicksilver]

It is clear from these facts that art imitates nature and creates its own peculiar products, some of them for use, and some only for show, such as wall-paints, and others for both purposes equally, such as quicksilver; for this has its use too. It is made when cinnabar mixed with vinegar is ground in a copper vessel with a pestle made of copper [mercury sulfide reduced to mercury by the copper]. And perhaps several other things of this kind could be discovered.

65–66 [gypsum used to make plaster]

65 The nature of *gupsos* is peculiar; for it is more like stone than earth, and the stone resembles *alabastrites*. It is not cut out in a large mass but in small pieces. Its stickiness and heat, when it is wet, are remarkable; for it is used on buildings and is poured around the stone or anything else of this kind needing fastening.

66 After it has been pulverized and water has been poured on it, it is stirred with wooden sticks (for this cannot be done by hand because of the heat). And it is wetted immediately before it is used; for if this is done even a short time before, it quickly hardens and it is impossible to divide it. Its strength, too, is remarkable; for when the stones are broken or pulled apart, the *gupsos* does not become loose, and often part of a structure falls down and is taken away, while the part hanging up above remains there, held together by the bonding. And it can even be removed and calcined and repeatedly rendered useable [also used to clarify wine: § 67 and Plutarch, *Natural Questions* 10 (914d-e)].

(Caley and Richards [1956] 45, 48, 50, 54–60)

Odors

8

For tastes and odors alike are derived from these two things: the method of the makers of spices and perfume-powders is to mix solid with solid, that of those who compound unguents or flavor wines is to mix liquid with liquid: but the third method, which is the commonest, is that of the perfumer, who mixes solid with liquid, that being the way in which all perfumes and ointments are compounded. Further one must know which odors will combine well with which, and what combination makes a good blend, just as in the case of tastes: for there too those who make combinations and, as it were, season their dishes, are aiming at this same object. So much for the ingredients and the methods whereby these arts attain their ends.

21–23 [use of bain-marie in making scents]

21　Of the spices used in making perfumes and their treatment. Almost all spices and sweet scents except flowers are dry and hot and astringent and mordant. Some also possess a certain bitterness, as we said above, as iris, myrrh, frankincense, and perfumes in general. However the most universal qualities are astringency and the production of heat; they actually produce these effects.

22　All spices are given their astringent quality by exposure to fire, but some of them assume their special odors even when cold and not exposed to fire; and it also appears that, just as with vegetable dyes some are applied hot and some cold, so is it with odors. But in all cases the cooking, whether to produce the astringent quality or to impart the proper odor, is done in vessels standing in water and not in actual contact with the fire [double-boiler or bain-marie, used especially by Maria, below Section 8]; the reason being that the heating must be gentle, and there would be considerable waste if these were in actual contact with the flames; and further the perfume would smell of burning. [21–22 are also in Humphrey, Oleson and Sherwood [1998] 388]

23　However there is less waste when the perfume obtains its proper odor by exposure to fire than when it does so in a cold state, since those perfumes which are subjected to fire are first steeped either in fragrant wine or in water: for then they absorb less: while those which are treated in a cold state, being dry, absorb more, for instance bruised iris-root.

(Hort [1926] 335, 347)

Fire

1 [troubling properties of fire]

Of the elemental substances fire has the most special powers. Air, water, and earth undergo only natural changes into one another, none of them is able to generate itself. But fire is naturally able to generate itself and to destroy itself; the smaller fire generates the larger, and the larger destroys the smaller. Moreover, most forms of generation of fire take place by force, as it were; for instance, that caused by the striking of solids like stones, and those caused by friction and compression, as in fire-sticks and in all those substances which are in process, such as those which are ignited and fused (in fact, it is from air that the clouds undergo their concentrations and compressions, for of course the motions by which fire-winds and thunderbolts are generated are forcible), and whatever other ways we have observed, whether above the earth, on it, or beneath it. Most of these appear to come about by force. [also in Humphrey, Oleson and Sherwood [1998] 38]

[3: fire differs in needing a substrate, fuel]

12 [concentrated fire]

Conversely, for the same or a similar reason burning bodies are consumed more quickly in winter than in summer. For summer makes fire weaker, just as the sun does, and fire itself does this to light. On the other hand, winter and the cold of the surrounding air concentrate fire. Anything that is concentrated has greater force, which is why the light of lanterns carries farther. Generally speaking, the strength of any body, just like its weight, is greater when collected and condensed. It is thereby increased, as it were.

17 [paradoxical heating power of cold]

That cold is a powerful force in collecting and gathering heat is shown by the molten and broken fragments of ductile and fusible metals. For tin and lead are said to melt on occasion in the area of the Euxine [Black Sea] when the frost and cold are intense, while bronze is broken [this "melting" is also known as "tin pest," a low-temperature structural change causing crumbling]. This clearly occurs because the moisture is turned into vapor as the heat is contracted and condensed; as the vapor rushes forth, it causes the breach.

30–33 [varieties of flame]

[30–31: purer flames lacking water and earth burn clearer]

   32   Given this variation, it is but natural that some substances, being less warm, nonetheless heat faster and more intensely. Example: the flame from reeds heats water and persons more than the flame from wood, although in general the hottest fire is that from the most solid fuels. Yet charcoal heats least though it is most solid. And the heat from the sun darkens the human body, that from fire does not.

33 Thus we may lay all these and like phenomenon to the fineness or the coarseness of the fuel. For example, the flame from reeds heats flesh and water most rapidly because it is fine and compact. It is fine because the reeds are light, compact because they are continuous. A fire that is fine can penetrate, and the heat is made possible by the continuous admixture of the fuel. The flame of reeds then is in general preferable to the heat of charcoal; its output is both greater and more concentrated.

59 [extinguishers]

The greater extinguishing power of liquids is due to their ability to penetrate to a high degree to the burning edge of a fire, as has been said of vinegar. If a viscous substance is admixed, this effect is increased. The vinegar makes its way by penetration, the viscous substance, so to say, puts an oily covering surface upon the material. Hence we are told that the most efficacious means is a mixture of white of egg with vinegar; part is viscous, part is penetrating. This is most helpful against the conflagrations caused by siege machines. [compare: Aineias, *Tactics* 34 (mix vinegar and birdlime), and Humphrey, Oleson and Sherwood [1998] 40]

(Coutant [1971] 2, 10, 12, 20, 22, 38)

9.2 Sotakos wrote about –300 and was used by Pliny; he is otherwise unknown.

Stones

(fr. in Apollonios, Marvels *36)*

The Karustian stone has woolly and downy excrescences, from which napkins are spun and woven. Wicks are also woven from it which when ignited glow and don't burn up. The washing of the dirty napkins does not occur by water, but brushwood is ignited and then the napkin placed thereon and the dirt flows off but the napkin becomes white and clean through the fire and useful for the same purposes again. And the wicks remain unconsumed the entire time they're burned with oil. The smell of the burning wick tests the "befallen ones" [epileptics]. [perhaps compare Theophrastos, *Stones* 17; paraphrased by Dioskourides 5.138 and Pliny 19.19–20 in Humphrey, Oleson and Sherwood [1998] 352]

*(Keller [1877] 52)

9.3 Epikouros of Samos (–340 to –269, and a citizen of Athens) adapted the theories of Demokritos, although his own goal was not explanation but peace of mind. Thus, his method was to allow for as many models as possible, so long as all were natural, and he rarely offers novel explanations, merely rendering existing ideas into atomist terms.

Letter to Herodotos

68–70 [nature of "physical attributes"]

68 Now as for the shapes, colors, sizes, weights, and other things predicated of a body as permanent attributes – belonging either to all bodies or to those which are visible, and knowable in themselves through sensation – we must not hold that they are *per se* substances: that is inconceivable. Nor, at all, that they are non-existent. Nor that they are some distinct incorporeal things accruing to the body. Nor that they are parts of it; but that the whole body cannot have its own permanent nature consisting entirely of the sum total of them, in an amalgamation like that when a larger aggregate is composed directly of particles, either primary ones or magnitudes smaller than such-and-such a whole, but that it is only in the way I am describing that it has its own permanent nature consisting of the sum total of them.

69 And these things have all their own individual ways of being focused on and distinguished, yet with the whole complex accompanying them and at no point separated from them, but with the body receiving its predication according to the complex conception.

70 Now there often also accidentally befall bodies, and ephemerally accompany them, things which will neither exist at the invisible level nor be incorporeal. Therefore by using the name in accordance with its general meaning we make it clear that "accidents" have neither the nature of the whole which we grasp collectively through its complex of attributes and call "body," nor that of the permanent concomitants without which body cannot be thought of. [Alexander of Aphrodisias, *Mixture* 2.2–3 describes Epikouros' explanation of mixed properties as arising from atomic juxtaposition]

(Long and Sedley [1987] 33–34)

9.4 Chrusippos of Soloi (*ca.* –280 to –206) wrote so comprehensively and authoritatively on the Stoic teachings that his interpretation became standard (however, all his writings are lost). He studied and wrote in Athens for about a quarter of a century, and then was head of the Stoic school from –231.

(title unknown)

(paraphrase in Ioannes of Stobi, Selections *1.129–130)*
["element" defined]

"Element" has three meanings. First, it means fire, because out of it the remaining elements are composed by alteration [condensation] and into it they get their resolution [by "diffusion"]. Secondly, it means the four elements, fire, air, water, earth, since all other things are composed of a particular one of these or more than one of these or all of these: all four in the case of animals and all terrestrial compounds, two in the case of the moon, which is composed of fire and air, and just one, in the case of the sun, which is composed of fire; for the sun is pure fire. On the third account, element is said to be that which is primarily so composed that it causes generation from itself methodically up to a terminus and from that receives resolution into itself by the like method.

(Long and Sedley [1987] 280)

(title unknown)

(paraphrase in Alexander of Aphrodisias, Mixture *3.3)*
[role of *pneuma* in mixture]

While the whole of substance is unified because it is totally pervaded by a *pneuma* through which the whole is held together, is stable, and is sympathetic with itself, yet some of the mixtures of bodies mixed in this substance occur by juxtaposition, through two or more substances being composed into the same mass and juxtaposed with one another "by juncture," and with each of them preserving the surface of their own substance and quality in such a juxtaposition, as, one will grant, happens with beans and wheat grains in their juxtaposition; other mixtures occur by total fusion with both the substances and their qualities being destroyed together, as happens with medical drugs in the joint-destruction of the constituents and the production of some other body from them; the third type of mixture occurs through certain substances and their qualities being mutually coextended in their entirety and preserving their original substance and qualities in such a mixture: this mixture is blending in the strict sense of the term.

(Todd [1976] 115, 117)

9.5 Bolos of Mendes (–180 ± 30) wrote a work in 4 books (gold, silver, gems, and purple) based on ancient artisanal formulas explained through his saying "Nature rejoices in nature, . . .," which was perhaps derived from Aristotle's "the qualities supersede one another" (fragment quoted by Zosimos 12.4 (p.150B.)), and which recurs in the astrologer Petosiris (fr.28-R.: Firmicus Maternus 4.16). He seems to have advertised his work as by Demokritos or at least it is often so cited (see also Chapter 10): see Wilson [1998].

Physical and Mystical Matters

3 [the key, revealed in a vision]

"Nature delights in nature [*sumpatheia*], and nature conquers nature [*antipatheia*], and nature masters nature." (We were entirely amazed that in one brief sentence the whole work was summarized.) I have come to Egypt bringing this book, so that you may disregard much quibbling and confused [literally: "poured-together"] matter.

4 [making gold]

Take quicksilver and fix it on the metal of Magnesia [one of various lead alloys, see Zosimos 28.5 (p.195B.)] or on Italian antimony or on unfired sulfur or on *aphroselênon* [a copper–silver alloy named from its metals' gods: Zosimos 6.7 (p.123B.)] or cooked *titanos* [gypsum or lime] or Melian alum or arsenic, or whatever you prefer. Project this white "earth" [because no longer "wet" and metallic] on copper and you'll have "shadowless" copper [brilliant and untarnished]. Add yellow silver [probably electrum] and you'll have gold on gold and it will be metallized gold-glue [usually malachite]. Yellow arsenic [orpiment] does the same, and prepared realgar, and completely everted cinnabar. But only quicksilver makes copper shadowless. For nature conquers nature.

[13, 20: from the book on silver translated in Keyser [1995/6] 213–214]

1 [purple-dye: originally from book 4, now the opening]

To make a pound of purple: two obols of iron slag in 7 drachmas of urine [ratio is 1:21], place on a fire until it boils. Take the decoction from the fire and pour into a vase (pour the purple in first and then pour the decoction onto the purple) and let it be steeped a day-and-night. Then take 4 pounds of sea-lichen [orchil, *Rytiphloea tinctoria*; compare Theophrastos, *Plant Researches* 4.6.5] and pour on water so that it's four *daktuls* [about 7.5 cm] above the sea-lichen; let it sit till it plumps, and strain it and heat the strainings, set out the wool and pour it on. Squeeze the loose parts so that the wash reaches the roots, and leave it for two day-and-nights. Next take it and dry it in the shade [to prevent sun-bleaching], and pour off the wash. Then put into the same wash two pounds of sea-lichen and add water to the wash so that it has the first ratio. And leave it thus until the lichen plumps. Then filter and add the wool as before, and do it for one day-and-night. Then take and rinse off in urine and dry in the shade.

*(Berthelot [1888/1967] 41–44)

(title unknown) [cited as "Demokritos" in "Zosimos";
compare Theophrastos, *Stones* 60]

Who does not know that the vapor of cinnabar is quicksilver, of which it is composed? So that if one grinds cinnabar in natron-oil [sodium carbonate in oil], thoroughly mixing and sealing it in a double-boiler, setting it upon an uninterrupted fire, then all the vapor will be captured fused onto metals. [Dioskourides 1.68.7 describes sublimation of frankincense; 5.95.1 distillation of mercury]

*(Berthelot [1888/1967] 123)

9.6 Poseidonios of Apamea (in Syria) became a citizen of the "free city" of Rhodes (a Roman protectorate), where he wrote and taught between about −100 and −50. He was widely traveled and well-connected. Much of what we know about his work comes from Strabo, who wrote in the generation after him (see below).

(title unknown)

(fr. 235 E–K; paraphrase in Strabo 7.5.8) [asphalt]

In the country of the Apollonians [modern Vlone, on Hipparchos' parallel of 15-hour longest days] is a place called Numphaion. It is a rock from which fire rises, and under it flow warm asphaltic springs, apparently produced from burning clods of asphalt. Nearby on a hill is a mine of the stuff. If a bit is cut, it fills out again in time, with the earth deposited in the dug hole changing into asphalt. "Vine-earth" (bituminous earth) mined in Seleukeia in Pieria [Nahr el ᶜÂsî at the mouth of the Orontes in Syria] is a cure for vermin-infected vines; if it is mixed with olive oil and smeared on the vines, the insects are destroyed before they reach the sprouts of the roots [Theophrastos, *Stones* 49; Dioskourides 5.160]. Similar earth was found also in Rhodes, when I was occupying the office of the *prutaneia* [town-assembly presidency, obtained in −90 ± 10 through his connections with wealthy Rhodian oligarchs: Strabo 14.2.5], but it needed a greater mixture of olive oil [for fluxing].

(Kidd [1999] 307)

9.7 Strabo of Amaseia (−63 to *ca.* 21) believed in the authority of Homer and the moral and political utility of history and geography (1.1.14); he studied under Xenarchos (see Chapter 3), and wished his work to promote mutual Greek and Roman understanding (1.1.16), as well as peace of mind through acceptance of the natural order governed by divine providence.

Geography

16.1.15 [asphalt]

Babylonia produces also great quantities of asphalt, concerning which Eratosthenes states that the liquid kind, which is called *naphtha*, is found in Susis [*ca.* 300 km east of Babylon in Persia], but the dry kind, which can be solidified, in Babylonia; and that there is a fountain of this latter asphalt near the Euphrates River; and that when this river is in flood at the time of the melting of the snows, the fountain of asphalt is also filled and overflows into the river; and that there large clods of asphalt are formed which are suitable for buildings constructed of baked brick.

Other writers say that the liquid kind also is found in Babylonia, and state the great usefulness of the dry kind in the construction of buildings, but they say also that boats are woven with reeds and, when plastered with asphalt, are waterproof. The liquid kind, which they call *naphtha*, is of a singular nature; for if the *naphtha* is brought near fire it catches the fire; and if you smear a body with it and bring it near to the fire, it bursts into flames; and one cannot quench these flames with water (for they burn more), unless with a great amount, though they can be smothered and quenched with mud, vinegar, alum, and bird-lime [compare Theophrastos, *Stones* 59]. It is said that Alexander, for an experiment, poured some *naphtha* on a boy in a bath and brought a lamp near him, and that the enflamed boy would have been nearly killed if the bystanders had not, by pouring on him a very great quantity of water, mastered the fire and saved him.

Poseidonios says of the springs of *naphtha* in Babylonia, that some send forth white naphtha and others black; and that some of these, I mean those that send forth white *naphtha*, consist of liquid sulfur (and it is these that attract the flames), whereas the others send forth black *naphtha*, liquid asphalt, which is burnt in lamps instead of oil.

(Jones [1930] 215, 217)

9.8 Maria the Jewess worked in the first century BCE, and her writings survive only in extracts in Zosimos and later writers (Patai [1982]). She is credited with the invention of the *tribikos* ("three-jar") still and the method of producing in a *kêrotakis* (encaustic-painters' wax-softener used as a reflux condenser) the golden mercury–copper amalgam (13% mercury); the "bain-marie" (double-boiler) is named for her. She made extensive use of sulfur ("divine": *theios*) water, and credits god with her inspiration.

(title unknown)

(p. 171B) [theory]

The sulfurous [*theiodês*] is mastered by the sulfurous – Nature delights and conquers and dominates nature. Just as a human is composed of elements, so also is copper; and just as a human is composed of liquids and solids and *pneuma*, so also is copper. (*Pneuma* is a cloud, as Apollo says in his oracles: "and *pneuma* darker, wet, unmixed.")

(p. 182B) [copper]

Copper is cooked with sulfur, reheated with natron-oil, and cleaned off; and repeating these steps many times it becomes fine shadowless gold. And god says this: "Let everyone know by experiment that heating the copper with sulfur does nothing; but if you heat the sulfur alone it not only makes copper shadowless, but also turns it toward gold." God granted me this, to know that copper is first heated with sulfur, then the metal of Magnesia is; and the vapor streams forth until the sulfurous escapes with the shadow, and it becomes shadowless gold. [sulfur purifies in *Iliad* 14.228 and *Odyssey* 22.481]

(pp. 192–193B, p. 198B) [*molubdochalk*, leaded-copper, as quality-less metal, black lead]

Without black lead the Magnesian metal which we have completed and perfected does not exist. . . . You will find black lead: take this, I say, after mixing it with quicksilver. . . . *Molubdochalk* is "cinnabar" or "lead" or "annual stone.". . . Which you are going to immerse and project onto it the limit of yellow realgar, so that it will no longer potentially but actually be purified gold.

If our lead is black, it has become that (for ordinary lead is black from the start). How did it become that? If you do not demetallize metals and do not metallize the unmetals and make the two one, nothing expected happens. And if all metals are not attenuated in the fire and the pneumatized vapor is not sublimed, nothing will be sublimed to the limit. . . . I don't mean with lead simply, but with our lead. Here's how they prepare the black lead, for it's cooked with ordinary lead, and ordinary lead is black from the start, but ours becomes black not having been so at first.

(p. 146B) [the *kêrotakis* (Figure 9.2)]

Take sulfur-water and a little gum and put them into hot ashes, and thus the sulfur-water is fixed. ... In the preparation of gold-film ["flower-of-gold," meaning a surface coating] let the sulfur-water and a little gum be placed on the platform of the *kêrotakis* so that the gold-film is fixed by them; and heat this a little while over a dung-fire. One part "our copper," one part gold, make a twice-smelted plate and

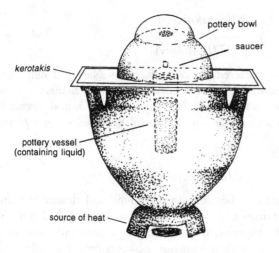

Figure 9.2

set it upon the suspended sulfur-water and let it cook for 3 day-and-nights until done. [compare Hopkins [1938]]

(p. 236B) [the *tribikos* (Figure 9.3)]

Make three tubes from beaten copper with a small flat end a little thicker than a pastry-pan, in length 1 + ½ cubits [*ca.* 65 cm]. So make such tubes, and make another having a width of about a palm [*ca.* 7.5 cm] with an opening fitted to a copper vessel. And let the three tubes have an opening fitted as a nail into the neck of a small jar; one the "inverse" [thumb] tube, so that the other two "index" [literally: "licking-finger"] tubes on the sides are fitted to the two hands. And about

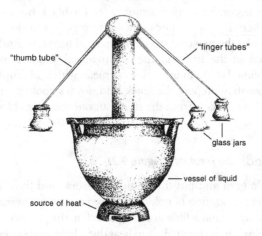

Figure 9.3

the base of the copper vessel the three holes are fitted to the three tubes and once fitted let them be soldered, receiving the *pneuma* from above. Set the copper vessel above the pottery dish holding the sulfur, lute the joints all around with bread dough [or "bear fat"?], and affix to the ends of the tubes large glass jars, thick enough that they won't break from the heat of the water entering their insides.

*(Berthelot [1888/1967] 146, 171, 182, 192, 193, 198, 236)

9.9 Anaxilaos of Larissa worked in Rome (–40 ± 10) whence he was banished in –27 by Augustus; he wrote works entitled *Paignia* ("*Recipes*") and *Baphika* ("*Tintings*"), and probably developed the distillation of wine (Dioskourides 1.72.3 distills pine-pitch; see Butler and Needham [1981] and Wilson [1984] 46–49).

Paignia

(*fr. in* P. Stockholm, *2*) [to make silver]

Anaxilaos traces back to Demokritos [i.e., Bolos?] also the following recipe [to make silver]. He rubbed common salt together with chopped alum in vinegar and formed small truncated cones from these and let them cool for three days in the bath chamber. Then he ground them, cast copper together with them three times and cooled, quenching in sea water. The outcome is the proof. [also in Humphrey, Oleson and Sherwood [1998] 230; another recipe of Anaxilaos is preserved in Pliny 19.19]

(Caley [1927] 981)

9.10 Dioskourides of Anazarbos (around 40 to 50) traveled widely collecting material for his pharmacy-book (*Medical Materials*), arranged by affinity of effect; book five concerned "minerals" (5.123–147 stones, 5.151–162 earths), of which *P. Leyden X* (below Section 17) excerpted the ten sections listed (See also Chapters 10, 12).

Medical Materials

5.74 Kadmeia [impure zinc oxide] [*P. Leyden* X 105]

5.75.3–4 Pompholux *furnace* [in Humphrey, Oleson and Sherwood [1998] 222]

5.75.5–6 [sublimation of *kadmeia*]

When the *kadmeia* is vaporised, the fine and light part is carried up to the upper part of the furnace and settles on its walls and roof, and being solidified by accumulation, at first it's like bubbles [*pompholuges*] atop water, but later, as the augmentation proceeds, it comes to resemble yarn-balls. But the heavier part moves downward and is spread around on the furnace and the floor and is considered worse than the light part, because it's earthy and full of dirt when collected. [compare Pliny 34.128]

5.76.1–2 [preparation of "burnt copper"]

Burnt copper [defined by process not composition] when good is red and when rubbed has the color of cinnabar, but when black is more burnt than necessary. It's prepared from ship's nails placed in an unfired pot, under-plastered with sulfur and salt in equal quantity, and in turn over-plastered therewith; the pot is covered and luted round with clay and placed in a furnace, until it's done cooking.

Some plaster with alum instead of sulfur; some place the sulfur and salt separately in the pot and fire for enough days; some use only sulfur, but they create soot. Others anoint the nails with chopped alum plus sulfur and vinegar and fire in an unfired pot; others sprinkle them with vinegar and cook in a bronze pot, and after the firing they do it again thrice, then set aside. [compare Pliny 34.106]

5.84.3 [treatment of stibnite]

Stibnite is smeared with fat and buried in coals until the fat glows, taken out, and quenched in the milk of a son-bearing woman or old wine. Then it's placed on coals, fired, and blown upon until enflamed (for if burned more it becomes lead [i.e., antimony]). Then it's washed like *kadmeia* [5.74.7: in mortar with water until no longer "oily"] and burnt copper; though some wash it like lead dross [5.82: in mortar repeatedly rinsing with water]. [compare Pliny 33.103–104 in Humphrey, Oleson and Sherwood [1998] 212]

Other extensive alchemical passages

5.89 *Chrusokolla* ("gold-glue") [*P.Leyden* X 106]; 5.94 Cinnabar [*P. Leyden* X 110]; 5.95 Mercury [*P.Leyden* X 111]; 5.96 Sinopian red ocher [*P. Leyden* X 107]; 5.98 *Chalkanthon* [copper sulfate; in Cohen and Drabkin [1958] 367]; 5.100 *Misu* [a shiny hard yellow copper ore, possibly chalcopyrite; *P. Leyden* X 104]; 5.104 orpiment [*P.Leyden* X 102]; 5.105 Realgar [arsenic sulfide; *P. Leyden* X 103]; 5.106 Alum [aluminum sulfates or any astringent; *P. Leyden* X 108]; 5.113 Natron [sodium carbonate; *P. Leyden* X 109]: 5.115 Quicklime [in Humphrey, Oleson and Sherwood [1998] 243]

*(Wellmann [1907/1958] 3.41–42, 45, 56)

9.11 "Isis" writes to her son "Horus" (probably in the first century CE), telling of the angel Amnaël who brought her these recipes. (Aischulos attributed all human arts and sciences to Prometheus, *Prometheus Bound* 445–506, but Plato and his followers – plus many alchemists – looked to Hermes: *Kratulos* 425, *Phaidros* 274, and *Philebos* 18.)

To Horus

7 [theory]

My son, now that you have heard this prologue, understand the whole creation and genesis of these things, and know that "human knows how to beget a human, lion a lion, and dog a dog." But if one of these happens to arise unnaturally, it is generated as a monster, and will not have structure [*sustasis*]. For "nature delights in nature and nature conquers nature."

9 [whitening metal]

Take quicksilver and fix it, either by clay or by Magnesian metal or by sulfur, and set aside (this is the "warm-freeze" [*chliaropagês*, amalgam]). Mixture of species: 1 part of warm-freeze lead, and two parts of marble, and 1 part of "massive" stone [or "pure" stone?], and 1 part of yellow realgar [orpiment, arsenic sulfide, see Theophrastos, *Stones* 40, 50], and 1 part malachite. Mix these with powdered lead and re-smelt thrice.

11 [softening metal]

Take copper and iron, and smelt them, and slowly mix in these powders: one part sulfur and 10 parts Magnesian until the iron is thoroughly softened, grind and set aside. Then take a carat [1/3-obol, *ca.* ½ g] of tempered copper, cast 4 parts from it, and mix with it 1 part of the ground iron, adding slowly and stirring, until the iron and copper are "co-unified."

13 [gold at last]

Prepare a wash of gold-plate or gold-leaf, without *chalkanthon* or casting dross, and pour the flakes into a glass; set aside 35 days until it's all macerated. Then extract and preserve.

17 [arsenic vapor for silvering]

Raise sublimed vapor thus: take orpiment, boil in water, put in a mortar, grind with vermilion and oil, and put in a pan or saucer. Set this above the door of the furnace in the coals, until the vapor sublimes. (Do the same with realgar.)

*(Berthelot [1888/1967] 30–32)

9.12 Kleopatra (perhaps first century CE) employed devices similar to Maria's, and is credited with a book on make-up and one on weights and measures. The diagram [Figure 9.4] here is all that survives of her work on *Chrusopoiia* (*Aurifaction* or *Gold-making*); the numbers in the figure refer to the notes below (unnumbered entities remain mysterious).

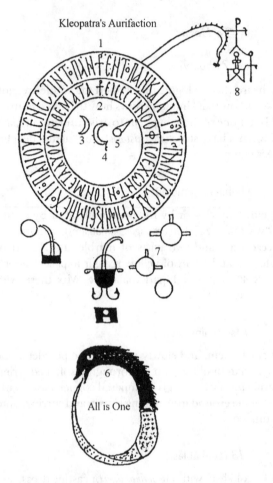

Figure 9.4

[1, very carefully penned Greek meaning "All is One, and All is through One, and All is for One, and if One does not contain All then the All is Nothing"; 2, very carefully penned Greek meaning "The snake is One that has the *ios* with two compositions" (*ios* means poison, rust, dye, transformation, and purple: Wilson [1984]); 3, symbol of mercury; 4, symbol of silver filings; 5, symbol of gold; 6, the Ouroboros ("tail-eating") snake symbolizing cyclic transformation (the Greek inscription which it encircles is translated); 7, two *tribikoi*, from above?; 8, a *kêrotakis*?.]

*(Berthelot [1888/1967] 1.132)

244

9.13 Menelaus of Alexandria, active 96 ± 2, wrote on mathematics and materials-science, and made astronomical observations at Rome. This work addressed to Domitian as patron of science is extant only in Arabic (and unprinted); Menelaus employs throughout the example-plus-proof style of Heron's *Metrika*.

Densities

["Archimedes discovered specific gravity and determined the gold–silver alloy of a crown, but we have nothing written by him on it; Mantias addressed to the emperor Germanicus" – either C. Julius Caesar Germanicus (**Caligula**) or Ti. **Claudius** Caesar Germanicus, so 45 ± 8 (compare Pliny 33.79: Caligula sought gold from orpiment) – "a book describing a method employing multiple balances and waters of differing densities" – which Menelaus quotes – "but my method is better":]

1–2 [introduction]

1 Let there be two bodies A and B of the same material in the same liquid. I say that the ratio of the volume of A to that of B is equal to the ratio of the weight of A to the weight of B in this liquid. [proof using Archimedes, *Floating Bodies* 1.Prop.5]

2 Given two bodies of two different substances, if the weight of one in the air equals the weight of the other, then if they are placed in any liquid whatever, their weight is not equal. The one of them made of the material whose nature is denser is heavier. [proof follows]

[3: reverse case]

4.1 [density of compounds]

Given two bodies of different substances which have the same weight in some liquid, if one places them in a liquid heavier than the first, then the body formed of the denser substance is heavier; and if in a liquid lighter than the first, then the body formed of the denser substance is lighter.

[example using fresh then salt-water]

Let the two bodies A and B have the same weight in freshwater and let body A be made of the denser substance, then the space occupied by A is less than that occupied by B. Thus the weight of the liquid volume equal to A is less than the weight of that equal to B. Let the weight of the liquid volume equal to A, in fresh water, be ΔΓ, and in salt-water be ΔE; and let the weight of the liquid volume equal to B, in fresh water, be HZ, and in salt-water HΘ. So ΔΓ < HZ and ΔE < HΘ (Figure 9.5). The ratio of the volume A to the volume B equals the ratio of the weight ΔΓ to the weight HZ [buoyant force in fresh water], and equals the ratio ΔE:HΘ [buoyant force in salt-water]. The ΓE and ZΘ are the two remainders in the same ratio, and so ΓE < ZΘ.

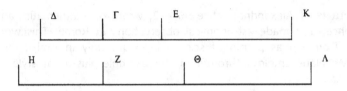

Figure 9.5

Further let the weight of body A be the magnitude ΔK and the weight of body B be the magnitude HΛ. Then the weight of body A in fresh water is ΓK and in salt-water EK; the weight of body B in fresh water is ZΛ and in salt-water is ΘΛ. From the previously-determined results it's known that ΓK = ZΛ, and we've shown that ΓE < ZΘ. Therefore, the remainder EK > ΘΛ.

[4.2: go from salt-water to fresh water; 4.3 ratio of immersed weights of A and B in fresh water is less than that in salt-water]

5 [the Menelaus equation]

Given two objects of equal volume that have any desired volume. Let one be pure gold and the other pure silver; we measure the weight of each one. We prepared a pure silver object equal in volume to the originally given object, which is composed of gold and silver. We measure the weight of this silver and that of the original object. I say that the ratio of the excess of the weight of pure gold over the weight of the same-volumed silver to the excess of the weight of the mixed body over the weight of the same-volumed silver is equal to the ratio of the weight of the pure gold to the weight of the gold mixed into the body composed of gold and silver. [proof omitted; this is the "Menelaus" equation for calculating the density of alloys whose constituents don't react; in modern notation:

$$(m_{gold} - m_{silver}) / (m_{alloy} - m_{silver}) = m_{gold}/m_{alloyed\text{-}gold}]$$

*(Wurschmidt [1925] 384, 386, 388–390, 392)

9.14 Plutarch of Chaironeia (*ca.* 50 to *ca.* 120), writing under the Flavian emperors, Trajan, and in the early years of Hadrian, was a Platonist teacher of philosophy; much of his copious output is in the form of moralizing essays or dialogues. Just as Theophrastos doubted fire was an element, so Plutarch wonders whether cold is merely the absence of heat.

On the Principle of Cold

1 (945f–946a) [is "cold" a principle?]

So is there, Fauorinus [of Arles, philosophical litterateur, lived 85 to 155], a primary and essential power of Cold (as fire is of Heat) through whose presence and participation everything else becomes cold? Or is coldness rather a privation of warmth, as they say darkness is of light and rest of motion? Cold, indeed, seems to be stationary, as heat is mobile; while the cooling off of hot things is not caused by the presence of any power, but by displacement of heat, for it can be seen to depart completely at the same time as the remainder cools off. The steam, for example, which boiling water emits, is expelled in company with the departing heat; that is why the amount becomes less by cooling off; for this removes the heat and nothing else takes its place.

15.2 (951c–e) [not air but water is cold]

In the first place, it is improbable that air, adjacent to the *aither*, touching and touched by the revolving fiery substance, should have a contrary power. For it's impossible for two substances whose boundaries touch and are continuous not to be mutually affected – and if affected, for the weaker not to be replete with the power of the stronger. Nor is it reasonable to suppose that Nature has placed side by side destroyer and destroyee, as though she were the creator of strife and dissension, not of union and harmony. She does, indeed, make use of opposites to constitute the universe; yet she does not employ them unmixed or repelling. She disposes them rather so that a space is skipped and an inserted strip placed whereby they will not destroy one another, but may enjoy communication and co-operation. And air occupies this, suffused as it is through a space under the fire and above water, and makes distribution both ways and receives contributions from both, being itself neither hot nor cold, but a blending and union of hot and cold. When these are so fused, they meet harmlessly and the fused matter gently rejects or receives the opposing extremes.

19.1 (953d–e) [Earth's core is icy]

Every power, presumably, whenever it prevails, by a law of nature changes and turns into itself whatever it overcomes. What is mastered by heat is enflamed, what is mastered by *pneuma* turns to air, and anything that falls into the water, unless it escapes, dissolves and liquefies. It follows, then, that whatever is completely frozen must turn into primordial cold. Now freezing is an excess of cooling, and terminates in a complete alteration and petrifaction when, since the cold has obtained complete mastery, the moist is frozen solid and the heat is expelled. This is the reason why the earth at its bottommost point is practically all solid frost and ice. For there undiluted and unmitigated cold abides at bay, thrust back to the point farthest removed from the *aither*. And visible features, cliffs and crags and rocks, Empedokles thinks have been fixed in place and are upheld by resting on the fire that burns in the depths of the earth [frr. B52, 62–DK]; but the indications are rather that all things from which the heat was expelled and evaporated were completely frozen by the cold.

21.1 (954d–f) [coldness of element earth]

We are, further, informed by physicians that generically earth is by nature astringent and cold, and they enumerate many metals that provide a styptic and staying power for medicinal use. The element of earth is not sharp or mobile or light or prickly or soft or ductile, but solid and compact like a cube [Plato's model]. This is how it has weight; and the cold, which is its power, by thickening, compressing, and expelling the humidity of bodies, induces their shivering and shaking through its inequality; and if it becomes complete master and the heat departs or is extinguished, it fixes the body in a frozen and corpselike condition. This is the reason why earth does not burn at all, or burns only grudgingly and with difficulty. Air, on the other hand, often shoots forth flames from itself and, turning into fire, makes streams and flashes of lightning. Heat feeds on moisture, for it is not the solid part of wood, but the damp part, that is combustible; and when this is distilled, the solid, dry part remains having become ashes.

(Pearson and Sandbach [1957] 231, 263, 275, 279, 281)

9.15 Alexander of Aphrodisias, professor of Aristotelianism from 203 ± 5 and intellectual opponent of Galen, wrote numerous commentaries and essays on Aristotle's thought, often imputing his own developments to Aristotle.

On Mixture

12.4 [how fire "mixes" with earthy materials]

In general, since the iron is kindled by a particular fire which depends on specific matter, then, if the fire goes through the iron, it must go through it while protecting the matter on which it depended when it was adjacent to the iron; but neither pieces of wood, nor coal, nor any other matter supporting the fire which burns the iron come to be within the iron. So we are left with the fact that fire must come to be in the iron when separated from its matter. But if it is separated, it must acquire new matter in the change, and there is none except the iron itself. For to the extent that there is moisture in it, it becomes matter for the fire; certainly the iron that is heated becomes harder after its extinction than before, since the moisture in it is expended by the fire, and it is kindled as long as there is some moisture in it (as also with pieces of wood).

15.3–4 [the separation of mixtures is not mechanical but qualitative]

3   For just as a heated stone cast into milk (a uniform body containing in potentiality something both moist and solid) separates each of them from it, and in some way creates them, making the one into cheese, the other into whey, not through separating a part actually inherent in the milk but by creating each of

them from every part, so must the action of the sponge dipped into the pitcher holding wine blended with water be understood; for by its own quality it creates from the whole mixture the water that can be easily purified by it, and reconstitutes it as separate through the extent and nature of its power.

4 For as the onset of a slight fermentation in must, which is a uniform body, creates and separates from the whole of it both air and wine (clearly these bodies were not in it in actuality beforehand, since it is impossible that air should actually be contained by water, as we said just above, but the product is separated out as a whole just when the fermentation takes place); so must it be understood that agents which separate constituents from which blends have been formed also do not separate what is actually inherent in blends but cause an alteration by a specific power, and actualize bodies that as a result of blending are present in them potentially.

(Todd [1976] 145, 147, 157)

9.16 Zosimos of Panopolis (around 250) addressed his encyclopedia of alchemy to his noble sister Theosebeia, for her further edification in the art. Each book was devoted to a letter, and was allegedly based in part upon the *Pneumatics* of Archimedes and of Heron (p. 237B).

On Sulfur ("Divine") Water

(pp. 138–139B)

My lady, once when I was at your place for instruction, I was amazed at the entire operation you called "*structor*" [from Latin for "architect(ure)" or "building"]; sufficient astonishment fell upon me at the work there that I was beside myself and venerated the "fixer" [*poxamos*]; and we considered in our own mind how each artisan took few starting points from their predecessors and themselves practiced better. This was what drove me into astonishment: the cooking of the "breathy" [or "perforated"] "bird," how it's refined sealed off from vapor and heat and the making of a wash, even if it lacks no tinting. Thus amazed, our mind reined in to our pursuit: whether our composition can be refined and colored by exhalation of vapor of sulfur-water. I looked to see if one of the ancients mentioned this device, and I found nothing. Disappointed at that, I consulted your books and found in the Jewish books [of Maria?] next to the device known to the art and called "*tribikos*," and this is the description of the device:

Take arsenic and whiten thus: make a fat paste flat in the shape of a small speculum, and bore small holes sieve-like, and set therein a small fitted dish, in which let there be one part sulfur, and into the sieve put just as much arsenic; then seal it over with another small dish and lute the joint all around, and after two day-and-nights you'll find white lead. Take ¼ mina [about 120 gm] of this and smoke it for a day, adding a little bitumen, and so on.

249

(unknown section) (pp. 129, 202–203B)

Fear not to heat, and "aquify" [lit.: "make watery"] metals many times, because the myriad heatings of copper make it more tintable. Evert nature and you'll find what you seek, for nature is hidden within; when nature is everted, it will no longer appear white according to the conspicuous quicksilvering, but yellow according to the advertised yellowing of the *ios* ["poison," "dye," and "transformation": Wilson [1984]]. Now where are those who say nature cannot be changed? See: nature has changed, becoming solid according to the quality of gold, and it will be precipitated to black. For if the liquidity of the quicksilvering weren't circulating through the earthiness and powder of the solid body, and dissolvings and aquifactions weren't circulating through the essential quality of the quicksilvering, our expectation would come to nothing; and if it weren't dissolved and aquified and heated, our expectation would come to nothing. And if it weren't dissolved and heated and chilled, our expectation would come to nothing. But if everything happens in proper order, our hope of the outcome is realized with divine foreknowledge.

(unknown section) (p. 208B)

My lady, not everyone thinks yellowing follows directly after whitening, for the white compound being refined generally turns into the yellow compound. . . . Some do something extra besides these, for they let it cool and then pour and smooth out the yellow sulfur-water in the sun, for the instructed number of days, and after that they boil and cook it. . . . The "destroyed" sulfur-water, which has two parts lime and one part sulfur, boiled in a pot and filtered off, and boiled again, this is the sulfur-water which is cast for two colors. [*P. Leyden* X 89 gives the same recipe]

Final Account 1 (pp. 209, 239–240B)

Zosimos greets Theosebeia.

My lady, the entire kingdom of Egypt depended on these two arts, *chronokratoric* tintings [or "mummifications"?; the *chronokrator* is astrological] and gold-winnings. For the divine art of transformations, i.e., the teaching which is the business of all those pursuing all handicrafts, and the arts (I mean the four which teach manufacturing [the four books of Bolos]) were given only to the priests. For the natural gold-winning belonged to their kings, so that if it chanced to be spoken to a wise priest, interpreted to them from their ancestors or predecessors, they received and retained it as inherited. And knowing these procedures, they did not practice their knowledge, for they were punished: just as the artisans who know how to strike royal coins don't strike their own since they'd be punished, so also under the Egyptian kings the artisans of smelting and those knowing about this procedure don't practice it for themselves, but they performed it for the Egyptian kings working for their treasuries. They had their own directors set over them, and great was the tyranny over smelting, and not only over that but over gold-winning (for if anyone was found mining he had to be authorized in writing). [compare Agatharchides, *On the Red Sea* book 5, frr.23–29]

*(Berthelot [1888/1967] 129, 138–9, 202–3, 208–9, 239–240)

(unknown section) [he asserts (10.3, p. 145B): "just as minute yeast transforms the entire loaf, so will a small bit of gold transform the whole"]

Take good clean barley and soak a day [in water]; drain and set to rest in a breezy spot until the following day, and again soak for 5 hours. Put into a shouldered [or "short-handled"?] perforated vessel, and soak; on the following day dry it out until it blisters, then dry it in the sun until it shrinks (for they are bitter). Grind the rest and make bread, adding bread-yeast, and cook it rather raw, and when it effervesces, pour off the sweet water, and filter it through a colander or fine sieve. Others bake bread and put it in a basket in water and boil it a little (so it does not seethe or even get warm), and drain and filter, and this preparation they heat and set aside.

*(Gruner [1814] 10–16)

9.17 The Leyden Papyrus "X" (*"P. Leyden X"*) was written around 250 and buried with its owner, probably near Egyptian Thebes. About three-quarters of its eight-dozen recipes involve alloys or imitations of gold and silver, corresponding to books 1 and 2 of Bolos (about a dozen describe how to make gold ink). (Other recipes are translated in Humphrey, Oleson and Sherwood [1998] 362, 383.)

3 Purifying tin for mixing with asemos ["uncoined," i.e., silver]

Take tin purified of any other substance, melt it, let it cool; after having well mixed and soaked it with oil, melt it again; then having crushed together oil, bitumen, and salt, rub it on the metal and melt a third time; after fusion, break apart the tin after having purified it by washing; for it will be like hard silver. Then if you wish to employ it in the manufacture of silver objects, so that it is undetected and has the hardness of silver, blend four parts of silver and 3 parts of tin and the product will become as a silver object. [also in Humphrey, Oleson and Sherwood [1998] 229; compare *P. Stockholm* 4]

6 Doubling of asemos

The doubling of *asemos* is done thus: One takes: refined copper, 40 drachmas; *asemos*, 8 drachmas; tin in buttons, 40 drachmas; one first melts the copper and after two heatings, the tin; then the *asemos*. When both [sic] are softened, re-melt several times and cool as in the preceding recipe [tin–mercury amalgam]. After having augmented the metal by these proceedings, clean it with talc. The tripling is effected by the same procedure, the weights being proportioned in conformity with what has been stated above. [compare *P. Stockholm* 7]

25 *Purifying silver* [cupellation]

How one makes silver pure and brilliant. Take 1 part of silver and an equal weight of lead, soften in a furnace, and melt until the lead has been consumed; repeat the operation often until it becomes brilliant. [Keyser [1995/6] 212]

31 *Recognizing whether tin is adulterated*

After having melted it, place some papyrus below it and pour; if the papyrus burns, the tin contains some lead. [also in Humphrey, Oleson and Sherwood [1998] 229; same method in Pliny 34. 163]

35 *Manufacture of* asemos *that is black like obsidian*
[Egyptian niello]

Asemos, 2 parts, lead, 4 parts. Place in a new earthen vessel, add a triple weight of unfired sulfur, and having placed it in the furnace, melt. And having withdrawn it from the furnace, beat, and make what you wish, whether you wish to make figured objects in beaten or cast metal. Then polish and cut. It will not rust. [Pliny 33. 131; compare Moss [1953]]

37 *For giving to copper objects the appearance of gold*

And neither fire nor rubbing against the touchstone will detect them, but they can serve especially for a ring. Here is the preparation for this. Gold and lead are ground to a fine powder like flour, 2 parts of lead for 1 of gold, then having mixed them, knead them up well with gum, and one coats the ring with this mixture; then it is fired. One repeats this several times until it takes the color. It is difficult to detect, because rubbing gives the mark of a gold object, and the heat consumes the lead but not the gold.

55 *Another preparation of gold*

To gild silver in a durable fashion. Take some mercury and some leaves of gold, and form a waxy mass; taking the vessel of silver, clean it with alum, and taking a little of the waxy material, lay it on with the polisher and let it freeze. Do this five times. (Hold the vessel with a clean linen cloth lest you spoil it.) Then taking some embers, prepare some ashes; then smooth it with the polisher and use as a gold vessel. Proven. [compare Vitruuius 7.8.4 and Pliny 33. 65, 125]

(Caley [1926] 1151–1152, 1155–1156, 1158)

9.18 The Stockholm Papyrus ("*P. Stockholm*") was written around 250 and buried with its owner, probably near Egyptian Thebes. Its fourteen-dozen recipes are almost evenly divided between altering stones and tinting cloth (corresponding to books 3 and 4 of Bolos). (For other stone recipes, see Humphrey, Oleson and Sherwood [1998] 383–384.)

43 (= 88) Preparation of smaragdos

Mix and put together in a small jar ½ a drachma of copper green [malachite or verdigris], ½ a drachma of Armenian blue [Dioskourides 5.90: probably azurite], ½ a cup of the urine of a virgin youth, and two-thirds of the fluid of a steer's gall. Insert stones each weighing about ½-obol. Lay the cover upon the pot, lute the cover all around with clay, and heat it for six hours with a gentle fire of olive wood. But if this sign appears, that the cover becomes green, then heat no more, but cool off and take the stones out. Thus you will find that they have become emeralds. (The stones are of crystal; all crystal, however, changes its color by boiling.) [compare Diodoros of Sicily 2.52 and Strabo 2.3.4]

61 Another recipe for whitening pearls

[prior recipe: feed it to a chicken and it emerges white; see also 10–13, 25]
Quicklime, which if not yet slaked in water after burning in the ovens, contains its own interior invisible fire, they dissolve in dog's milk – from a white bitch, however. They then knead the lime and rub it around about the pearl and leave it 1 day in this manner. After they have wiped off the lime, they find the pearl has become white.

74 Preparation of verdigris for smaragdos from solid Cyprian copper

Clean a sheet using pumice with water, dry, and smear it very lightly with a very little oil. Fold it up and tie a cord around it. Then hang it in a vase with sharp vinegar so that it does not touch the vinegar, and carefully close the vase so that no evaporation takes place. Now if you insert it in the morning, then scrape off the verdigris carefully in the evening, but if you insert it in the evening, then scrape it off in the morning, and suspend it again until the sheet is used up. As often as you scrape it off, again smear the sheet with oil as explained previously. (The vinegar is unfit for further use.) [compare Theophrastos, *Stones* 57; Dioskourides 5.91]

109 Collection of woad

Cut off the woad and put together in a basket in the shade. Crush and pulverize, and leave it a whole day. Air thoroughly on the following day and trample about in it so that by the motion of the feet it is turned up and uniformly dried. Put together in baskets and lay it aside. (Woad thus treated is called "charcoal.")

110 *Woad dye* [familiar to Disokourides 2.185 and Pliny 20.59]

Put about a talent [*ca.* 26 kg] of woad in a *pithos* [large storage jar], which stands in the sun and contains not less than 15 *metretês* [about 450 liters], and pack it in well. Then pour urine in until the liquid rises over the woad and let it be warmed by the sun, but on the following day get the woad ready in such a way that you can tread around in it in the sun until it becomes well moistened. One must do this, however, for 3 days together.

118 *Another recipe for purple* [a costly color often imitated: Pliny 9.125–141; "authentic" purple was derived from the murex shellfish: see Thompson [1947] 209–218]

Dyeing in purple with herbs. Take and put the wool in the juice of henbane and lupines boiled sour in water. This is the preliminary mordant. Then take the fruit clusters of buckthorn [*Rhamnus infectorius*], put water in a kettle and boil. Put the wool in and it will become a good purple. Lift the wool out, rinse it with water from a forge, let it dry in the sun and it will be first quality.

135 (= 155) *Another recipe for purple*

Cold-dyed purple. Pulverize quicklime in cistern water. Pour the lye off and mordant what you wish therein from morning until evening. Then rinse it out in fresh water and color it in the first place in an extract of orchil. Then add *chalkanthon.*

(Caley [1927] 988–990, 993, 995–996)

10

BIOLOGY:
BOTANY AND ZOOLOGY

We live, but not alone: our life is deeply enmeshed with the life around us. Plants grow and some offer food; animals move and often act almost human. Our earliest art depicted these companions and victims, admired for their wit and exploited for food, clothing, and tools. Soon we tamed those we could (sheep, goats, and cows; later dogs, horses, and a few others), and admired those we never could (lions were brave, foxes clever, rabbits swift). Different peoples identified themselves by the animals they hunted, herded, or worshipped; people everywhere learned their local flora, their uses and habits. Field and fountain, lake and ocean were full of marvelous beasts and monsters, which every culture somehow classifies.

In Mesopotamia early legend created human–animal hybrids (fish–men, bird–men, cat–men, and the like), some representing forces of nature. The Egyptians set animal heads upon human forms to depict their gods (the god of death was envisaged with the head of the corpse-eating jackal), and saw some animals as sacred. The Cretan civilization delighted in depicting sea creatures, notably the polyp. In the earliest Greek literature, humans are often likened to animals: warriors prance like horses, lions, or rams (*Iliad* 6.506–511; 5.135–143; 3.195–198); they guard like hounds (*Iliad* 10.183–187); they fight at bay like boars, donkeys, lions, or stags (*Iliad* 11.414–418, 558–562, 544–555, 473–481); and attack like eagles, wasps, or wolves (*Iliad* 15.690–692; 16.259–263, 352–355). Sometimes we are radically distinguished from animals (Hesiod, *Works and Days* 276–284), or even thereto transformed (*Odyssey* 10.200–400, 19.518). Animals play humane roles in the numerous Greek fables of human behavior (Hesiod, *Works* 204–215, etc.), and other mythic tales told of whispering trees, powerful herbs, and hybridized or chimerical beings like minotaur, griffon, or sphinx.

The earliest Greek writers who we know to have sought causal explanations within Nature for the nature of living beings suggested that, just as plants (and perhaps sometimes small pesky animals) spring up from the moist and fecund Earth, so originally did all life. Anaximander of Miletos may have proposed that life originated in the sea or in watery containers (–580 ± 30, perhaps). Empedokles of Akragas (–455 ± 20) certainly hypothesized that not beings but parts thereof arose from the earth and linked together in chance ways; only those that could survive did, while mixed and monstrous types (such as ox-headed humans and human-faced oxen) perished from internal disharmony. Analogy of parts (plant-seeds are like animal eggs) and harmony of material (sight happens through innate ocular fire) were prominent as explanations. Plants grew up because of air, down because of earth, and would fruit year-round in milder climes (as preserved in Theophrastos, *Plant Etiology* 1.12.5, 1.13.2). Anaxagoras of Klazomenai, working in Athens (also –455 ± 20), suggested that every kind of substance sprouted from invisible seeds (as plants and animals did): every kind of thing had its seed, and every thing – not just fecund Earth – contained the seeds of every other kind of thing; he re-asserted the idea that all life emerged originally from muddy Earth.

Pythagoras of Samos (perhaps –530 ± 30) declared instead that the human soul is immortal and is reborn ever and again, sometimes in animal bodies. That explained our close kinship with the animals, while plants contained power to restore balance (as recorded by Ailian, *Historical Miscellany* 4.17). Alkmaion of Kroton (perhaps –500 ± 20), described as a follower, asserted the earth was mother and the sun father of plants (Lebedev [1993]); another of Pythagoras' followers applied his notions of the balance of opposites to explain plant growth and habit: Menestor of Subaris (–400 ± 40).

Demokritos of Abdera (writing *ca.* –410 ± 30) is attested to have composed much on biology, but little survives and most is mixed with material from Bolos and others. He is quoted by Theophrastos as explaining plant growth by hypothesizing differently-sized internal passageways for fluids (*Plant Etiology* 2.11.7–9) and various tastes via the manifold shapes of aggregated atoms (6.1.6–6.2.4). In roughly the same era, an Androtion wrote a book on agriculture exploiting notions of plant-sympathy (still available six centuries later to Athenaios), and Kleidemos composed a work touching on figs, olives, vines – and lightning. Herodotos of Halikarnassos (–435 ± 10) recorded many marvelous plants and animals from the extreme parts of his flat earth. In the medical library ascribed to Hippokrates the work *On Regimen* (2.46–49) classifies animals by domicile (land/sea/air), relation to humans (wild/tame), and edibility properties (dry/moist, heavy/light, strong/weak); another work in the same library, *Nature of the Child* (22–27) draws extensive parallels between plant growth and human embryonic development. On the comic stage, Aristophanes of Athens spoofed the avian ways of fellow-citizens in his *Birds* (–413).

In his dialogues, Plato of Athens (writing –365 ± 20) sought to turn traditional implicit classifications into explicit taxonomies, in two *ad hoc* and incompatible ways: in the *Sophist* (220–223) animals are classified by how they are hunted (land/sea, then sea-creatures are winged/watery, etc.), while in the *Statesman* (263–267) the division starts with tame/wild, of which the tame are land/sea, then land animals are winged/terrestrial, etc. The *Timaios* instead commences with humans and our anatomy (69–90: liver and heart and brain, bones and blood and flesh, etc.), and derives from us all other living beings by karmic degeneration (90–92): birds from mental lightweights, quadrupeds from the passionate, snakes from fools, and so on down to shellfish. A little later, Mnêsitheos of Athens (–350 ± 20) returned to the taxonomic question (again on a dietetic basis), as did Plato's nephew, student and successor Speusippos of Athens (head of the Academy from –346 to –338) in his (lost) *Similars*, on the basis of multiple characteristics.

Aristotle of Stageira, (writing –335 ± 10) wrote more on the natural philosophy of animals than on any other single subject. "We ought to study their nature as joyfully and zealously as we do mathematics and astronomy, since Nature's purposive operation is here most clearly visible and beautiful" (*Parts of Animals* 1.5 [644b22–645a37]). Plants, animals, and ourselves are part of one continuous series wherein Nature proceeds from "soulless" beings up to plants then testaceans, other sea-creatures with rudimentary sensation, and finally the mobile and sensing animals (*Animal Researches* 7[8].1 [588b4–589a9]). Within the animals, increasing hotness explains the increasing perfection of reproductive mode: chilly insects lay a larva which grows into an egg (the chrysalis); slightly warmer fishes, crustaceans, and cephalopods lay imperfect eggs; birds and reptiles manage a perfect egg; some are quite warm and lay internal eggs (ovoviviparous such as the sharks); and hottest and best of all are the viviparous, who produce perfect offspring (*Generation of Animals* 2.1 [733a18-b18]). Generation normally takes place from seed, which is derived not as some earlier thinkers had said from the whole body, but is a kind of "residue" left over from nutriment (*Generation of Animals* 1.17–18 [722a1–726a25]). The spontaneous generation of some living beings (plants and insects mostly) does occur but is imperfect, resulting in a half-formed type of creature (*Animal Researches* 5.1 [539a16–b14]); some fish are formed from mud and sand, especially eels, but also testaceans and sponges, etc. (*Animal Researches* 6.15–16 [569a10–570a25], 5.15 [547b18–34], 5.16 [548a22–29]). Aristotle's system of *differentiae* rejects division by one attribute at each step, instead simultaneously employing multiple attributes (the *differentiae*) for each division (*Parts of Animals* 1.2–3 [642b5–644a12]). He appears to have invented or more closely recognized several categories still in use (testaceans, cephalopods, ovoviviparous), and his system was sufficiently flexible to accommodate fuzzy borders between classes (apes greatly resemble humans but are sub-human; seals are land *and* sea creatures; bats are land *and* air beings; and sea-anemones even straddle the plant–animal divide). He carefully observed the life-cycles of animals, from well-known chickens in their eggs (*Animal Researches* 6.3 [561a5–562a22]) to the peculiar selachian "dog-fish" whose fry have "umbilical cords" and placentas (*Animal Researches* 6.10 [565b2–17]). Outdoor natural history of high quality peppers his writings, e.g., *Animal Researches* 6.14 (568a22–b23) on the small Greek catfish, the *glanis*, or *Animal Researches* 4.5 (530a32–531a8) on the five "lantern-like" jaws, and other parts, of the sea-urchin. Animal habits too were observed and often assimilated to human behavior (*Animal Researches* 8[9].3–7 [610b20–613b6], 8[9].44–48 [629b5–631b5]).

10.1 Theophrastos of Eresos (–370 to –286) succeeded Aristotle as the head of his Lukeion "school," and wrote widely, amassing data questioning Aristotle's system (Keyser [1997c]). The works here excerpted were part of a set of works on natural philosophy (see also Chapters 8, 9, 11, and 12). Theophrastos' careful and precise distinctions are noteworthy.

Plant Researches

1.3.1 [categories]

Now since our study becomes more illuminating if we distinguish different kinds, it is well to follow this plan where possible. The first and most important classes, which comprise all or nearly all plants, are: tree, shrub, under-shrub, herb. A tree is a thing which springs from the root with a single trunk, having knots and many branches, and it cannot easily be uprooted; for instance, olive, fig, vine. A shrub is a thing which rises from the root with many branches; for instance, bramble, Christ's thorn. An under-shrub is a thing which rises from the root with many trunks as well as many branches; for instance, savory, rue. An herb is a thing which comes up from the root with its leaves and has no main stem, and the seed is borne on the stem; for instance, grain and vegetables.

[1.13.3: distinguishing epigynous, hypogynous, and perigynous flowers]

2.1.1–2 [how plants start to grow]

1 The ways in which trees and plants in general originate are these: spontaneous growth, growth from seed, from a root, from a piece torn off, from a branch or twig, from the trunk itself; or even from small pieces of cut-up wood (for some trees can be produced also in this manner). Of these methods spontaneous growth is perhaps primary, but growth from seed or root would seem most natural; indeed these methods too may be called spontaneous; so they are found even in wild kinds, while the remaining methods depend on human skill or at least on human choice.

2 However all plants start in one or other of these ways, and most of them in more than one. Thus the olive is grown in all these ways, except from a twig; for an olive-twig will not grow if it is set in the ground, as the fig will from a sprig, or the pomegranate from a branch.

4.4.8 [cotton]

The trees from which the Indians make their clothes have a leaf like the mulberry, but the whole tree resembles the wild rose. They plant them in the plains in rows, wherefore, when seen from a distance, they look like vines.

. . .

4.7.7

The island of Tulos [Bahrain] also produces the "wool-bearing" tree in abundance. This has a leaf like the vine, but small, and bears no fruit; but the vessel in which the wool is contained is as large as a spring apple when closed, but when ripe it unfolds and puts forth the wool, of which they weave their fabrics, both cheap and costly. [for this "tree" and its "wool," Strabo 15.1.21 cites Aristoboulos of Kassandreia, who wrote around −305 ± 10]

4.6.1–4 [maritime plants]

1 However, the greatest difference in the natural character itself of trees and of tree-like plants generally we must take to be that mentioned already, that some plants, as some animals, are terrestrial, some aquatic. Not only in swamps, lakes and rivers, but even in the sea there are some tree-like growths, and in the outer sea [Atlantic: see Chapter 5] there are even trees. In our own sea all the things that grow are small, and hardly any of them rise above the surface; but in the ocean we find the same kinds rising above the surface, and also other larger trees.

2 Those found in our own waters are as follows: most conspicuous of those which are of general occurrence are seaweed [*Posidonia oceanica*] "oyster-green" [*Ulva lactuca*] and the like; most obvious of those peculiar to certain parts are the sea-plants called "fir" [*Cystoseira abies-marina*], "fig" [?], "oak" [*Cystoseira ericoides*], "vine" [*Fucus spiralis*], "palm" [*Callophyllis laciniata*]. Of these some are found close to land, others in the deep sea, others equally in both places. And some have many forms, as seaweed, some but one. Thus of seaweed there is the broad-leafed kind, ribbon-like and green in color, which some call "sea-leek" and others "girdle-weed." This has a root which on the outside is shaggy, but the inner part is made of several coats, and it is fairly long and stout, like *kromuogeteion* [a kind of onion].

3 Another kind [*Cystoseira foeniculum*] has hair-like leaves like fennel, and is not green but pale yellow; nor has it a stalk, but it is, as it were, erect in itself; this grows on oyster-shells and stones, not, like the other, attached to the bottom; but both are plants of the shore, and the hair-leafed kind grows close to land, and sometimes is merely washed over by the sea; while the other is found further out.

4 Again in the ocean about the pillars of Herakles there is a sea-weed of marvelous size, they say, which is larger, about a palm's-breadth [*Laminaria saccharina*]. This is carried into the inner sea along with the current from the outer sea [compare Straton in Section 5.3], and they call it "sea-leek"; and in this sea in some parts it grows higher than a man's waist. It is said to be annual and to come up at the end of spring, and to be at its peak in summer, and to wither in autumn, while in winter it perishes and is thrown up on shore. Also, they say, all the other plants of the sea become weaker and feebler in winter. These then are, one may say, the sea-plants which are found near the shore. But the "oceanic seaweed," for which sponge-fishers dive, belongs to the open sea.

(Hort [1916] 23, 25, 105, 317, 329, 331, 333, 343, 345)

6.3.1–3 [silphion, now probably extinct]

1 The silphion has a big thick root; its stalk is like ferula in size, and is nearly as thick; the leaf, which they call *maspeton*, is like celery: it has a broad seed, rather leaf-like, called the "leaf." The stalk is annual, like that of ferula. Now in spring it produces this *maspeton*, which purges sheep and greatly fattens them, and makes their flesh wonderfully delicious; after that it sends up a stalk, which they say is eaten in all ways, boiled and roast, and this too, they say, purges the body in forty days.

2 It has two kinds of juice, one from the stalk and one from the root; so the one is called "stalk-juice," the other "root-juice." The root has a black bark, which is stripped off. They have regulations, like those in use in mines, for cutting the root, in accordance with which they fix carefully the proper amount to be cut, depending on previous cuttings and the supply of the plant. For it is not allowed to cut wrongly or more than allowed; for, if the juice is kept and not used, it goes bad and decays. Those conveying it to Peiraios [port of Athens] deal with it thus: having put it in vessels and mixed meal with it, they shake it a long time, and from this process it gets its color, and this treatment makes it thereafter keep without decaying. This is how things are with the cutting and treatment.

3 The plant is found over a wide tract of Libya, they say for a distance of more than four thousand stades [about 750 km], but it is most abundant near the Surtis, starting from the Euesperidês islands [renamed Berenikê by the Ptolemies and now called Benghazi]. It is a peculiarity of it that it avoids cultivated ground, and, as the land is brought under cultivation and tamed, it retires, plainly showing that it needs no cultivation but is a wild thing [Herodotos 4.169]. The people of Kurênê say that the silphion appeared seven years before they founded their city; now they had lived there for about three hundred years before the archonship at Athens of Simonides [in –309].

[8.2.1–2: some trees and all legumes are dicots; plants such as grasses are monocots] [9.5.1–3: marvelous tales regarding cinnamon and cassia; compare Herodotos 3.110–111]

(Hort [1926] 15, 17, 19)

Plant Etiology

1.5 [spontaneous generation]

Cases of spontaneous generation occur in the smaller plants (broadly speaking), especially in annuals and herbaceous plants. They nevertheless sometimes also occur in larger plants, either after spells of rain or when some other special condition has arisen in the air and the earth. For it is thus that silphion is said to have come up in Libya, when the water was "pitchy" and thick, and the forest now existing there is said to have come from another such cause, not having existed before. [*Plant Researches* 3.1.6] But rainy spells not only cause decays and alterations, the water penetrating far and wide, but they can also feed what is formed and make it grow larger, while the sun warms and dries it, this being also how most authorities account for the generation of animals as well.

2.4.4–5 [soil types]

Among soils good for trees some are better suited to one tree, some to another, and such are the distinctions that the agriculturists make. For instance stony soil, and still more white soil [chalky], is a good producer of the olive, since it has moisture and a good deal of *pneuma*, and the olive requires both. Meadow land and sandy soil are good producers of the vine, and so in general is any soil that is open, light, lean, and with a water table within easy reach of the rain, since the vine requires plenty of food because it is hot, open, fluid and an abundant bearer (in fact it is perhaps these very features that make it an abundant bearer). Furthermore, the roots of fluid trees do not easily decompose, like those of dry trees, and so are able to attract and retain and transmit the food. So too with the rest: for every tree there is a soil appropriate to its nature, and the same soil is more appropriate to some, less so to others.

2.13.1–2 [effects of locale]

When the crop changes with the country the mutation evidently follows the air and the soil, since all plants get their food through these and from these, and the food has a strong effect in producing similarity, seeing that in animals too it is through the food that similarity to the females comes about [Aristotle, *Generation of Animals* 2.4 (738b25–36): female produces body, male produces soul of offspring]. Not only seeds, plants and trees are observed to change but also animals, and animals in a way even more, since assimilation to the regional character affects even their shapes [Aristotle, *Animal Researches* 7[8].28 (606a13–b3)], whereas this is not so noticeable in crops. Still the change occurs in crops too, most noticeably in color, size and flavor: in color, as grains, from black to white [*Plant Etiology* 3.21.3, barley] and from white to black, and in flavor, fruit too [*Plant Etiology* 2.13.4], whereas a change in its color is either not noticeable or not so frequent, except in the instances where the whole tree is changed, with the result that a white variety comes from a black, a thing that occasionally occurs in trees growing from seed. [accounts of changes: *Plant Researches* 2.4]

(Einarson and Link [1926] 33, 35, 225, 227, 305, 307)

3.2.6 [best time to plant]

So one must always plant and sow when the earth is in heat, since then the sprout that comes forth is best, just as in animals when the seed enters a womb desiring it [Aristotle, *Generation of Animals* 2.4 (739a31–35)]. The earth is in heat when it is moist and warm and the weather temperate, since then it is loose and sends the shoot up quickly and is in general nutritious. This receptivity occurs chiefly, at least for trees, at two seasons, spring and autumn, these being in fact the times when more planting is done, and for a greater number of different trees it is done in spring, for the earth is then soaked through and the sun by its warmth brings about growth and the air is mild and dewy, so that all this combines to rear the slips and make them sprout well.

3.6.1 [fertilizer]

It is evident that manure not only gives the earth a loose texture but also warms it through, both of which lead to rapid sprouting. But there is a dispute about how it is to be used, and not all the experts apply it in the same way. Some mix it directly with the earth and put the mixture around the slip. Others put the manure in a layer between the earth that makes the bottom of the hole and the fill of earth at the top, since if put below it moves up after a rain (the liquid solution that moves up being, they say, the best part), but if put on top it loses its moisture to the sun, and when it rains the liquid does not reach the bottom. All however agree to this extent: that the manure should not be pungent and strong but light, and this is why they chiefly employ that of pack-animals, since pungent and strong manure heats too thoroughly, and some of it also dries the slips [*Plant Researches* 2.7.4].

3.18.1 [plant gender]

That the fruit does not remain on the female date-palm unless you shake the flower of the male over it together with the dust [pollen: *Plant Researches* 2.8.4] (this too being reported by some) occurs only in the date-palm, but is similar to the caprification of fig-trees [Herodotos 1.193; *Plant Etiology* 2.9.15]. From these instances one would be most inclined to infer that even a female tree cannot by itself bear completely formed fruit; except that this should hold not of just one or two female trees but of all or most of them, since this is how we decide the nature of the class of females. And in the cases before us that of the date-palm is very strange indeed, since caprification is considered to have a clear explanation [*Plant Researches* 3.8.1, 3.9.1–6: male/female in another sense: Negbi [1995]].

(Einarson and Link [1990a] 19, 41, 135, 137)

5.17.1 [girdling]

Stripping the bark all around ["girdling"] is a form of killing (discussed earlier [*Plant Researches* 4.15.1–4]) effective with all or most trees. For whether the stripping is of parts that control life (as some assert), death would be reasonable; or whether the parts exposed get thick and the thickening spreads to the whole tree, death is reasonable with this process too.

(Einarson and Link [1990b] 181)

On Fire

60 [the peculiar powers of the salamander]

If along with a liquid of this kind [a cold liquid, like vinegar] coldness be present by nature, this contributes to the extinction [of the fire], as is the case with the salamander. This animal is cold by nature, and the liquid flowing from it is viscous and at the same time contains a very penetrating secretion. The evidence: waters and fruits become poisonous when the secretion is mixed with them, especially from a dead salamander.

(Coutant [1971] 38, 40)

10.2 Bolos of Mendes (–180 ± 30) wrote on the powers and nature of living things (he also composed a work on alchemy: see Chapter 9.5). He either advertised his work as by Demokritos or at least it is often so cited: Vitruuius 9.pr.14, 9.2–3; Columela 1.1.7, 7.5.7, 11.3.2,64; Pliny 24.160–166.

Farming

(fr. 81 paraphrased in Geoponica *15.2.21–38)*

[bovine apifaction; other sources record that "Egyptians bury a bull up to its horns to get bees (likewise scorpions come from dead crocodiles and wasps from dead horses)"; Nikandros of Kolophon (below Section 4) probably versified this (he refers to wasps from horse-corpses in *Theriaka* 741), and certainly Vergil did: *Georgics* 4.295–314]

21–26 This is how to make bees: let your house be ten cubits high and ten wide (and the other sides equal). Make one entrance and four windows, one in each wall. Lead a thirty-month-old cow into this, well-fed and very fat, and station around it many young men and let them strike it hard, beat it with clubs, and slaughter it, grinding together the flesh and bones. (Let them take care not to draw blood from the cow – for the bee is not begotten from blood – which is possible if at first they don't strike violently.) Then immediately, using fine clean linens anointed with pitch, stuff every opening of the cow, i.e., eyes and nostrils and mouth and anus. Then spreading around much thyme and turning the cow upside down, leave the house and immediately seal the door and windows with waterproof cement, so that there's no entrance or draft for air or wind.

27–28 In the third hebdomad [week] open up all sides and let in light and clean air, except whenever there's a strong breeze (if this happens you must close the entrance). And when the ensouled material seems to have drawn in self-sufficient *pneuma*, you must immediately close up the house with cement as before. On the eleventh day after this, open up and you'll find it full of "swarming bees" [*Iliad* 2.89] gathered upon each other, but only the horns and bones and hair of the cow remain and nothing else.

. . .

30–33 From the head the "kings" [queens] are generated, and they excel the

others in size and strength. And then you'll observe the first change and transformation of the flesh into animals, and as it were a begetting and generation:
when the house is opened they are small and white in appearance and all alike
and incomplete [larvae], and you won't yet see the animals multiplying all over
the calf; they are all motionless and slowly growing. You may see their winged
parts being articulated, and their taking on their proper color, and then sitting
around the king and flying towards him; and briefly you'll see their trembling on
their wings through inexperience with them, and the shakiness of their limbs.

34–38 They sit on the windows buzzing, shoving and pushing each other in
lust for the light (it's better to open and uncover the windows by day). Take care
lest the nature of the bees now changes, due to being closed up too long and not
breathing air in time, so that they suffocate. Let the beehive ["bee house"] be
nearby and when the windows are opened and they fly out, fumigate with thyme
and white *kneôros* [*Daphne oleoides*, the fragrant herb described by Theophrastos, *Plant
Researches* 6.2.2]: for you'll draw them with the smell to the apiary, being cultivated by the smell of the flowers, and by fumigating you'll make them willing to
enter the apiary.

*(Wellmann [1921] 57–58)

10.3 Agatharchides of Knidos (*ca.* –215 to *ca.* –140) worked in Alexandria
under Ptolemy V "(Epiphanes")" and Ptolemy VI ("Philometor") (a time of strife
and chaos). He wrote several works of history enriched with anthropology and
geography, and towards the end of his life, fleeing the chaos in Alexandria, he
published *On the Red Sea*, which is lost but paraphrased by Diodoros of Sicily,
the first-century BCE historian, and more closely but less completely by Photios
the ninth-century bishop of Constantinople (material found only in Diodoros is in
italics).

On the Red Sea

Book 5, fr. 72 [the rhinoceros and its battles with elephants]

The rhinoceros ["nose-horn"] is *similar in courage and strength* to the elephant,
although it is not as tall. Its color is similar to that of common boxwood [yellowish:
from its mud-bathing] and the texture of its skin *is the toughest*. At the end of its
nostrils it has a snub horn that is as hard as iron [African rhinos are two-horned,
Indian rhinos one-horned], and it sharpens its horn (thrusting forward with its
chest) against a convenient rock, but if it meets an elephant in battle (with which
it always contends over grazing-rights) it slips under the elephant's belly and rips
open the encircling flesh *using its horn like a sword*. This immediately drains the
elephant of blood, and many have been killed thus. But if the rhinoceros fails to
hit the elephant's belly, then the elephant first grabs the rhinoceros with its trunk,
and instead the rhinoceros is rendered powerless and killed by repeated blows of
the trunk and tusks since there's a great difference in their strength and power.
[Diodoros of Sicily 3.35.1–3]

Book 5, fr. 77 [carnivorous bulls: Aristotle, *Animal Researches* 8(9).45
(630a18–b18), the aurochs; Ailian 17.45]

Of all the animals known, however, the wildest and altogether hardest to master is the carnivorous bull [if not mythical, the African buffalo, *Syncerus caffer*]. In mass it is larger than tame bulls, and in speed *it approaches the horse*. (It is fiery red in color.) Its mouth opens as far as its ears, and its eyes are more brilliant than those of a lion *and they shine at night. The horns are peculiar:* mostly the animal moves them like its ears [Ailian 2.20 says the same of the oxen of Eruthrai, near Chios], but when it gets into a fight, he holds them rigid. The direction of its hair-growth is opposite to other animals. *Outstanding in bravery and power,* the beast attacks the boldest beasts and hunts other animals, *getting its food by eating the flesh of its victims.* But it destroys the herds of the natives, *and engages in mad combat with whole troops of shepherds and packs of dogs.* It alone cannot be wounded by lance or bow, so no-one has succeeded in subduing it, though many have tried; but if it falls into a pit or *is captured by a* similar trap, then it is quickly choked by its own rage, *and in no way does it exchange freedom for the kindness of domestication.* The Trogodutes rightly judge this animal *the best, since Nature gave it* the courage of a lion, the speed *of a horse,* and strength of a bull, and to be invulnerable to iron. [Diodoros of Sicily 3.35.7–9]

Book 5, fr. 99 [fragrant plants in Saba]

A natural fragrance pervades the whole coast of Saba [South Yemen], *because almost everything that excels in scent grows there unceasingly,* providing a pleasure to visitors that is greater than what can be imagined or described. Along the coast balsam grows in abundance and cassia and another sort of plant which *has a peculiar nature:* when fresh, it's very delightful to the eye but suddenly fades (so that the usefulness of the plant is blunted before they can send it to us).

In the interior there are large dense forests, in which grow tall trees: myrrh and frankincense, cinnamon, [date-]palm and *kalamos* [a reed of the genus *Cymbopogon*] and other such trees *with similar sweet scents; and one can't count their peculiar properties and natures because of the excessive quantity of the scent collected from all of them.* So what is experienced by those who have tested it with their senses seems *divine and* inexpressible. It is not pleasure from old stored spices nor that of a plant separated from what bore and nourished it, but that of one blooming at its divine peak and giving off its own natural wondrous scent. Thus many come to forget mortal blessings and think they have tasted ambrosia (seeking a name for the experience proper to its extraordinariness). [Diodoros of Sicily 3.46]

Book 5, fr. 110 [maritime olive trees and the sea fan]

In this strait [Bab al Mendeb, between Yemen and Ethiopia], a strange thing happens to olive trees. When the tide is high they are all covered, but they bloom all during the sea's ebb tide. [compare Pliny 13.141 describing the tree as having evergreen leaves like bay, scent like violets, and berries like olives, probably following Nearchos of Crete, −305 ± 10] There is a kind of plant that grows there underwater in rocky areas, like black rush, and the natives call it "Isis' tresses" seeking to add simple faith to a mythical fiction. When this plant is struck by a wave, it bends flexibly since its whole stalk is soft like that of other plants. But if a person cuts off a piece and brings it up to the air, it immediately becomes harder than iron [a sea-fan of the genus *Gorgonasia*; compare Iuba of Mauretania in Pliny 13.142].

*(Henry [1974] 174–176, 183, 188)

10.4 Nikandros of Kolophon, a priest of Apollo (around −130 ± 20), versified contemporary learning on venomous beasts, as well as on poisons (see Chapter 11), bee-keeping (lost), hunting (lost), etc. His colorful depictions deploy more recondite verbiage than instructive particulars.

Theriaka

157–189 [the asp (lines 159–160 excised as spurious)]

Next to consider is murderous asp who is covered all over with
Dryness of scales, and is slowest and dimmest of serpents that strike; well, her
Form is a terror to look at and slithers in motion that's sluggish and
Weighty to rise up from *writhing in* coils, and she's always appearing to
Gaze at you sleepily looking askance and she's peering so fixedly.
Whenever *suddenly* thuds on her ears any noise or else brightness she
[165] Spies, then *it's surely* she sloughs off the dullness of sleep from her carcass,
Whirling her circular coil on the ground *where she crawls in her track*, and
Banefully bristling she raises her *sorry old* head from the middle.
Stretched out in length she will measure a fathom, most *shameless and* horrid of
Earth's teeming creatures; the thickness *she girdles around* is the width of a
[170] Spear that was fashioned by spear-maker *skillful in* making a spear *to be*
Deadly for hunting *and slaughter* of bulls and the deep-throated lions.
Sometimes the color of powdery dust is what spreads on her back, and
Sometimes she's shiny and quincey her yellow, and otherwise ashen, and
Often when darkened with dirt of Aithíops, instead she is smoky and
[175] Brown like the sludge which the manifold mouths of the Nile when it's flooding
Pour in the sea, where its stream is colliding with waves of the deep.
Over the eyes there appear what can seem like callosities marking the
Brow and beneath these the eyes *as if shadowed* are gleaming with redness,
Standing up over her coil, as she puffs out her dust-colored neck while
[180] Ceaselessly hissing *confusion* when fastening death from the depths onto

Travelers meeting this beast that is boiling with grudges of rage.
Fourfold in number the fangs that she has which are hollow below, and
Curving and long they are rooted and set in her jaws very firmly for
Casting her poison, then shrouded with membranes to cover them deeply:
[185] Therefrom she belches out pitiless poison on bodily members.
Would that such terrible monsters assail just inimical heads, for
Flesh is unmarked by their bite and moreover *at strike-point* appears nothing
Swollen or deadly inflamed, but the victim is painlessly dying, and
Slumberous lethargy brings on destruction and close of his life.

715–737 [deadly spiders]

Next to consider with care are the deeds of the ravening spider, and
Symptoms attending its bitings: the "Grape"[a] with a hue that is smoky is
Also called "pitchy"; its feet they go creeping one after another, and
Right at its center his belly is hardened with teeth for destruction.
Horrible creature *advancing to* strike: still your skin is remaining
[720] Just like unwounded; up higher the lights of your eyes are empurpled, a
Shudder is pressing upon your poor carcass and instantly skin and the
Masculine genitals tauten down under, the member projecting and
Dripping with mixture of ichors; the very same moment chilblain is
Darting upon you, prostrating your hips and support of your knees.
[725] Let yourself learn of another, the "Starlet"[b] whose back is adorned with
Bands that are edged with a color that's bright as it gleams on its skin; in
Biting she casts upon victims a shivering *most* unexpected, with
Cranial torpor descending, then knees are unfastened beneath them.
Darting about while aloft is another, the "Blue" *as it's called,* who is
[730] Downy; by this one a terrible pricking of skin is inflicted on
Whom *he is hobbling,* and heavy the heart in *the breast of* that victim, as
Night is descending upon him while spiderous vomit he's spewing, a
Ruinous web from his throat: his proximate death is allotted.
Huntsman's another, a spider who's like unto "Wolf"[c] in his shape, the
[735] Slayer of carrion flies who is stealthily trapping the bees, the
Wasps in the fig, the gadflies, and whatever may come to his trap; yet on
Man he inflicts with his strike a mere burdenless trifle of sting.

[[a] the "Grape" is *Latrodectus mactans*, also mentioned by Ailian 3.36; [b] the "Starlet" is a color-variant of *L. mactans*; [c] the "Wolf" is the webless class of spiders: Aristotle, *Animal Researches* 8(9).39 (623a1–8); Beavis (1988) 47–54]

(Gow and Scholfield [1953] 39, 41, 77: versified)

10.5 Leonidas of Buzantion (*ca.* −100 ± 20) wrote a book on *Fishing*; a few extracts survive in paraphrases by Ailian (see below) – see also: 2.50 killer sting-ray, 3.18 the *Tetrodon* puffer fish, and 12.42 bait for the parrot wrasse.

Fishing

(fr. in Ailian, Animals' Characters 2.6) [dolphins]

While sailing past Aiolis I saw with my own eyes at the town called Poroselenê [largest of the Hekatonnesi ("hundred-isles" or "Apollo's-isles"), between Lesbos and the mainland "Aiolis"; normally named Pordoselenê, a spelling avoided by the fastidious as resembling "fart-moon": Strabo 13.2.5–6] a tame dolphin which lived in the harbor there and behaved towards the inhabitants like personal friends. Moreover, an old woman and old man dwelling there fed this foster-child, offering it the most alluring baits. What is more, the old couple's son was brought up with the dolphin, and the pair cared both for the dolphin and their son, and somehow by dint of being brought up together the man-child and the fish gradually and imperceptibly fell in love, and, as the song has it, "a super-reverent counter-love was cultivated" among them. So then the dolphin came to love Poroselenê as his homeland and grew as fond of the harbor as of his own home, and what is more, he repaid his care-givers their raising expenses. And this was how he did it. When fully grown he had no need of being hand-fed, but would now swim further out, and as he ranged abroad in search of marine quarry, would keep some to feed himself, and the rest he would bring to his relations. And they understood this and even gladly awaited his tribute. That then was one gain; another was this. As to the boy so to the dolphin his foster-parents gave a name, and the boy, bold due to their common upbringing, would stand upon some spot jutting into the sea and call the name, and with the calling would say sweet things. The dolphin, whether he was racing with an oared ship, or plunging and leaping about the fish roaming in shoals about the spot, or was hunting under stress of hunger, would rise to the surface with all speed, like the great wave of a ship driving onward, and drawing near to his playmate would play and leap about at his side; now swimming close by the boy, now seeming to challenge him and even induce his favorite to race with him. And what was even more astounding, he would at times even decline the winner's place and actually swim second, as though presumably he was glad to be defeated.

[Herodotos 1.23–24 on the dolphin of Arion of Methumna on Lesbos; Aristotle, *Animal Researches* 8(9).48 (631a8–631b5); Thompson [1947] 52–56]

(Scholfield [1958a] 93, 95, 97)

10.6 Poseidonios of Apamea (in Syria) became a citizen of the "free city" of Rhodes (a Roman protectorate), where he wrote and taught between about −100 and −50. He was widely traveled and well connected. Most of what we know about his book *On the Ocean* comes from Strabo, who wrote in the generation after him (see below); these extracts may derive from that work or another.

(title unknown)

(fr. 241 E–K from Strabo 3.5.10) [unusual trees]

Poseidonios also reports a tree at Cadiz with branches bending to the ground, and often with sword-like leaves a cubit in length, but four fingers in breadth [*Draco dracaena*, native to the Canary and Madeira Iss.]. And he says that at New Carthage [Cartagena] there is a tree that yields bark from the thorns, from which very fine woven fabrics are made [the dwarf palm].

Well, I too know a tree in Egypt like the one at Cadiz in the bending down of the branches, but unlike it with regard to the leaves, and too in having no fruit [a willow?]; Poseidonios says that the Cadiz one has fruit. Thorn fabric is woven in Kappadokia too, but it is not a tree that produces the bark-bearing thorns, but a low-lying plant. With regard to the tree at Cadiz, the following has also been reported, that if a branch is broken off, it oozes milk; if a root is cut, red moisture seeps up [the sap runs whitish but dries to red].

(title unknown)

(fr. 245 E–K from Strabo 17.3.4) [apes resemble people]

Poseidonios, when he was sailing from Cadiz to Italy, was carried near the Libyan shoreline and saw a thicket which projected into the sea full of apes; some were on the trees, some on the ground, and some with young and suckling them. He said the sight made him laugh: the heavy breasts of the mothers, some males bald, others ruptured, and exhibiting other such ailments and afflictions. [compare Section 11.19 (Galen)]

(Kidd [1999] 313–314, 316)

10.7 Damostratos, a Roman senator around –40 ± 20 (possibly C. Claudius Titianus "Demostratus"), wrote a work in 20 books entitled *Fishing*; a few extracts survive (two verbatim) in Ailian (see below); the others are: 15.4 the blue "moon fish" of Cyprus whose power to wax and wane trees and wells depends on lunar phase; 15.19 the tricky mating of the tortoise.

Fishing

(fr. in Ailian, Animals' Characters 13.21) [a "real" triton]

Damostratos in his treatise on Fishing says that at Tanagra he has seen a preserved Triton [human head and torso; fish tail: originally a god (Hesiod, *Theogony* 930–934 and Apollonios the Rhodian, *Argonautika* 4.1551–1619), whom Herakles fought, and who represented the dangers of the sea].

It was in most respects as portrayed in statues and pictures, but its head had been so marred by time and was so indistinct that it was not easy to make it out or recognize it. And when I touched it there fell from it rough scales, quite hard and resistant. And a member of the Council, one of those chosen by lot to regulate Greece and entrusted with the rule for a single year, intending to test and examine the nature of what he saw, removed a small piece of the skin and put it on a fire; whereupon a heavy smell from the burning object assailed the nostrils of the bystanders. But we were unable to conclude whether the animal was terrestrial or marine. The experiment however cost him dear, for shortly afterwards he lost his life while crossing a small, narrow strait in a short, six-oared ferry-boat. And the Tanagrians maintained that he suffered this because he profaned the Triton, and they declared that when he was taken lifeless from the sea he disgorged a fluid which smelt like the hide of the Triton when the man ignited and burnt it.

(fr. in Ailian, Animals' Characters *15.9)* [the fifteen-foot long "crane fish,"
perhaps *Regalecus banksi* or *Nemicthys scolopaceus*]

I saw the fish and was filled with astonishment, and I was anxious to preserve it so others could see. And when the cooks got to work and opened it up, I myself inspected its internal organs and I saw spines on both sides of the chest cavity which met and turned their points towards one another; they were triangular like the three-sided law-tablets, and imbedded in them was a liver of considerable length, and below that was a gall-bladder, with a long tube as in skin-bags. You would have said on seeing the gall-bladder that it was a damp bean-pod. So both gall-bladder and liver were extracted, and the latter swelled up to equal the liver of the largest fish, whereas the gall-bladder, which happened somehow to have been placed on a stone, caused the stone to melt away and then it disappeared from sight. [Thompson [1947] 43]

(Scholfield [1959] 119, 223, 225)

10.8 Strabo of Amaseia (–63 to ca. 21) believed in the authority of Homer and the moral and political utility of history and geography (1.1.14); he studied under Xenarchos (see Chapter 3.13), and wished his work to promote mutual Greek and Roman understanding (1.1.16), as well as peace of mind through acceptance of the natural order governed by divine providence. He eschewed marvels and myths, useful only to impress children and the ignorant (1.2.8).

Geography

15.1.21 [amazing trees in India]

[the amazing banyan tree, *Ficus benghalensis*; compare Theophrastos, *Plant Researches* 1.7.3, 4.4.4; *Plant Etiology* 2.10.2]
In truth, India produces numerous astounding ["paradoxical"] trees, among which is the one whose branches bend downwards and whose leaves are no smaller than a shield. Onêsikritos [of Astupalaia, writing –305 ± 10], who even in

rather superfluous detail describes the country of Mousikanos (which he says is the most southerly part of India), relates that it has some huge trees whose branches have first grown to the height of twelve cubits [about 5 m], and then their remaining growth is downwards, as though bent down, till they have touched the earth; and then distributed along the ground they have taken root like rooting branches, and then sprouting have formed trunks; and that the branches of these trunks again, likewise bent down in their growth, formed another rooting branch, and then another, and so on successively, so that from only one tree there formed a vast sunshade, like a tent with many poles. He says also of the size of the trees that their trunks could hardly be embraced by five men [i.e., about 3 m diameter]. Aristoboulos also, where he mentions the Akesinê and its confluence with the Huarôtis [the Chenab and Ravi tributaries of the Indus River], speaks of the trees that have their branches bent downwards and of such size that fifty cavalry (according to Onêsikritos, four hundred) can pass the noon in shade under one tree. (Aristoboulos mentions also another tree, not large, with pods, like the bean, ten *daktuls* [*ca.* 18 cm] in length, full of honey, and says that those who eat it cannot easily be saved from death.) But the accounts of all writers of the size of these trees have been surpassed by those who say that there has been seen beyond the Huarôtis one which casts a shade at noon of five stades [about 1 km; a note about cotton follows: see Theophrastos, *Plant Researches*, above].

(Jones [1930] 33, 35)

10.9 Alexander of Mundos (around 15 ± 25) wrote a work on animals (book 2 covered birds), of which a few extracts survive in later writers; see also Ailian 3.23 (the filial piety of storks), 4.33 (the tricky chameleon), Athenaios 9(388a, 391b, 391f) (the francolin, owls, domestic and wild sparrows), and Plutarch, *Marius* 17.3 (semi-tame vultures). He also wrote on dream-interpretation (Artemidoros 1.67, 2.9, 2.66).

Animals

Book 2 (fr. in Athenaios, Deipnosophists 9[392c]) [quail]

The female quail is slender-necked compared to the male, and lacks the black region under the chin. When dissected, it is seen not to have a large crop, but it has a large heart with three lobes. It also has the liver and the gall-bladder attached to the intestines, a spleen small and hard to find, and testicles [ovaries: compare Herophilos in Section 11.3] under the liver, like roosters. [Thompson [1936] 215–219; Pollard [1977] 61–62]

Book ? (fr. in Athenaios, Deipnosophists 5[221b–d])

[the "gorgon": possibly the gnu; compare Ailian 7.5]

The gorgon is the animal which the Numidians in Libya, where it occurs, call

271

"down-looker" [*katoblepon*]. As most say, based on its skin, it's like a wild sheep; but some say it's like a calf. They say that it has a breath so strong that it destroys everyone encountering the animal. And it carries a mane hanging from its forehead over the eyes; whenever it manages to shake this aside, despite the mane's weight, and catches sight of anything, it kills the thing it sees, not by its *pneuma* but by the influence which naturally emanates from its eyes, and makes it a corpse. It came to be known like this. Some soldiers in the expedition of Marius against Iugurtha [–106; Sallust, *Iugurthine War* says nothing about gorgons] saw the gorgon, and thinking that it was a wild sheep, since its head was bent low and it moved slowly, they rushed forward to get it, thinking that they could take it with their swords. But the creature, being startled, shook aside the mane which lay over its eyes and immediately made those rushing upon it into corpses. Again and again other people did the same thing and became corpses; and since all who attacked it at close quarters always died, some inquired of the natives about the nature of the animal; and some Numidian cavalry, at the command of Marius, lay in ambush for it at a distance and shot it; they then returned with the animal to the commander.

(Gulick [1930] 275; [1927] 501, 503)

10.10 Philon of Alexandria, a Jewish philosopher and politician (active 30 ± 25); most of his copious output consists of allegorizing commentaries upon Hebrew scripture (Torah and Prophets); his outlook is strongly influenced by Platonism and Stoicism. This work (composed around 50) is constructed on the pattern of Plato's *Phaidros* 227–236, and attacks Philon's nephew Ti. Claudius Alexander, a convert to paganism and defender of women's rights.

Animals

20–21 [intelligence of bees]

[from the argument by "Alexander" that animals are partly rational; compare Aristotle, *Animal Researches* 5.21–22 (553a16–554b21), 8(9).40 (623b3–627b22), and Ailian 1.59]

20 The intelligence of the bee is hardly distinguishable from the contemplative ability of the human mind. During the spring, when every fertile plain and mound is in bloom, swarms of bees fly over orchards, gardens, arbors, and green fields, and hover over the sweet-smelling blossoms, florets, and buds to suck the dripping dew (especially that of thyme and vines called honeysuckle). By the wonderfully-fashioned mechanism of their bodies, they transform dew into natural honey. It happens thus: the bee receives the dew like a seminal substance [Stoic "disseminate rationality"], becomes full, and hurries to pour it where the efflux will not be wasted. When it is ready to pour, it makes cells which are compact, of suitable size, and hard enough. This dual construction of this animal has a dual significance: there is something like bodily substance (wax),

and something like indwelling soul (honey). The wax after being filled with dew is tightly sealed to protect the incomplete work from exposure and to shield it against the attacks of inherently malicious animals. The hive is like a walled palace near a city, thickly walled all around. Since it would be easy to attack an exposed site, the inner sections are made with narrow and intricate passages that are inaccessible and cannot easily be attacked.

21　It is not only important to enclose the hive tightly, but behind the mighty wall the bee takes charge as captain of the guard and keeper of the wall. I think it holds a leading position, for it is very clearly seen waiting at the gates and looking all around like one watching from an observation post. If its adversaries are inactive, it is likewise quiet; but when they attack, it's immediately incited to avenge. If the structure needs reinforcement, it adds buttresses from within, for fear that enemies might break in unexpectedly. When the bees are aroused from their hiding place, they immediately come buzzing, ready to attack. With their stings raised they terrify their enemies because they employ all their defenses when it becomes necessary to avenge.

92 [Philon's "refutation"]

The bee is dutiful when it goes about its work: it gathers flowers and creates a beautiful honeycomb which it fills with honey in a most amazing manner. Yet I repeat, because it should be emphasized, that these are not accomplished through the animals' foresight, but are to be ascribed to Nature who manages all. They do nothing by thought; their work is to attack various created things and to seize from everywhere whatever is found until their proper work is completed.

(Terian [1981] 75–76, 106)

10.11 Dioskourides of Anazarbos (around 50 ± 10) traveled widely collecting material for his pharmacy-book (*Medical Materials*), arranged by affinity of effect, and drawing on Krateuas (–80 ± 20) and Theophrastos. (See also Chapters 9.10 and 11.13.)

Medical Materials

1.8.1–2

In the Ligurian alps, Keltic nard [valerian] grows, known locally as "saliunca"; it also grows in Istria. It's a little shrub (made up into bundles with its roots) and has longish yellowish leaves and a quince-yellow flower. Only the stems and roots have any use or fragrance, so you must sprinkle the bundles a day before; clean off the dirt, and set on a floor covered with damp parchment, and wash off the next day: for the usefulness won't be washed off with the foreign and chaffy stuff because of the "elasticity" of the juice. It's adulterated when a similar herb is plucked with it, which because of its stench they call "goat." The distinction is easy: that herb is stemless, and has shorter whiter leaves, and does not have a bitter root or an aroma like the real plant. [compare Theophrastos, *Plant Researches* 9.7.4; Pliny 12.45–46, 21.43]

1.77.1

The *kedros* [*Juniperus oxycedrus*, cedar] is a tall tree from which one gathers the so-called *kedria* [cedar-pitch]. It has a fruit like the cypress, but much smaller. Another smaller *kedros* also grows, spiny and bearing a fruit like the *arkeuthos* [*Juniperus macrocarpa*, juniper: Dioskourides 1.75], the size of the myrtle and rounded. The best *kedria* is thick and shiny, firm, with a heavy smell, shed in drops, not scattered. [compare Theophrastos, *Plant Researches* 3.12.3; Pliny 13.52–53]

2.106.1–2

The "Egyptian bean," which some call the Pontic bean, mostly grows in Egypt, and is found in Asia and Kilikia on lakes [*Nelumbium speciosum*]. Its leaves are as large as a wide-brimmed hat, and it has a cubit-long stem, a finger thick, and a rose-colored flower twice as large as a poppy, which after it blooms bears a little seedpod very like a wasps'-nest, in which the bean barely overshoots the cover like a bubble. It's called "capsulet" or "boxlet" since it's planted set in muck and thus its seeds get expelled into water. Its root is thicker than a reed's, and is eaten boiled or raw, called *kolokasios*. [Herodotos 2.92; Theophrastos, *Plant Researches* 4.8.7]

2.159.1–2

The pepper tree is said to grow in India, producing from its twigs a fruit as long as string-beans, which is the "long pepper," and having inside a seed like millet, which in the end will be the pepper, the fruit when unfolded in its season produces clusters, bearing shriveled berries, some being like unripe grapes, which are the white pepper, useful especially for eye-salves and antidotes for poisons. It's "long" because of its unlimited suitability for antidotes for poisons, and the black pepper is sharper than the white and tastier and more aromatic through being ripe, and is more useful for seasonings, but the white is unripe and weaker in all those ways. Choose the densest, full, black pepper, not too shriveled, fresh and not branny. (An ill-fed thing is found among black peppers, empty and light, called *bregma*.) [Theophrastos, *Plant Researches* 9.20.1]

3.5.1

"Sweet-root" [licorice (Figure 10.1)], which some call Pontic root, some gentian-root, some Skuthian-root, and some "quench-thirst" root, grows mostly in Kappadokia and Pontos. It's a little shrub with two cubit stalks which are thick

Figure 10.1

with leaves, like mastich [*Pistacia lentiscus*], oily and sticky to touch; the flower is like hyacinth [*Scilla bifolia*], the fruit is the size of the berries of the plane [*Platanus orientalis*] but rougher, with pods like the lentil, small and red; the roots are long, like boxwood or gentian, somewhat astringent, sweet, and juicy like buckthorn. [Theophrastos, *Plant Researches* 9.13.2; Pliny 22.24, 25.82]

3.148

Cannabis: a plant useful in daily life for the plaiting of the most resilient ropes (Figure 10.2). It bears leaves like the ash [*Fraxinus ornus*] but bad-smelling, has long hollow stems, and a round edible fruit which if eaten in quantity quenches sexual desire [or: "fertility"]; pureed when green it's good for analgesic ear-drops. [Herodotos 4.74–75 knows another use; compare Pliny 19.173–174, 20.259]

Figure 10.2

3.149

Wild cannabis [*Althaea cannabina* (Figure 10.3)] bears shoots like those of the elm but darker and smaller; it's a cubit tall; the leaves are like the tame sort, rougher and darker, the flowers reddish like rose campion [*Lychnis coronaria*: Dioskourides 3.100], and the seeds and roots like mallow [*Althaea officinalis*].

Figure 10.3

4.20.1

Sword-plant [*Gladiolus segetum* (Figure 10.4)] which some call blade- or dagger-plant because of the shape of the leaf: it's like iris, smaller and narrower, pointed like a blade and fibrous; it puts forth a cubit-long stem on which are purple

Figure 10.4

flowers arranged in rows, and a round fruit; it also puts forth two roots like small bulbs, one lying atop the other; the lower one is emaciated and the upper bulkier. It grows especially in fields. [Theophrastos, *Plant Researches* 6.8.1, 7.12.3; Pliny 25.138]

 *(Wellmann [1907/1958] 1.12–13, 76, 180, 224–225; 2.8–9, 157, 184–185)

10.12 Plutarch of Chaironeia (*ca*. 50 to *ca*. 120), writing under the Flavian emperors, Trajan, and in the early years of Hadrian, was a Platonist teacher of philosophy; much of his copious output is in the form of moralizing essays or dialogues.

Natural Questions

26(918b–e) [animal instinct]

Why do animals, when suffering some malady, seek out and pursue the things that have helpful properties, with frequent benefit from their use? Thus dogs eat grass in order to vomit up their bile; pigs go after river-crabs, since by eating them they get relief from headache; when a tortoise has eaten the flesh of a viper it proceeds to feed on marjoram; and they say that the bear, when suffering from nausea, gets rid of it by picking up ants with its tongue and swallowing them. But these animals have neither been taught nor tried and experienced these remedies. [compare Aristotle, *Animal Researches* 8(9).6 (612a1–8, 24–33); Philon, *Animals* 39]

 Is it really true that, just as honeycombs excite bees by their scent and attract them from a distance, and carrion has the same effect on vultures, so crabs act on pigs and marjoram on the tortoise, while ants' nests draw a bear to them by odors and effluxes that are conducive and proper to its well-being, their perceptions guiding these animals without any calculation of advantage?

 Or are these appetites induced in the animals by the bodily constitutions [*kraseis*] brought about by their diseases, which give rise in them, through changes in their fluids, to various pungencies, sweetnesses, or certain other unusual and abnormal qualities? There is a clear example of this in the case of pregnant women, who even eat stones and earth. This is also why clever physicians know in advance from the appetites of the sick which cases are hopeless and which may recover. For example, Mnêsitheos [of Athens] records that a patient who in the initial stages of disease of the lung has an appetite for onions recovers, while one who wishes for figs dies, the reason being that their appetites follow the constitution of their bodies and their constitutions follow the disease. It is plausible then that such animals also as are overtaken by diseases that are not completely destructive or fatal acquire just that bodily condition and constitution which leads and guides each of them by way of its appetites to the things that are its salvation.

(Pearson and Sandbach [1965] 205, 207)

Animals' Cleverness

10.3(966e–7a) [spiders' webs: Beavis [1988] 34–37, 40–42 –
and thirsty animals]

There is more than one reason for admiring spiders' webs, the common model for both women's looms and fowlers' nets; for there is the precision of the thread and the weaving, which has no disconnected threads and nothing like a warp, but is wrought with the continuity of a thin membrane and a binding from a stickiness inconspicuously worked in. Then too, there is the dyeing of the color that gives it an airy, misty look, the better to hide it; and most notable of all is the art itself, like a charioteer's or a helmsman's, with which the spinner handles her artifice. When some prey is entangled, she perceives it, and uses her wits, like a skilled handler of nets, to close the trap suddenly and make it tight [Aristotle, *Animal Researches* 8(9).39(622b28–623b2); Philon, *Animals* 18–19; Ailian 1.21, 6.57]. Since this is daily under our eyes and observation, the account is believable. Otherwise it would seem a fiction, as I used to think the tale of Libyan crows which, when thirsty, throw stones into a pot to fill it and raise the water until it is attainable; but later when I saw a dog on board ship, since the sailors were absent, putting pebbles into a half empty jar of oil, I was amazed how it knew that lighter substances are forced upward when the heavier settle to the bottom.

13.1(968f–9a) [reconnoitering foxes]

Currently the Thrakians, whenever they propose crossing a frozen river, use a fox as an indicator of the solidity of the ice. The fox moves ahead slowly and lays her ear to the ice; if she perceives by the sound that the stream is running close underneath, judging that the frozen part is not deep, but is only thin and insecure, she stands still and, if she is permitted, comes back; but if she is reassured by not hearing noise, she crosses over. And let us not declare that this is an irrational precision of perception, but a syllogism based on perception: "What makes noise must move; what moves is not frozen; what is not frozen is liquid; what is liquid gives way." [*Principle of Cold* 12(949d); Ailian 6.24]

16.1(971a–d) [a clever mule: same tale in Ailian 7.42]

For one of the mules that were used to carry salt, on entering a river, accidentally stumbled and, since the salt melted away, it was unburdened when it got up. It recognized and remembered the cause, so that every time it crossed the river, it would deliberately lower itself and immerse the bags, crouching and leaning to both sides. When Thalês [a Greek wise man from Miletos around –650, often credited with any clever discovery] heard of this, he ordered the bags filled with wool and sponges instead of salt, and the mule driven thus laden. So when it did its usual thing and soaked its burden with water, it came to know that its cunning was unprofitable and thereafter was so attentive and cautious in crossing the river that the water never touched the burdens even by accident. [Aristogeiton's horse similarly malingers in Philon, *Animals* 40]

17.1(972b) [smart elephants; also Philon, *Animals* 28 and Ailian 2.11: a literate elephant]

Iuba [the second, king of Mauretania for almost fifty years, –24 to 23, and very learned] says that elephants exhibit social capacity with intelligence. Hunters dig pits for them, covering them with slender twigs and light rubbish; so when any elephant of a number travelling together falls in, the others bring wood and stones and throw them in to fill up the hole so that their comrade can easily get out [same trick in Ailian 8.15]. He also relates that, without any instruction, elephants pray to the gods, purifying themselves in the sea and, as the sun rises, worshipping it by raising their trunks, as if they were hands. [Ailian 7.44; but they revere the moon in Ailian 4.10]

30.2(980b–c) [sponges: compare Aristotle, *Animal Researches* 5.16 (548a10–19); same tale in Ailian 8.16]

The sponge is driven by a little creature not resembling a crab [as in the prior marvel], but like a spider. Now the sponge is no lifeless, insensitive, bloodless thing [*contra* Aristotle]; but it clings to the rocks, as many other animals do, and it has a peculiar movement outward and inward which needs some admonition and supervision. For otherwise it is open and its pores are relaxed because of its sloth and dullness; but when anything edible enters, the guard gives the signal, and it closes up and consumes the prey. Moreover, if a person approaches or touches it, informed by the scratching of the guard, it shudders and so closes itself up by stiffening and contracting that it is not an easy, but a difficult, matter for the hunters to undercut it.

(Cherniss and Helmbold [1957] 365, 367, 377, 389, 391, 395, 397, 447)

10.13 Dionusios of Philadelphia (or "Periegetes," the guide) around 125 ± 15 wrote a poem on birds and fowling, of which this later paraphrase survives (attributed to Oppian, see Section 16). Book 1 treats land-birds, 2 water-birds, and 3 fowling; the poem may have been used by Ailian. (Many other passages are translated by Pollard [1977].)

Birds

1.6 [falcons and hawks]

There are many species of *hierax* and they are swift for hunting, being especially destructive to doves and pigeons [see *Birds* 3.5]. Some snatch smaller birds, and others are very lazy and timid of flight, so that they want to be fed by others and make their strike too late and are only set upon frogs. And others share the hunt with people, having bonds and fearing the prey of birds. They are enemies of larks and swallows, so that one might say they are kin to Tereus [turned into a hoopoe which attacks such birds]. They are weak-sighted above every other race of birds, so learn the treatment of this suffering: it's the milk of wild lettuce which it is usual to cut up as medicine for falcons. [Thompson [1936] 114–118; Pollard [1977] 80–81]

1.32 [phoinix]

I have heard of an Indian bird which exists without seed or sex, called *phoinix* [Herodotos 2.73; Tacitus, *Annals* 6.28: one was seen in 34 CE; Ailian 6.58], and they say it lives a long time and with utter fearlessness, since nothing can harm them, neither arrow nor stone nor spear nor snare of men attacking them. And their death is the origin of their life: for if a *phoinix* growing old ever sees that it is more sluggish in flight, or that the beams of its eyes are dimmer [see Chapter 7], he gathers twigs-and-straw [*karphos*: "nesting material" and also "kindling"] upon a high rock, making a pyre of death as well as a nest [*kalia*: or "shrine"] of life; the *phoinix* sits amid this and the heat of the sun's rays ignites it. The *phoinix* is destroyed, but another young one is immediately created from the ashes and behaves just like its "father," so that this bird is created without father or mother but only by a sunbeam. [end of book 1; Thompson [1936] 306–309; Pollard [1977] 100]

2.1 [introduction: water fowl]

We must also speak about "amphibious" birds, such as enjoy the sea and rest upon crags and make their nests along shores, of which there are many varied species (for the sea supports no fewer birds than the land). Now the nature of the sea- and land-birds is not the same: with land-birds the feathers are set more widely on their bodies, usefully for swift flight, but dense and naturally-unwettable feathers fence in the "amphibious" birds all around so they can swim unsodden. They also have broad passages behind their mouths through which they easily send a captured fish down to their bellies. Most also have "solid" [webbed] feet which they use like oars to swim on the waters.

2.7 [pelican: Aristotle, *Animal Researches* 8(9).10 (614b27–30); Ailian 3.23 describes its piety: also in medieval Christian bestiaries, such as the "*Physiologus*," which drew heavily on Greek theories of animal behavior]

No smaller food-lust grips the pelicans, who have the longest necks, but they do not dive completely submerged, like the shearwaters [the previous bird], since as they dive they continuously submerge their necks (which are a fathom long) but show their backs above the sea, and consume every fish they encounter, receiving them into their huge chasm of a mouth. There's a pocket fitted to the front of their chests into which for awhile they cram all their booty, not even refusing scallops or mussels [Aristotle, *Animal Researches* 4.1 (525a22–29): floating testaceans such as *Ianthina*; and 5.15 (547b11–18): coastal testaceans]; for awhile they swallow whatever they find with these aims, then vomiting up again everything, now dead, they eat they flesh and discard the shellfish – these are closed while alive but after death they open and separate. [Thompson [1936] 231–233; Pollard [1977] 75]

*(Garyza [1955/7] 205, 215–216, 222)

10.14 Arrian of Nikomedia (*ca.* 86 to *ca.* 160) studied philosophy and fancied himself a latter-day Xenophon of Athens; in his mid-twenties he made friends with the future Emperor Hadrian, who appointed him to various offices. See also Chapter 5.18.

Hunting

4–5 [good dogs, and the best of the dogs]

4.1 I too will say what criteria one should use to judge which hounds are quick and well bred, and what a person should pay attention to if he wished to distinguish those that are poor and slow. 2. First they must stand long from head to tail; in considering every type of dog, you could not find any single mark of speed and breeding as accurate as the length of the body, and conversely shortness of body indicates slowness and poor quality. So much so that I have often seen dogs with many other defects, but, because they happened to be long, they were quick and spirited. 3. Moreover, bigger dogs, if in other respects similar, are better than smaller ones by reason of their size itself. The poor examples of big dogs are those that have loose and uneven limbs; so that in this state they would be worse than the little ones if they shared other defects equally. 4. They should have light and well-knit heads; if they are hook-nosed or blunt-nosed, this will not make a great difference; nor even should it greatly matter whether they have the parts below their foreheads sinewy; the only bad ones are those with heavy heads, and those with thick noses not coming to a point but ending flatly. 5. Let the eyes be large, set high, clear and splendid, astonishing the viewer. Best of all are those that are fiery and very bright, like those of leopards, lions or lynxes; second to these are those that are black, if they are wide open and fierce looking: third are those that are gray; for indeed the gray ones also are not bad, nor an indication of poor quality hounds, provided that they are clear and fierce looking.

5.1 For I myself reared a hound with the grayest of gray eyes, and she was fast and a hard worker and spirited and agile, so that when she was young she once dealt with four hares in a day. 2. And apart from that she is most gentle (I still had her when I was writing this) and most fond of humans, and never previously did any other dog long to be with me and my fellow-huntsman Megillos as she does. For since she was retired from the chase, she never leaves us, or at least one of us. 3. If I am indoors she stays with me, and accompanies me if I go out anywhere; she escorts me to the gymnasium, and sits by while I am exercising, and goes in front as I return, frequently turning round as if to check that I have not left the road somewhere; when she sees I am there she smiles and goes on again in front. 4. But if I go off to some public business, she stays with my friend, and behaves in the same way to him. If one of us is ill, she does not leave him. If she sees us even after a short period of time, she jumps up in the air gently, as if welcoming him, and she gives a bark with the welcome, showing her affection. When she is with one of us at dinner she touches him with her paws alternately, reminding him that she too should be given some of the food. And indeed she makes many different noises, more than any other dog that I think I have seen; and she shows audibly what she wants. 5. And because when she was being trained as a puppy she was punished with a whip, if anyone even to this day should mention a whip, she goes up to the one who has said it and crouches down like one beseeching, and fits her mouth to his mouth as if she is kissing, and jumps up and hangs from his neck, and does not let him go until the angry one gives up the threat. 6. And so I think that I should not hesitate to write down the name of this dog, for it to survive her even in the future, viz. that Xenophon the Athenian had a dog called Hormé, very fast and very clever and quite out of this world.

7 [dogs' character]

7.1 Also the temperament of the hounds will provide the sensible observer with no less evidence both ways [good and bad character]. For a start, those that regard everyone in a hostile manner are not good; but if you find some that are unfriendly to those they don't know, but friendly to the one who feeds them, this is good rather than bad. 2. I knew a hound which was gloomy at home and took no pleasure in anyone who approached her, but when she was taken out to the hunt she was overjoyed, and smiling and fawning on anyone who came near showed clearly that staying at home was boring for her; this also is a good sign. 3. Best, though, are those which are most fond of humans, and do not feel that the sight of any man is something alien. But those which are afraid of humans and frightened by a sudden noise and get into a state of confusion and are easily disturbed by a host of things – these features show irrational and foolish animals – just like humans if they are cowardly and foolish, hounds of that kind could never be of high quality. ... 7. The best of all have a frowning appearance, and appear proud, and their step is light and quick and delicate; and they twist their bodies both ways, and stretch up their necks, as horses do when they are proud.

16 [hares worth hunting]

16.1 The best hares are those that have their seats in clear and open ground; through self-confidence they do not hide themselves, but, as it seems to me, they challenge the hounds. 2. And these same hares, when they are pursued, do not run to wooded valleys or groves of trees, even if they chance to be very close at hand so that they could easily remove them from the danger, but they press on to the plains, competing with the hounds. If the following hounds are slow, they run as far as the pursuit lasts; if they are fast, as far as they are able. 3. And often it happens that when they have turned away onto the plains, if they sense a good hound keeping up with them to the extent of being overshadowed by it, they frequently wrong-foot it by their twists and turns, and themselves return to the wooded valleys or if they know a hiding-place somewhere. 4. One must treat this as proof that the hound beat the hare; true huntsmen do not take out their hounds to catch the creature, but for a trial of speed and a race, and they are satisfied if the hare manages to find something that will rescue her. 5. Sometimes if she takes refuge in a small clump of thistles, if the huntsmen see this and that she is frightened and exhausted, they call off their hounds, particularly if there has been a good contest; so that I myself, when accompanying the chase on horseback and arriving as the hare has been caught, have often snatched her away alive, and having got hold of her and put the hound on a leash, I have let the hare run to safety; and if I arrived too late to save her, I have hit myself on the head, because the hounds had killed a worthy adversary. [Plutarch, *Animals' Cleverness* 15.3(971a): the hounds kill only for victory and not for mere food]

(Phillips and Willcock [1999] 95, 97, 99, 101, 109, 111)

10.15 Ailian (Claudius Ailianus), a Roman freedman of Praeneste (lived from 168 ± 3 to 233 ± 3), never left Italy but learned Greek and wrote many works to show the operation of divine providence and justice; unspecified citations and quotations herein refer to this work (his *Historical Miscellany* is cited in this chapter's prolegomenon).

Animals' Characters

2.16 [Skuthian elk?]

Whenever a flush or a pallor arises upon our smooth and hairless skin, there's no wonder: but the animal *tarandos* transforms itself hair and all, and accomplishes a myriad color-changes so as to bewilder the eye. It is Skuthian, and its hide and size resemble a bull's. The Skuthians consider this creature's hide a good ward against spears, and cover their shields with it. [pseudo-Aristotle, *Marvels* 30 (832b8–17)]

2.17 [remora]

There is a fish whose province is the open sea, black in appearance, as long as an eel of moderate size, and deriving its name from what it does: with evil purpose it meets a vessel running before the wind, and biting into the tip of the prow as if vigorously curbing with bit and tightened rein a wild and violent horse, it checks the vessel's onrush and holds it fast. In vain do the sails belly in the middle, to no purpose do the winds blow, and depression comes upon the passengers. But the sailors understand and realize what the ship suffers. And therefrom the fish has acquired its name, for experts call it the "ship-holder" [*Echeneis remora*: Thompson [1947] 67–70 (Figure 10.5)].

Figure 10.5

2.19 [bear]

The Bear knows not how to produce a cub, nor would anyone allow, on seeing its offspring after labor, that it was a live birth. Yet the Bear has been in labor, though the lump of flesh is unmarked, unformed, and shapeless. But the mother loves it and recognizes her child, keeps it warm beneath her thighs, smoothes it with her tongue, and fashions it into limbs, and little by little brings it into shape; and when you see it you will say that this is a Bear's cub. [compare 6.3: bears' hibernation]

3.16 [partridge]

When partridges are about to lay they make themselves what is called a "threshing-floor" out of nesting-material [*karphos*]. It is plaited, hollow and well-suited for sitting in. They pour in dust and construct as it were a soft bed; they enter and after screening themselves over with nesting-material to hide from birds of prey and human hunters, they lay their eggs in complete tranquillity. Next, they do not entrust their eggs to the same place but to another, emigrating as it were, because they are afraid that they may sometime be detected. And when they hatch their hatchlings they heat the tender beings, and warm them with their wings, enveloping them in their feathers, like swaddling-clothes. They do not however wash them, but render them brighter by putting dust on them.

If a partridge sees someone approaching intending harm to itself and its young, it thereupon rolls about in front of the hunter's feet and fills him with the hope of seizing it as it whirls around. And he bends down to catch his prey, but it eludes him. Meantime the young run off and get away. So when the partridge is aware of this, it takes courage and releases the fowler from his fruitless business by flying off, leaving the man gaping. Then when the mother-bird is secure and well placed, she calls her young, and recognizing her voice they flutter towards her. [Aristotle, *Animal Researches* 8(9).8 (613b15–614a35); Philon, *Animals* 35; Plutarch, *Animals' Cleverness* 16.2(971d); Thompson [1936] 234–238; Pollard [1977] 50–51]

3.41 [rhinoceros? – and possibly the origin of the unicorn fable]

India produces one-horned horses, they say, and the same country fosters one-horned asses, and from these horns they make drinking-vessels. And if someone puts a deadly poison in them and anyone drinks, the plot will not harm him. For it seems that the horn both of the horse and of the ass is an antidote to the poison. [compare Agatharchides above, Section 3; rhinoceros horn is an aphrodisiac in *Koiranis* 2.34 (around 100 ± 50)]

(Scholfield [1958a] 115, 117, 119, 173, 175, 201)

9.24 [angler fish]

There is a species of [fish called] frog which bears the name of "angler," and is so called from what it does. It possesses baits above its eyes which one might describe as elongated eyelashes, and at the end of each one is attached a small spherule. It is aware that nature has equipped it and even stimulated it to attract other fish by these lures. Accordingly it hides itself in muddier spots and those filled with more slime, and quietly extends those hairs. Now the tiniest fishes swim up to these eyelashes, imagining that the round, swinging objects at the end are bait; meanwhile the Angler lies in ambush, never stirring, and when the little fishes are near to him, he withdraws the hairs towards himself (they are drawn in by some secret and invisible means), and the little fishes, made neighbors by gluttony, provide a meal for this frog. [*Lophius piscatorius*: Aristotle, *Animal Researches* 8(9).37 (620b10–19); Thompson [1947] 28 (Figure 10.6)]

Figure 10.6

(Scholfield [1958b] 245)

16.2 [marvelous birds of India]

I learn that in India there are parrots, and I have also mentioned them earlier on [13.18], but this seems a most fitting place to relate what I did not relate on the former occasion. I hear there are three kinds, and all learn like children and become talkative in the same way and speak human speech. In the forests, however they utter the notes of birds, and do not produce intelligible and distinct speech, but are unlearned and cannot talk as yet. [Philon, *Animals* 13; Dionusios 1.19; Thompson [1936] 335–338]

There are also peafowl in India, larger than anywhere else [Aristotle, *Animal Researches* 6.9 (564a25–564b13); Ailian 5.21; Thompson [1936] 277–281; Pollard [1977] 91–93], and doves with green plumage: anyone seeing them for the first time and not possessing a knowledge of birds, would say that they were parrots not doves, but they have beaks and legs the same color as those of partridges in Greece [perhaps *Crocopus chlorogaster*]. And the roosters there are of immense size, and their combs are not scarlet like those of our country, but of varied hue like flower-garlands. And their tail-feathers are not arched or curved in a circle but flat, and they trail them, just as peacocks do when not raising them aloft. And the wings of Indian roosters are golden with the dark gleam of an emerald [*smaragdos*: see Chapter 9].

(Scholfield [1959] 261, 263)

10.16 Oppian of Anazarbos around 175 ± 5 versified the struggle between human and sea-beast, two orders of being which he assimilated, and dedicated his work to emperor M. Aurelius Antoninus (1.3).

Fishing

5.62–108 ["guide fish"]

All of the overgrown beasts of the Ocean (aside from the fish called the
Dog) are proceeding on *wearisome* weighty-limbed ways of unease, since
They are perceptive in vision of nothing that's far off at sea, and they
[65] Never would travel on limbs of a giant across the *wide* Ocean,
Massively rolling along, very tardily making their way. So
Therefore with all of them swims a companion, a *travelling* fish[a] that is
Dusky to eye, and is long in the body and thin in the tail; it
Swims out in front, very prominent, showing their way through the sea,
    and
[70] Pointing it out; for this reason they call him "Hegétor," the Guide;
    he is
Wondrously favored by Whales as a *friendly* companion, and also as
Leader and guard, since he easily bears the cetacean wherever he
Wishes to go; for alone of all fishes he follows that guide, while
Trusting his thought to a *comrade* that's faithful; who's wheeling about
    in the
[75] Closest proximity, *stretching* his tail-fin extended to eyes of the

Whale, thus explaining each point to the beast, whether presence of prey
   to be
Seized or else pain that is threatening close *and is coming upon him*, or
Sea that is shelving to shallows, a depth it were better for whale to
Flee and avoid; as if speaking with voice of a human, the tail is
[80] Rightly declaring it all, and this weight on the waters obeys. And
So you can see that he serves as the champion, plus as the eyes and the
Ears of the beast; and the whale is enabled to see and to hear, the
Reins of his life for safe-keeping entrusted to *smallest of fishes*; as
Sons are embracing their aged begetters with kindness and love, they are
[85] Thoughtfully tending their years to repay them the price of their
   nurture, and
Zealously aiding and cherishing fathers, now weakened *and strengthless* in
Eyes and in bodily members, by holding out arms for support in the
Street as they walk, and attending them closely in all that they do: for
Sons will provide to their aging old fathers a youthful new strength.[b] And
[90] Likewise the guide-fish embraces with fondness the fang-bearing fish
   of the
Deep, as he steers it along like a rudder that's guiding a ship. Well,
Surely they shared an ancestral relation of blood at their birth, or
Willing and free did the *little* one make the cetacean companion.
Clearly not ever does valor or beauty confer any boons as does
[95] Elderly wisdom; and strength without sense can be nothing but
   vanity.
Look you: a man who is little but offers good counsel can sink or bring
Safety to *great* men *and* mighty, for even unconquered cetacean, who's
Boundless of body, procures as a friend just a slip of a fish. And
Therefore the first task *to finish* is capture of Guide who is scouting for
[100] Whale by entrapping that fish with some bait and the power
   of fishhook;
Never while guide-fish is living can anyone master and conquer the
Monster; but after removing the fish, his destruction comes swiftly. Then
Knowing no longer securely the pathways empurpled of *deep* sea, nor
Able to *run or* escape from the pain that is coming upon him,
[105] Just like a mercantile *sea-going* vessel whose steersman has perished,
   he's
Wandering wildly, defenseless and helpless, whichever direction the
"Grey-water"[c] drives him, and only on shadowed unguessable paths is he
[110] Carried, a widow bereft of his succouring charioteer.

[a] *Naucrates ductor*: Plutarch, *Animals' Cleverness* 31.2(980f–981b); Ailian 2.13;
Thompson [1947] 208 (Figure 10.7); [b] Oppian was exiled with his father whom he
tended (compare also Pindar, *Olympians* 8.70–71); [c] to Greeks the sea was *glaukos*,
meaning gray or blue]

Figure 10.7

*(Mair [1928] 464, 466, 467)

10.17 Anonymous This writer from Apamea (2.125-7, 156-7) whose name is no longer known to us (his work is attributed to Oppian) composed this poem dedicated to the emperor Caracalla around 215 ± 5. In this extract he describes subspecies of lions around the Mediterranean, now all extinct, and rare in his day.

Hunting

3.20–62 [lions: compare Aristotle, *Animal Researches* 8(9).44(629b5–630a8)]

[20] Each of the tribes of the lions possesses a form that is unique: so
Those who lie down near the waters of Tigris, a broad-flowing river with
Far-sounding stream, are the ones that are bred in Armenia, mother of
Archers[a]; indwellers of land of the Parthians, fertile for grazing, do
Come into being with bravery less *than imposing*, and yellow's their
[25] Hair, while their neck is expanded *for bearing* a head that's enlargèd, and
Brightly are shining their eyes under eyebrows most bushy and deep, those
Brows overhanging the nose *they encompass*; and out of the neck and the
Cheeks there are growing symmetrical tufts of luxuriant hair. But the
Bountiful plowland of Great Eremboi[b] (they're a people we mortals will
[30] Designate "fortunate") nourishes *different* lions, possessing a
Neck that is shaggy with hair that is flowing, and chest of a like guise; a
Twinkling of fire from their eyes *is emitted like* flashes of lightning, and
Second to none is the masculine beauty they bring into being: this
Family of lions the infinite Earth has begotten as scanty.
[35] Thronging *in masses* the lions of Libya, whose soil is so loamy, do
Roar in that waterless country, *revealing* their might by the sound, but
Hairy they're not any longer and thin is the sheen on their hides, and it's
Fearful to gaze on their visage or throat, and on all of their members the
Color that's found is a blackening tint which is mingled with deep-blue;
[40] Strength without limit is found in the limbs of these Libyan lions,
Lording it over all lions *no matter how* lordly those others.
Once from the *land of* Aithíop there came to the Libyan country a
Marvel magnificent: Black-colored lion, and comely his mane, with
Broadness of head, who was hirsute of foot and resplendent of eye, and
[45] Showing some reddening only adjacent to yellowy mouth-parts.

288

Once in my life I have held in my vision the murderous beast (*which is
Better* than reading), transported for show in imperial spectacle.
Feedings aren't needed on every day by the tribe of the lions, but
One day's devoted to meals, and the next one is given to labors[c];
[50] Nor is the lion found sleeping surrounded by rocky recesses: he
Sleeps out in plain sight, with bold-heated soul that is never at rest, re-
clining where powerful Night at her evening may happen to find him.
This I have heard from the vigorous keepers of lions *in cages*, that
Under the paw of his forefoot (the right one), the reddish-brown lion
[55] Carries the swiftest benumbing to bind up the knees of his quarry. The
Lioness loosens her midriff for bearing her progeny five times:
Empty in truth the oracular tale of her singleton birthing[d].
Five in her litter the first time, then fourfold the kits from her following
Labored delivery, next in her sequence of birthings *it happens that*
[60] Three are the kittens that leap out, and twins are the fruit of the fourth
   of her
Labors for young, and then last from her womb does the mother of noblest
Animal offspring deliver the glorious king of the lions.

[[a] Armenians and Parthians were renowned for archery; [b] the Eremboí in *Odyssey* 4.84
became identified with Arabians (Strabo 1.2.34); [c] lions' feeding habits from Aristotle,
Animal Researches 7(8).5 (594b18–28); [d] lions' birthings from Aristotle, *Animal
Researches* 6.31 (579b8–11) and *Generation of Animals* 3.1 (750a21–b1), compare also
Ailian 4.34]

*(Mair [1928] 114, 116)

11

MEDICINE

Our bodies function in manifold ways whose purpose and manner at first elude our grasp or even escape our notice: we breathe, our chest thuds, we need air and drink and food and sex, and excrete substances with peculiar powers. Excursions from these norms excite our wonder (and a desire to restore our balance): blood flows from wounds, diseases weaken or kill us, and some foodstuffs transform our body for a time, or are even fatal. Every human culture consumes certain foods not for thirst or hunger but to achieve certain effects, especially the relief of pain or the alleviation of malfunction. The unseen origins of diseases are countered in attempts to restore balance (cooling fevers, warming chills, evacuating foreign or improper material, etc.). Displaced or broken bones can be returned to their place, and wounds can be closed so as to speed healing; many cultures successfully practiced trepanation (the removal of sections of the skull when damaged or to relieve perceived pressure).

The early medical tradition of Mesopotamia records that to treat affections whose focus was seen as internal, drugs were mashed into beer as a potion, while visibly external ills were plastered with soapy lye washes, spiced plant resins or oils, mineral solutions, mud or even dung; those recipe books arrange their prescriptions by affected part from the head down. Doctors also engaged in incantations, prognoses, and legally-regulated surgery; many ills were attributed to unseen spirits, some were believed to be infectious. Egyptian medical texts record an equally-wide range of recipes (often employing honey, malachite, myrrh, opium, or fresh meat) and contain the earliest known contraceptive prescription. Doctors were credited with specialties (eyes, teeth, anus, etc.), and practiced a form of triage based on prognosis: give up, or contend with the ill, or promise a cure. Bodily ills were seen as caused by "sufferingness" (wkhdw), a mobile morbid principle or pain matter, which moved through the body along 22 tubes (joining heart, anus, and other vitals), normally full of life-giving fluids (blood, water, air, etc.).

The earliest Greek literature shows that wounds and ailments were treated with simples (not compounds) such as fig-tree sap to clot (Iliad 5.900–904), or "bitter root" to ease pain (Iliad 11.842–848), or Egyptian nepenthes to drive away sorrow (Odyssey 4.219–234, probably opium, smoked in Minoan Crete).

Diseases seemed often the work of angry gods (Apollo in the *Iliad*, or Zeus in Hesiod, *Works and Days* 100–104), yet surgery and internal medicine were also divinely granted (Asklepios' two sons, Machaon and Podaleirios). Hesiod offers an anthropocentric explanation of swollen feet in winter (starvation: *Works and Days* 496–497); while Pindar describes medicine as *discovered* by Asklepios, thereupon rewarded with divine honors (*Pythians* 3.47–53). Malaria, known as a recurring fever (*kausos*), and tuberculosis, known as wasting (*phthisis*), were frequent, almost endemic (Grmek [1989] 177–197, 245–283).

   Just before the Persian invasion of Greece, Demokedes of Kroton is recorded as successfully practicing medicine around the Aegean – he was in turn hired by the cities of Aigina and Athens, and then by the tyrant of Samos – and in Persia (Herodotos 3.125, 129–138, who elsewhere gives a Greek merchant's view of contemporary Egyptian and Babylonian medicine). Around the same time, Alkmaion of Kroton (perhaps –500 ± 20) taught that health was a balance of bodily powers and illness a tyranny caused by environmental factors or misuse of the body (as by exhaustion). He may have been the first to postulate three or more juices – i.e., "humours" – blood, bile, phlegm, and perhaps *melancholê* (black bile) and others. He also made extensive contributions to the physiology of sensation (see Chapter 12). Empedokles of Akragas (–450 ± 20) claimed to be able to cure all diseases, including old age (perhaps a reference to his belief in reincarnation). He also speculated that both males and females contributed seed to form an embryo which grew male on the hotter right-hand side of the womb, its growth being by accretion of like to like. Anaxagoras of Klazomenai (also –455 ± 20) advocated a theory of matter in which every sensible quality derived from infinitesimal "seeds" of each kind of stuff; this he employed to explain how out of bread our bodies produce blood and bone. Diogenes of Apollonia (–430 ± 10) attributed life to *pneuma* (approximately "living air" or "spirit"); he described the human body as irrigated by twin veins running from head through neck to spleen or liver then legs.

   The medical library attributed to Hippokrates provides extensive data on Greek medicine between –440 and –350, which even to summarize adequately would need a book of its own. Certain practices and concepts stand out. The body is composed of the "humours," a small number of fundamental liquids: lists vary but four became standard: hot wet blood, hot dry bile, cold wet phlegm, and cold dry *melancholê* (black bile). Health is a balance of those, maintained by regimen (diet plus activity), and when lost restored primarily by adjustment of regimen (a theory often attributed to Herodikos of Selumbria, –440 ± 20). Beyond regimen, doctors resorted to drugs, then surgery, then cautery (to dry and heat body parts too cold or wet or soft). Drugs, usually simple, restored the humoural balance more rapidly and precisely than regimen, particularly when toxic *hellebore* was used ("white" *hellebore* was *Veratrum album*, "black" was the Christmas rose, *Helleborus niger*). Surgery consisted primarily of extraction of arrows, reduction of fractures and dislocations, trepanning, and excision of stones (mostly bladder stones). A fifth of the about five dozen works may be mentioned. *Airs, Waters,*

Places explains health and disease on environmental principles. *Aphorisms* (beginning "life is short, *technê* long, opportunity brief, experiment slippery, judgment hard") summarizes medical observation and practice. *Arthroi* (i.e., *Joints* or *Dislocations*) and its kin *Fractures* and *Mochlikon* form a set of practical guides by an experienced and opinionated craftsman describing reductions and settings of bones, some by simple machines. The seven books of *Epidemics* are case notes from various times and places recording the progression of mostly fatal ailments, emphasizing the *krisis* (decisive paroxysm of the disease). The three books of *Gunaikia* ("women's matters") treat female bodies as essentially differing (moister, e.g.) and reproduction as the sprouting within them of male seed; many drug remedies are prescribed (including contraceptives: 1.78). *Nature of the Child* and its companions *Seed* and the mistitled *Diseases 4* describe reproduction as due to both male and female seed, and discuss heredity and sex differentiation. Polubos may be the author of (part of) *Nature of People* which explains the body in terms of the standard four humours, and health as their balance. The *Oath* is probably the work of a Pythagorean around –350: it could not have been widely followed since it prohibits surgery (even for bladder stones). *Prognosis* asserts the leading role of predicting the course of a disease, both to acquire a reputation and to decide correctly how and which diseases to treat; many examples are given of observation-based forecasting. Several works on regimen form part of the library: *Regimen in Health* describes diet as the best preventative, *Regimen in Acute Diseases* suggests that sick people recover best by eating barley gruel, while *Regimen* (in four books) advocates restoring health by balancing input (food) and output (exercise), diagnoses being based on the symptoms of a human body composed of fire and water. The *Sacred Disease* explains that epilepsy is not caused by any divinity or *daimôn* but by entirely natural causes, i.e., a curable excess of phlegm on the brain. *Wounds in the Head* is another practical work by a confident and experienced craftsman, advocating that depressed skull fractures will often heal naturally, but an injured skull without a hole should be trepanned.

In the same period, Thoukudides ("Thucydides") of Athens (–400 ± 5) describes an unidentifiable plague in medically precise terms, including that it was infectious (2.47–54). Plato of Athens (writing –365 ± 20) describes the doctor Eruximachos as a respected member of the Athenian elite at this time (*Banquet*) – not all doctors were itinerant craftsmen. In Plato's own summary and probable account of medicine in the *Timaios*, he is primarily concerned with "psychology" (see Chapter 12) but notes that the heart is the font of blood, cooled and padded by the lung (a single organ), and the soft liver is easily affected, so the spleen serves as its cleanser (70–72). The compositions of marrow including brain, and of bone, are described in terms of his four elements, while flesh is padding, sinews are bonds, and the skin a bag (73–76). His model of the veins resembles Diogenes' (77–78), and that of breathing somewhat recalls Empedokles' (78–79). Old age and death arise from the weakening, imbalance, and dissolution of the elements composing the body (81–86).

Reproduction involves seed from the spinal marrow sown on the womb and there nurtured (91).

Mnêsitheos of Athens (around –350 ± 20) wrote *Foodstuffs*, carefully distinguishing them and their putative effects (e.g., in Athenaios, *Deipnosophists* 2[54bd], 3[92b, 121d], 8[357ab], etc.), the *Structure of the Body*, and on *Heavy Drinking* (in Athenaios, *Deipnosophists* 11[483f]). Diokles of Karustos (perhaps –340 ± 20) is credited with the first anatomical handbook, an early (perhaps the earliest) Greek herbal, as well as many other works on all aspects of medicine. He invented a kind of ladle for extracting arrowheads without causing further harm.

Aristotle of Stageira (writing –335 ± 10) is the most likely author of the tenth book of the *Animal Researches*, which describes the operation of the womb (10.1 [633b10–4b26]) and its diseases (10.4 [636a26–b11]), especially the chorionic mole (10.7 [638a11–b38]); infertility of couples is discussed (10.5 [636b12–7b7]). Aristotle also wrote an illustrated *Anatomy*, to which he frequently refers, but it has perished.

11.1 Theophrastos of Eresos (lived –370 to –286) succeeded Aristotle as the head of the Lukeion "school," and wrote widely, amassing data questioning Aristotle's system (Keyser [1997c]). The works here excerpted were part of a set of works on natural philosophy (see also Chapters 8–10 and 12).

Plant Etiology

6.13.1–4 [medicinal plants]

1 We must however study the question in the case of medicinal plants and in general those that work by their potencies: here too the potencies of all the parts are neither the same nor equal, beginning with the roots, nor yet again are they the same or equal in all the upper parts (as leaves, twigs, and fruit). Now the difference in degree has a certain reasonableness; one is more likely to be surprised at another difference: some parts quite lack the power to do what other parts do (thus the seeds and stalks lack the potency of the roots, or again the fruit and roots the potency of the leaves).

2 Here too the reasons must be sought in the preceding discussion: each part has its special tempering and nature, and the tempering and nature make the potency differ as well, so that some parts bring about colliquescence and separation, others do not; and some do it to a greater, some to a lesser degree; and so with heating, concocting, chilling, drying and the rest.

That the fruit differs most of all from the rest in this way is not unreasonable, since the whole nature of root and fruit and the other parts is also dissimilar, some parts being quite unconcocted, others concocted; all moreover are composed of different constituents, a fact that makes them differ in flavor and in potency. For we observe this distinction in both the wild and cultivated kind:

plants with bitter roots that are full of fig-like sap have fruit that is sweet, and this suggests that the fruit is concocted from something unconcocted.

3 We must suppose that the same variation occurs in the potencies of medicinal plants as well, so that it is not unreasonable that the root is stronger in some directions, another part in others. In fact in plants of the same kind one root differs greatly from another, and so with the seeds and other parts, owing to the weather of the different countries when the plant grows: so cereals and other seed-crops differ in the matter of indigestibility and digestibility by reason of the differences in their food.

4 For this reason the excellence of a drug varies with the region where it grows, though one region may be at no great distance from the other: so the hellebore of Mt. Oitê is better than that of Parnassos, which is considered too strong to be suitable for use. So too with the grains strength arises from a similar cause: their heaviness comes from the harshness of the air and the abundance of their food, which lead to a large earthy component, as is true of the grains of Boiotia; and with medicinal plants similar causes apply.

5 Different countries are suited to the production of different powers, as with the seed-crops; so some countries do not even bring medicinal powers to full concoction. Thus both black hellebore and other medicinal roots are found in many places, but are of a dull and ineffectual sort. Hence it would appear that drugs require a type of air that is not only cold but in movement, and again the right amount of food and no more; at all events we see that most drugs are produced on mountains, and especially on the highest and greatest.

(Einarson and Link [1990b] 365, 367, 369, 371)

Plant Researches

9.8.2–8 [collecting botanicals]

[2: extracting juices; compare Disokourides, *praef*. 9, below Section 13]
[3–4: root-cutting]

5 Further we may add statements made by druggists and herb-diggers, which may either be apropos or exaggerated. Thus they enjoin that in cutting some roots one should stand to windward (for instance in cutting *thapsia* [*Thapsia garganica*] among others), first being anointed with oil: since one's body swells if one stands the other way. Also that the fruit of the wild rose must be gathered standing to windward, since otherwise the eyes are endangered. Also that some roots should be gathered at night, others by day, and some before the sun strikes on them, for instance those of the plant called honeysuckle.

6 These and similar remarks may well seem to be not off the point, for the properties of these plants are hurtful; they are said to grip like fire and burn; for hellebore also soon makes the head heavy, and one cannot dig it up for long; wherefore they first eat garlic and drink unmixed wine. On the other hand the following ideas may be considered far-fetched and irrelevant: for instance they say that the peony (which some call *glukusidê*), should be dug up at night, for if one does it by day and is observed by a woodpecker while gathering the fruit, the eyes are endangered; and if while cutting the root, the anus prolapses.

[7–8: further customs]

9.16.4–5 [wolf's bane: see Nikandros, *Alexipharmaka* 12–73]

4 Wolf's-bane [*Aconitum anthora*] grows in Crete and in Zakunthos, but is most abundant and best at Herakleia in Pontos. It has a leaf like chicory, a root in shape and color like a prawn, and in this root resides its deadly power [*dunamis*], but they say that the leaf and the fruit do nothing. The fruit is that of an herb, not that of a tree. It's a low-growing herb and shows no special feature, but is like grain, except that the seed is not in an ear. It grows everywhere and not only at Akonai, from whence it gets its name (this is a village of the Mariandunoi) and it specially likes rocky ground. Neither sheep nor any other animal graze it.

5 It is said that: to be effective it must be compounded a certain way, and not everyone can do this; and so physicians, not knowing how to compound it, use it as a septic and for other purposes; and, if drunk mixed in wine or a honey-drink, it produces no sensation; but it can be compounded so as to kill at a certain moment (which may be in two, three, or six months, or in a year, or even in two years); and the longer the time the more painful the death, since the body then wastes away, while if at once, death is easy. It is also said that no antidote has been discovered, like the natural antidotes to other poisonous herbs we hear of; but the natives can sometimes save one with honey and wine and such, but only occasionally and with difficulty.

9.17.1–2 [acquired resistance to drugs]

1 The powers of all drugs become weaker, and sometimes entirely ineffective, to those who are accustomed to them. Thus some eat enough hellebore to consume whole bundles and yet suffer no hurt; this is what Thrasuas [of Mantineia] did (who seems to have been very clever about herbs). And it appears that some shepherds do the like; wherefore the shepherd who came before the vendor of drugs (marveled at because he ate one or two roots) and himself consumed the whole bundle, destroyed the vendor's reputation: it was said that both this man and others did this every day.

2 For it seems that some poisons become poisonous because they are unfamiliar, or perhaps it's more true to say that familiarity makes poisons non-poisonous; for, when the constitution has accepted and mastered them, they cease to be poisons, as Thrasuas also claimed; for he said "the same thing was a poison to one and not to another"; thus he distinguished between different constitutions, as he thought right; and he was clever at distinguishing. Also, besides the constitution, it is plain that use has something to do with it.

(Hort [1926] 253, 255, 257, 299, 301, 305, 307)

11.2 Praxagoras of Kos (–300 ± 25) taught Herophilos, was the first Greek to call attention to the pulse as a diagnostic tool, distinguished veins from arteries, and suggested that arteries originate at the heart, contain only *pneuma*, and dwindle to tendons which control movement. His works are preserved only in quotations (those here are from Galen) or paraphrases (in *italics*).

Anatomy

(fr. 10S)

Behind the position of the tongue, the position of the uvula lies above, at the end of the palate. Behind it are the trachea and the esophagus. The trachea is ventral, the esophagus is dorsal and is attached to the cervical vertebrae. The trachea leads into the lung [considered unitary], the esophagus into the belly. Between the trachea and the tongue is the epiglottis covering the entrance to the trachea.

Associated Symptoms, Book 2

(fr. 90S) [diarrhea]

The excreta of food must pass quickly in people who suffer from diarrhea, since their intestines and especially the duodenum are lubricous.

. . . whatever passes through quickly, must be completely undigested. The rapid passage of food is the specific characteristic of this disease and the undigested state of the food must be reckoned among its necessary consequences. For not because the food remains completely undigested, is it quickly discharged, but because it is quickly discharged, it is not digested at all.

(title unknown)

(fr. 22–25S) [humours]

Praxagoras named the humours in his own way, calling them sweet, equally mixed, and vitreous. These belong to the genus of phlegm. Others are called sour, sodic [or "soapy": nitrôdes], salty and bitter. These are differentiated according to taste. Others are called leek-green because of their color, others yolk-like because of their thick consistency. Another is called corrosive humour, because it has the quality of corrosiveness, and another is clotting because it remains in the veins and does not pass through into the flesh, through being thin and venous. As a rule Praxagoras calls every liquid a "humour" [churos].

The tongue is sensitive to sweet phlegm, which he calls more specifically sweet humour.

The salty humour is engendered by everything that is strongly heating.

Not only fish having tough flesh but all other tough food also, when cooked excessively, produces the salty humour. He calls this humour not only salty or briny but also sodic. When one boils the juices alone, they first become more salty and then even bitter.

(title unknown)

(fr. 27S) [the pulse]

Praxagoras assigns the pulse to the arteries, just as he considers palpitation, tremor and spasm as affections of the arteries. The pulse is a natural event, while palpitation, tremor, and spasm differ from one other in intensity, but are unnatural movements.

(title unknown)

(fr. 64S) [causes of diseases]

The cause of jaundice: a cooling of the innate heat and of the humours of the body takes place, a condition which is also like a step toward dropsy. For jaundice, if intensified, turns into dropsy. This is confirmed because: jaundice is produced in winter, it attacks older people more often, the patients drink vinegar and use spicy condiments (the bile being abnormal), and they are neither feverish or thirsty.

(Steckerl [1958] 49, 59–61, 63, 77, 79)

11.3 Herophilos of Chalkedon (–280 ± 25) was the first Greek physician to perform human dissection and even vivisection, on which he based his anatomy, in which the brain was the seat of thought. He first described the nerves as organs distinct from tendons, connected to the brain, and of two types, motor and sensory. Paraphrases are in italic type.

Anatomy

Book 2?, (fr. 60vS) [an accurate description of the human liver]

The liver of humans is of a good size, larger than in certain other animals which match humans in size. And where it touches against the diaphragm, it is convex and smooth, but where it touches against the abdominal cavity and against its lump, it is concave and uneven. Here it may be likened to a certain fissure by which in embryos, too, the vein from the navel naturally extends into it.

The liver is not similar in all, but different in different creatures, in breadth, length, thickness, height, number of lobes, and in the irregularity both at the front where it is thickest and at the circular parts at the top, where it is thin. In some, it does not even have lobes, but is completely round and unarticulated, whereas in others it has two lobes, in still others more, and in many also four lobes.

Book 3 (at outset; fr. 61vS) [ovaries]

Two "testicles" ["twins": ovaries] are also attached to the uterus on the sides, one on either part, and they differ only a little from the testicles of the male. [contrast Aristotle, *Generation of Animals* 1.4 (717a12–b14), 5.7 (787b20–8a15), for whom male testicles were counterweights]

. . .

In females the two "testicles" are attached to each of the two shoulders of the uterus, one on the right, the other on the left, not both in a single scrotum but each of the two separate, enclosed in a thin, membranous skin. They are small and rather flat, like glands, sinewy at their surrounding covering but easily damageable in their flesh, just like the testicles of males. (In mares they are also quite sizable.) And they are attached to the uterus with no small number of membranes and with a vein and an artery implanted from the uterus into these "testicles." You see, the attachment is from the vein and the artery that go to each of the two "testicles," a vein from the vein and an artery from the artery.

The spermatic duct from each "testicle" is not very apparent, but it is attached to the uterus from the outside, one duct from the right, the other from the left [fallopian tubes]. Like the seminal duct of the male, its anterior part is also convoluted, and almost all the rest up to its end looks varicose. And the spermatic duct from each testicle grows into the fleshy part of the neck of the bladder, just like the male duct, being thin and winding in its anterior part where it touches the hipbones. Here it also terminates, like the pudendum penetrating to the interior from either side.

Eyes

(fr. 260vS) [drug to improve vision]

For those who cannot see in the daytime, twice daily rub on an ointment of gum, the manure of a land-crocodile, vitriolic copper, and the bile of a hyena made smooth with honey; and give the patient goat-liver to eat on an empty stomach.

Midwifery

(fr. 193vS, paraphrased) [uterus: compare Soranos 3.1–5]

The uterus is woven from the same things as the other parts, is regulated by the same faculties, has the same material substances at hand, and is caused to be diseased by the same things, such as excessive quantity, thickness, and disharmony in similars. Accordingly, says Herophilos, there is no affection peculiar to women, except conceiving, nourishing what has been conceived, giving birth, "ripening" the milk, and the opposites of these.

(fr.196vS) [difficult labor]

Difficult labor accordingly occurs because a woman has had a troublesome pregnancy with many fetuses, for example with three to five, as was observed by Simon the Magnesian [around –300 ± 20]. Difficult labor also occurs when the fetus is born in an oblique position or when the neck of the uterus or also its orifice is not distended sufficiently, or when the membrane containing the fetus, in which the water collects, is too thick and not capable of being broken before the birth.

Fetuses have been seen to issue forth without the membrane being broken; but such fetuses are also born with difficulty. Difficult labor also occurs because the uterus or its orifice is slack. And that the uterus is slack is a problem with the body.

But also because of external things – things that happen to one, things consumed, things done – and because too much blood-like moisture is excreted from the body, difficult labor occurs. Difficulty also arises because the uterus is distended by the fetus through pains during birth, and because of cold or heat or a tumor or an abscess in the intestines, in the upper abdominal cavity.

When a concavity arises in the loin and spine, it, too, becomes a cause of difficult labor. On account of fat in the upper abdominal cavity and in the hips, difficult labor also occurs as if the uterus is squeezed, and because the fetuses are dead.

[compare Demetrios of Apamea, fr.21vS in Soranos 4.2; he adds as a cause "hysterical suffocation," a mental disorder attributed to a malfunctioning uterus: Hippokrates, *Nature of Woman* 3, 14, *Diseases of Women* 2.123, Soranus 3.26–28]

Pulses

(fr. 177vS, paraphrased) [rhythms of pulses]

You see, the first pulse found in newborn children will have the rhythm of a short syllabled metrical foot, since it is short in both dilation and contraction, and it therefore is conceived of as consisting of two short time-units [⌣⌣], whereas the pulse of children who are growing is analogous to the metrical foot known as trochee [–⌣]. This pulse consists of three time-units, holding its dilation for two time-units, but its contraction for one. And the pulse of those in the prime stage of their lives is equal in both, that is, in dilation and contraction, and it is compared to the foot called spondee [– –], which is the longest of the disyllabic feet. It is actually composed of four time-units. This pulse Herophilos calls "in equal quantity." The pulse of those who are beyond their prime, and almost old, is itself also composed of three time-units, holding its contraction for twice as long as its dilation and longer [iambic: ⌣–].

(title unknown)

(fr. 259vS) [ointment for the anus]

Mix two drachmas dry or fresh roses, two drachmas white lead, two drachmas *pompholux* [zinc oxide], two drachmas saffron, two drachmas washed litharge [lead monoxide], two drachmas melilot, one drachma rush, one drachma opium, one drachma wool-grease, one roasted egg yolk, an adequate amount of rose-oil, two ladles [about 75 ml] of plantain [*Plantago maior*] juice.

(von Staden [1989] 183, 185–186, 321, 351, 365, 367–368, 423–424)

11.4 Erasistratos of Keos (–280 ± 25) employed mechanical explanations (such as that the heart is a pump of blood and *pneuma*), and sought the causes of diseases in an excess (*plêthos*) of blood, leaking from veins into arteries through tiny openings; he taught that veins, arteries, and nerves were woven together throughout the body. (All quotations herein are from Galen, as are paraphrases – in *italics*.)

Fevers

Book 1 [inflamed wounds]

Systems of treatment follow these principles to keep all wounds free from inflammation. The drugs that are rubbed into the surrounding healthy parts prevent, by their styptic and astringent action, the development of pressure by the blood poured out from above *on* the wounded parts. In the unaffected parts, on the other hand, there comes about an interchange between the many arteries and veins that have inosculations in the same places, transferring to the veins some of the blood that had gone across to the arteries. The practice of not giving food to wounded patients, during the time when inflammation is occurring, is also consistent with these principles; for the veins, when emptied of nutriment, will more readily receive back the blood that has gone across to the arteries, and when this happens the inflammation will become less.

(Brain [1986] 20)

Book 1 [heart as pump]

There are at the mouth of the vena cava three membranes, in arrangement very like arrow-barbs [tricuspid valve]. *The membranes of the "vein-like" artery are very similar in shape to those, but unequal in number, for to this mouth alone, only two membranes attach* [mitral valve]. *Each of the other two mouths has three membranes, all crescent-shaped* [semi-lunar valves]. *Each of the two mouths are exits, the one evacuates blood to the lung, and the other pneuma into the whole animal. These membranes perform a reciprocal service for the heart, alternating at the appropriate times – those which attach to the vessels which draw matter into the heart from the outside rebound at the entrance of the material and, falling back into the cavities of the heart, by opening their mouths, give an unimpeded passage to what is being drawn into that cavity. For material does not rush in spontaneously, as into some lifeless container; but the heart itself, dilating like a smith's bellows, draws the material in, filling itself by expansion* [diastolê: the heart actually <u>relaxes</u> to allow influx]. *The other membranes attached to the vessels which lead material out of the heart behave oppositely. For they incline outwards from within and, rebounding at the material passing out, open their mouths for as long as the heart is supplying material. But for all the rest of the time they firmly close their mouths not allowing any of the emitted material to return. So, too, the membranes attached to the vessels which lead material into the heart close their mouths whenever the heart contracts not permitting any of the in-drawn material to flow back out again.* [compare below Section 19 (Galen)]

(deLacy [1984] 397)

General Principles

Book 2 [blood vessels]

In the ultimate simple vessels which are thin and narrow, nourishment is drawn through the sides of the vessels and is deposited in the empty spaces left by the material which has been carried away.

(Longrigg [1998] 97)

Paralysis

Book 2 [nature of scientific research]

Those who are completely unused to inquiry are, in their first attempts, blinded and dazed in their understanding and straightway leave off the inquiry from mental fatigue, and are no less incapable than those who enter races without being used to them. But the man who is used to inquiry tries every opening as he conducts his search and turns in every direction and so far from giving up the inquiry in the space of a day, does not cease his search throughout his life. Directing his attention to one idea after another that is germane to what is being investigated, he presses on until he arrives at his goal.

(Lloyd [1973] 86)

(title unknown) [brain]

I examined also the nature of the brain. It was divided into two parts, like that of other animals, and has elongated ventricles lying there [right and left cerebral ventricles]. These two ventricles were connected by a passage where the two parts are joined. From here the passage led into the so-called *epenkranis* [cerebellum], where there was another small ventricle [modern name: "third ventricle"]. Each of the parts is divided off by the meninges; for the cerebellum was partitioned off by itself, and also the cerebrum, which is similar to the jejunum and has many folds. The *epenkranis* was furnished even more than the cerebrum with many varied convolutions. So the observer learns from these that, just as in other animals, the deer, the hare, or any other that far excels the others in running, is well provided with muscles and sinews useful for this, so in humans too, being far superior to other animals in intellect, this organ is large and very convoluted. All the nerves grow out of the brain, and on the whole the brain seems to be the source of bodily activity. For the sensation from the nostrils opened onto it as did those from the ears. And outgrowths from the brain led also to the tongue and the eyes.

(deLacy [1984] 441, 443)

(title unknown) [animals give off emanations]

If one were to take a creature, a bird for example or something similar, and place it in a vessel for some time without giving it any food, and then weigh it together with the excrement that has visibly been passed, one will find that there has been a great loss of weight, clearly because a considerable emanation, perceptible only by reason, has taken place.

(title unknown) ["do not feed a fever!" – compare also *Fevers 3*, in Brain [1986] 20]

For when the nutriment that is being distributed is neither fully digested nor elaborated according to what is customary for each individual, nor secreted in some other way, of necessity the veins are filled by digestion and distribution operating in accordance with their natural processes. When, on the one hand, the nutriment already in the veins is not consumed in any way, and, on the other hand, more nutriment continues to come into being from food, the veins, too, in the body are stretched to an even greater extent. When the veins can no longer accept further accumulation and other nutriment is being conveyed from the stomach, the nutriment already in the veins rushes into the arteries that lie alongside them.

(Longrigg [1998] 97, 116–117)

11.5 Andreas perhaps of Karustos (assassinated –216: Polubios 5.81.1–7) was the personal physician of Ptolemy IV, a competitor of Eratosthenes of Kurênê, and a member of the "school" of Herophilos; his works included *Poisonous Animals*, *False Beliefs*, and the *Casket* (on drug remedies). He was renowned for inventing an apparatus for reduction of dislocations, partially described by Oreibasios 49.4.8–13, 19–20, 50–51 (von Staden [1989] 472–477).

(fr. 45vS) [morays: compare Thompson [1947] 162–165]

Poisonous Animals   Only those morays have a fatal bite which come from a viper, and they are smaller and round and speckled.

False Beliefs [did Andreas change his mind?]   It is not true that the moray moves into lagoons and mates with the viper, for vipers don't feed in lagoons, preferring sandy deserts.

(Gulick [1929] 403, 405)

Casket?

(fr. 31vS) [cream for running sores, and slow-healing or bloody wounds, to prevent inflammation]

3 pounds litharge, 3 ounces each of copper scale [probably copper oxide], *chalkitis* [usually copper sulfate], powdered verdigris [*ios*: compare Chapter 9], 1 *kotulê* [about 220 ml] of vinegar, 8 pounds of aged oil; prepare in the usual way.

(fr. 32vS) ["rose-compound," good for great pain, fluxes great and small, blisters, and prolapses]

4 drachmas of rose-petals without the hips, 2 drachmas of saffron, 1 obol [1/6 drachma] of opium, 1 obol of Indian nard, 3 drachmas of gum acacia [Galen notes that some manuscripts have "1½ drachmas"]; the mixture dissolved in rain-water, apply with the cupping-glass.

*(Kühn vol. 13 [1827] 735, 765–766)

(title unknown)

(fr. 41vS) [hair loss]

There is bread in Syria made with mulberries, the eating of which causes hair-loss.

(Gulick [1927] 43)

11.6 Glaukias of Taras (–175 ± 20), an early member of the "Empiric" sect, wrote on Hippokratic exegesis, pharmacology, and bandaging (our numbering is for convenience).

[bandages]

1 The "beehive"-bandage [*tholos*] is done like this: leave a moderate length of the band on the face, unroll it up to the bregma, crown, and occiput, and order the attendant to hold it fast there; overlap the roll, and render it a forehead-bandage, then in the same way bring up the end at first left along the face to the bregma, crown, and occiput, and tie it off at the occiput.

. . .

2 The "strap" [broad bandage] having been taken by its two ends, place its middle upon the occiput. Then lay two diagonals under the ear-lobes to the eyes and up to the bregma, and there make a cross, then two cheek-bandages from occiput round to occiput, and again two circular cheek-and-forehead-bandages from occiput round to occiput; and again let the knot be above the forehead. Thus the binding is suitable for bandaging the jawbone and eyes.

*(Deichgräber [1965] 169)

11.7 Agatharchides of Knidos (*ca.* −215 to *ca.* −140) worked in Alexandria under Ptolemy V ("Epiphanes") and Ptolemy VI ("Philometor") (a time of strife and chaos). He wrote several works of history enriched with anthropology and geography. Towards the end of his life, fleeing the chaos in Alexandria, he published *On the Red Sea* (fragmentary).

On the Red Sea

(fr. in Plutarch, Table-Talk *8.9 [733bc])*

People around the Red Sea got sick with new and unrecorded symptoms, including the following: little worms would eat their way through the shin or arm and burst out. When they were touched, they went back in, and produced an intolerable inflammation, as they encased themselves in the muscular tissues. No one knows of this disease ever occurring before, or of its afterwards ever attacking anyone else, but this people alone. [guinea worm; a similar account in fragment 59, preserved by Photios and by Diodoros of Sicily 3.29.5–7]

(Minar [1961] 197)

11.8 Demetrios of Bithunian Apamea (date very uncertain: −140 ± 20), a member of the school founded by Herophilos, wrote especially on pathology (*Diseases* and *Semiotics*), but survives only in a few paraphrases, mostly of his gynecology (see von Staden [1989] 506–511).

(fr. 19vS; Soranos 3.19) [inflammation of the uterus]

The pain is located on the side directly involved. For it is not plausible that what's near the inflamed part should not be painful, while the unaffected parts feel pain (as some believed considering only the passing of the inflammation to the opposite parts and attacking that area). For it's much more logical that these parts too become sensitive only after the side directly involved in the inflammation has undergone much tension, because the opposite side gets the diversion and the sensation.

(fr. 17vS; Soranos 3.43) [uterine flux]

The flux is a "flow of fluid through the uterus for an extended time," since the flux may not be sanguineous only, but different at different times. The

304

differences lie in color and power [*dunamis*]. In color: for one kind is white (like barley juice), another watery, another red, another black, another slightly bloody (like washed meat), another is of uneven color, and another is pale. In power: one kind is inactive and occurs without irritation or pain, but another occurs with irritation and erosion and brings on a painful sensation at the time of its excretion. One kind comes from the whole body, another from the uterus, and another from some other part. And the white flux is said to be more stubborn than the red, since it has to pass through narrower ducts.

(Temkin [1956] 145, 165–166)

11.9 Nikandros of Kolophon, a priest of Apollo (around –130 ± 20), versified contemporary learning on venomous beasts, as well as on poisons (see Chapter 10), bee-keeping (lost), hunting (lost), etc. His colorful depictions deploy more recondite verbiage than instructive particulars.

Alexipharmaka

186–206 [hemlock and remedy; compare opium, 433–464]

Cognizant ought you to be of the damaging draught of the hemlock, a
Murderous beverage smiting the noggin with *harmful* disaster, and
Shadowy darkness of nighttime conferring, and whirling of eyes, so
Victims are going on staggering feet as they wander the streets, or
[190] Even they crawl upon hands; and a terrible choking obstructs and
Fills to repletion both throat at its bottom and narrowing windpipe;
Coldness descends on extremities, arteries stout in the limbs are con-
tracted within one, and breathing but little he gasps for his air very
Like to one swimming; his spirit, it gazes on *infernal* Hades.
[195] Give to the patient a surfeit of oil or of wine undiluted,
Till he can vomit, expelling the woeful destruction of poison; or
Else you may fashion an enemic cleansing and insert the implement;
Make him to drink very often of wine undiluted, or cut from and
Offer the twigs of the laurel of Témpê[a] or *daúkos*[b] of Crete (which
[200] First was the crown to be placed on the Delphian locks of lord
    Phoibos), or
Grind up some pepper to powder and mix it with seeds of the nettle, and
Give it, then bitters of sílphion poured into fruit of the vine.
Sometimes you'll offer a measure of fragrant perfume of the iris, and
Sílphion shredded to pieces then sopped into glistening oil, and
[205] Make him to drink of the honey-like sweetness of grape-juice, or
    vessels of
Milk that is foaming with warmth from the mildest of fires you may
    give out.

[a Tempê is the valley between Olumpos and Ossa; b Cretan *daukos* is the umbellifer *Athamata cretensis* (mainland *daukos* is parsnip)]

Theriaka

921–933 [snakebite remedies]

Truly, applying to deadly *and damaging* snakebite the cupping
Vessel of bronze you may empty from body both blood and the poison, or
Pour *upon puncture* the milky-white sap of the fig, or make use of an
Iron that's heated *to heal* in the heart of a burning-hot furnace.
[925] Sometimes the hide of a goat that once grazed you will fill up with
  wine and
Put into service whenever the wound is in hand or an ankle:
Shove in the sack up to midpoint of arm or of ankle the laboring
Patient, and draw up the fastening cords at the crotch *or the armpit*,
Till all the strength of the wine may extract from the skin all the pain.
[930] Other occasions will call for the leeches[a] to gorge on the wounds, or
Drip from an onion its juice *on the snakebite*, and sometimes the lees of the
Wine you should pour upon droppings[b] of sheep (or else vinegar), making a
Paste for a plaster, and wrap the fresh dung all around the *dread* wound.

[a earliest recorded medicinal use of leeches, later also employed by Themison of Laodikaia (–50 ± 30), a pupil of Asklepiades also working in Rome, and discussed by Galen in *Leeches, Diversion, Cupping-glasses, and Scarification* (Kühn vol. 11 [1826] 317–322); b because feces have an expulsive power, compare Galen, *Powers of Simples* 10.2.18–20 (Kühn v. 12 [1826] 290–295)]

(Gow and Scholfield [1953] 89, 91, 107: versified)

11.10 Asklepiades of Bithunia, a physician working in Rome (–100 ± 20), prescribed a mild regimen and few drugs, and developed a theory that disease is caused by irregularities in the healthy free motion through bodily pores of microscopic divisible corpuscles of which every body and the world are composed. (We have only paraphrases of extracts of his works; our numbering is for convenience.)

1 [corpuscles]

Asklepiades posited atoms as first principles, corpuscles perceptible to the intellect and without any normal quality, gathered together from the beginning and in constant motion. When they strike each other with mutual blows as a result of their particular kind of conflux, they are dissolved into innumerable fragments of parts differing in size and shape. They come together again, and through their addition and conjunction create all sensible things. They have in them the power of change in respect of their size, number, shape, and arrangement. It is not illogical that bodies with no quality should make up the sensible world. For one thing is true of the part, and another of the whole. So it is that silver is white, but a sliver of silver is black; goat's horn is black, but a shaving of it is white.

2 [fevers]

First, there are intelligible pores in us, differing in size from each other, second, parts of moisture and *pneuma* are gathered together from all sides out of intelligible corpuscles which are in permanent motion, and third, there are continuous emanations from us to the outside world, which vary according to the prevailing condition.

3 [assimilation of food]

Furthermore, there is no such thing as digestion in us, but a raw solute of food called *leptomeres* ["fine"] forms in the belly, and it passes through the individual parts of the body, apparently penetrating all the fine pores.

4 [bladder]

The liquid that we drink is dissolved into vapors and these pass into the bladder. The vapors then re-condense and resume their original form, becoming water once more from vapor. (In effect the bladder is a sort of sponge, or piece of wool, and not the entirely solid and watertight body it is, possessed of two very tough coats.)

5 [pneuma]

The heart and the arteries are dilated when they are filled with *pneuma* which flows into them because of the *leptomereia* ["fineness"] which they contain. When they are full, and the influx ceases, the coat contracts into its former natural state.

6 [lung]

The lung is like a funnel: the cause of respiration is the fineness in the chest, towards which the thick air flows from outside. It is pushed back again when the chest is unable to receive more or contain it. A small amount of fineness always remains in the chest (for it is not all excreted) and it is towards this which remains inside that the weight from the outside is borne back in again. Voluntary respiration occurs when the finest pores in the lung are gathered together and the bronchial passages are narrowed. For these things obey our will.

(Vallance [1990] 20, 26–27, 55, 81, 83)

11.11 Herakleides of Taras (–75 ± 20) was a student of Mantias (a Hero-philean pharmacologist, –125 ± 35, see von Staden [1989] 515–518), but left the school for the Empiric sect; he wrote on pharmacy and on Hippokratic exegesis; see also fragments of his *Banquet* in Athenaios, *Deipnosophists* 3 (79–80, 120).

Exterior Therapy [reduction of dislocation of the thigh]

Some suppose that the thigh doesn't stay reduced because the attaching tendon pulls the thigh towards the cup of the hip, but they are mistaken about the general procedure, and offer a mere excuse. For although Hippokrates [*Joints* 70] and Diokles didn't write about insertions, still Phulotimos [of Kos, –280 ± 20, loyal student of Praxagoras and writer on diet], Euênor [of Akarnian Argos, –320 ± 20, wrote on therapy and gynecology], Neileus [around –250 invented the winch-assisted reduction], Molpis [otherwise unknown], Numphodoros [–220 ± 20 invented a reduction machine described by Galen, *Use of the Parts* 7.14], and others did. And on two youths we accomplished the end proposed. For indeed often, and especially in adults, the joint slips out again. Now one ought not to judge the affair by theory (but since sometimes it does stay, one ought to consider that there is not always a separation due to the tendon, but that it slacks off and congeals again), since it's useful but not completely required to investigate this.

*(Deichgräber [1965] 176)

11.12 Apollonios Mus ("mouse" or "mussel"; –20 ± 30) wrote a history of his Herophilean school, on *Perfumes and Unguents*, and a *Ready Remedies*, all lost. Galen preserves some extracts (see von Staden [1989] 540–554).

Perfumes and Unguents

(fr. 8vS: Athenaios, Deipnosophists *15.38[688e–689b])*
[good and better supplies]

The best iris-root is that grown in Elis and in Kuzikos; the best rose-perfume is obtained in Phaselis, also from Naples and Capua; the best crocus-perfume, in Kilikian Soloi and Rhodes; the best spikenard, in Tarsos; the best drop-wort [*Spiraea filipendula*] is from Cyprus and Adramuttênê; the best marjoram and quince from Kos. Of henna the Egyptian is judged the best, next is the Cyprian and the Phoenician, especially that from Sidon. That called Panathenaic is made in Athens; the *metopion* and the Mendesian are made best in Egypt (the *metopion* is made with the oil obtained from bitter almonds).

The suppliers, the material itself, and the manufacturers, not the locales, make the perfume the best. For example Ephesos in earlier times excelled in perfumes, particularly in the kind called *megalleion*, but not now. Again, those of Alexandria used to be superior because of the city's wealth and the zeal of Arsinoë and [her mother] Berenikê [queens around –316 to –269]. In Kurênê, too, the rose-perfume was best when Berenikê the Great lived [queen there –245 to –220]. Drop-wort perfume in Adramuttion was once mediocre, later it became first quality due to Stratonikê, [from –187 to –171] the wife of Eumenes [II of Pergamon]. Syria in ancient times supplied all perfumes of excellent quality, especially that from fenugreek, but not now. And in Pergamon, in earlier times but not now, after a certain perfumer had worked hard at it, there was excellently manufactured what had never been made by anyone before, perfume from frankincense.

(Gulick [1941] 187, 189, 191)

Ready Remedies

Book 1 (fr. 11vS: Galen, Compound Drugs by Site 1.8)
[for dandruff]

1. Anoint the head with rose-oil; second, do the like with beet-juice; third after these do the same thing with tortoise-blood; fourth anoint the head with laurel oil; anoint similarly with crocus-perfume.
2. Rub with bull's urine and then after that use camel urine the same way. One must employ these for many days. [Galen remarks that one could hardly do that for one day; he then quotes and criticizes three more recipes]

Book ? (fr. 14vS: Galen, Compound Drugs by Site 2.1) [for hangover]

One must employ all recorded aids for headaches from fever, and especially for worsening headaches; [besides those:]

1. One must anoint the head with leaves of rue ground with vinegar and rose-oil to the consistency of gum;
2. one must anoint the head with bitter nuts [i.e., almonds?] ground with vinegar and rose-oil to the consistency of gum;
3. in the same way employ dry iris[-root?] and dry agnus-vitex, ground with vinegar and rose-oil;
4. employ bay-berry and leaves of rue, ground with vinegar and rose-oil.

Sleep, fasting, quiet, not moving, drinking warm water or honey-drinks, and evacuation are helpful.

*(Kühn vol. 12 [1826] 475–476, 514)

11.13 Dioskourides of Anazarbos (around 50 ± 10) traveled widely collecting material for his pharmacy-book (*Medical Materials*), arranged by affinity of effect, and drawing on Krateuas (–80 ± 20) and Theophrastos. (See also Chapters 9.10 and 10.11.)

Medical Materials

Preface. 5–9 [collecting and storing botanicals]

5   I now encourage you, and any who may chance upon my book, not to look at my verbal facility but at my careful practical experience. For I have exercised the greatest precision in getting to know most of my subject through direct observation, and in checking what was universally accepted in the written records and in making inquiries of natives in each botanical region. Furthermore I shall endeavor to use a different arrangement and describe the classes according to the properties of the individual drugs. It is, I suppose, obvious to everyone that pharmacology is a necessity, closely linked to the whole art of medicine and forging with its every part an invincible alliance. It can also continue to extend its range of preparations and mixtures and its trials on patients, for the knowledge of each individual drug has a great deal to contribute.

6   I shall also include the common and familiar *materia medica* in order to make my work complete. Before anything else, it is appropriate to consider the storage and collecting of individual drugs in their proper seasons, for these matters in particular determine the weakness or efficacy of drugs. For example, herbs should be gathered when the weather is excellent, for it makes a great difference if the collecting is done after recent droughts or heavy rains. Similarly, sites are important, whether they are in the mountains, high up, windswept, cold and arid, for the properties of such plants are stronger. Those of plants from flat and wet localities, in the shade and not open to the wind, are generally weaker, especially when plants are gathered in the wrong season or when they are decayed through some weakness.

7   One should not fail to note that plants often ripen neither sooner or later according to the specific character of the country and the climate. Some, according to their own particular nature, bear flowers and leaves in the winter, others produce flowers twice a year. Anyone wanting experience in these matters must encounter the plants as shoots, newly emerged from the earth, plants in their prime, and plants in their decline. For someone who has come across the shoot alone cannot know the mature plant, nor if he has seen only the ripened plants can recognize the young shoot as well. Great error is occasionally committed by those who have not made an appropriate inspection, as a result of the changes in the form of the leaves, the varying sizes of stems, flowers and fruits, and some other characteristics.

8   Indeed, precisely for this reason, some authorities have been deceived into saying that some plants bear neither flowers nor stem, nor fruit, like dog's tooth

grass, coltsfoot, and cinquefoil. But anyone who has seen these plants often and in many places will gain a particularly precise knowledge of them. Moreover, one must realize that some medical plants keep for many years, like white and black hellebore, and that the rest are useful for up to three years. On the other hand, one should gather the medicinal plants which are like young sprouts – French lavender, wall germander, felty germander, shrubby wormwood, Gallic wormwood, marjoram, and the like – when they are swollen with seeds, and their flowers before they fall off, their fruits while they are ripe, and the seeds when they are beginning to become dry before they drop off.

9   Extract juices from plants by infusion when the stems are recently sprouted, similarly with leaves; but to gain juices and drop-like gums by tapping, take the stems and cut them while in their prime. Gather roots for laying up in storage, as well as roots for juices and root barks, when the plants are beginning to shed their leaves. The clean roots should be dried out immediately in areas free from moisture, but roots with earth or clay adhering should be washed with water. Flowers and such parts that have a sweet smelling fragrance should be laid down in small dry boxes of limewood, but occasionally they can be serviceably wrapped in papyrus or leaves to preserve their seeds. As for moist drugs, any container made from silver, glass, or horn will be suitable. An earthenware vessel is well adapted provided that it is not too thin, and, among wooden containers, those of boxwood. Copper vessels will be suitable for moist eye-drugs and for drugs prepared with vinegar, raw pitch or juniper-oil. But stow animal fats and marrows in tin containers.

(Scarborough and Nutton [1982] 196–197)

1.30.1–4 Olive oil

1   The best oil for use in health is the raw-pressed which they call *omphakinon.* And of this, fresh oil, fragrant and not bitter, excels (also useful for preparing perfumes). It is also good for the stomach because it binds, and if soaked into wool and held in the mouth it can strengthen the teeth and stop pustules. The older and fatter oil is suitable for use in slackening drugs. Generally all oil is warming and softens the flesh, keeping bodies hard to chill and making them readier for activity.

[2: two oil-based recipes, an emetic, and a treatment for worms and constipation]

3   But the oil from the wild olive is more astringent and second-rate for use in health. It's better than rose-oil for those suffering from headaches. It wards off pustules and hair-loss. It cleanses dandruff and scabs and *lepras* [eczema: Grmek [1989] 165–168]. And if smeared on the hair every day, it slows graying.

[4: how to whiten oil]

1.78 Laurel

One type is thin leafed and the other has broader leaves. Both types are warming and softening, so that a decoction of them in a sitz-bath is good for conditions of the bladder and womb. The green leaves are somewhat and gently astringent. Ground into a plaster they help against stings of wasps and bees, and as a poultice with wheat and barley are able to soothe every inflammation, and if drunk, they dull the stomach and are emetic. But the bay-berries are more warming than the leaves. Ground into lozenges with honey or sugar they act against consumption and *orthopnoia* [being able to breathe only in an upright position] and the fluxes of the chest. Drunk with wine they aid against scorpion stings, and they cleanse *alphós* [loss of skin pigmentation]. Expressed laurel juice mixed with old wine and rose-oil helps ear-aches and ringing in the ears and hardness of hearing. It is mixed with refreshing and warming salves and with perspiration-inducers. But the bark of the root breaks [kidney or bladder] stones and kills embryos and is an aid against liver ailments when three obols of it are drunk with fragrant wine.

1.113 Cherries

Cherries if taken green are good for the belly, but if dry they bind the belly. Gum of cherries heals a chronic cough if taken with diluted wine, and it promotes a good complexion and sharp sight and the appetite. If drunk with wine, it helps those suffering from [kidney or bladder] stones.

2.82.1–4 Honey

1 The best is Athenian honey, and of this that called Humettion, next is honey from the Kuklades islands and after that honey from Sicily, called Hublaion. The most approved honey is the sweetest, and sharp, rather fragrant, yellowish, not watery, but sticky and elastic, and when drawn out returns back to the finger. Honey has a purgative power, opening the pores, drawing out moisture so that, if infused, it closes dirty and hollow wounds.

2 Boiled and applied, honey glues separated bodies and boiled with liquid alum and anointed cures rashes. When infused lukewarm with ground roasted salt, it cures noises and pains in the ears. Smeared on, it kills lice and nits. It restores the lack of a foreskin, if not lost by circumcision: the penis is softened with honey, especially by bathing, for thirty days. It cleans away what darkens the pupils, and when smeared on and gargled it heals inflammations around the throat and sore throats.

3 It also moves urine and is good for a cough and for those bitten by snakes. And if taken warm as a drink with rose-oil it is a good antidote for opium, and if licked [or "made into lozenges"] or drunk it is good for growths and those bitten by mad dogs. But indeed raw honey causes flatulence in the belly and provokes a cough so one must use it once it's been despumated [had the froth boiled off]. Springtime honey is best, next is summer honey, but winter honey being thicker is worse and makes bee-bread.

4 Honey made in Sardinia is bitter because the bees feed on wormwood, but smeared on the face it is good against freckles and blemishes. . . .

*(Wellmann [1907] 33–34, 78, 106, 165–166)

11.14 Aretaios of Kappadokia (either about 70 ± 20 or perhaps as late as 170 ± 20: Oberhelman [1994]) was a Pneumaticist who wrote on *Fevers*, on *Female Disorders*, and other works besides the one here excerpted.

Acute and Chronic Diseases

2.1 (pp. 15–16 Hude) [respiration]

Animals live by two principal things, food and *pneuma*; of these by far the most important is the respiration, for if it stops, one will not endure long, but die immediately. Its organs are many, the origin being the nostrils; the passage, the trachea; the container, the lung; the protection and receptacle of the lung, the chest. Now the other parts minister only as instruments to the animal; but the lung also contains the cause of attraction, for in its midst is seated a hot organ, the heart, the origin of life and respiration. It imparts to the lungs the desire of drawing in cold air, for it inflames them; but the heart attracts. So if the heart suffer primarily, death is not far off. [compare Herophilos, fr. 143vS on the mechanism of breathing, a natural dilation/contraction pumping air; and see Galen, *Use of Breathing* in Furley and Wilkie [1984]]

[symptoms of *peripneumonia* (lung inflammation) and of *phthisis*; the signs of a fatal and a successful *krisis*; how *peripneumonia* transits to *phthisis*]

3.1.1–2 (p. 36 Hude) [treatment of chronic diseases]

1 In chronic diseases the pain is great, the period of wasting long, and the recovery uncertain. For either they are not dispelled at all, or the diseases relapse on any slight error. For neither have the patients resolution to persevere to the end; or, if they do persevere, they commit blunders in a prolonged regimen. And if there's also the suffering from a painful treatment (of thirst, of hunger, of bitter and harsh medicines, of cutting or burning – of all which there is sometimes need in protracted diseases), the patients secretly evade it, truly preferring even death itself.

2 Hence, indeed, is developed the talent of the medical man, his perseverance, his diversity, and conceding pleasant things that are harmless, and in giving encouragement. But the patient also ought to be courageous, and co-operate with the physician against the disease. For, taking a firm grasp of the body, the disease not only wastes and corrodes it quickly, but frequently disorders the senses and even deranges the soul by the distemper of the body. Such we know mania and melancholy to be, which I will discuss later [see Chapter 12.9]. Now I'll give an account of *kephalaia* [headache and migraine, which transit to "vertigo": dizziness and tinnitus].

4.5.1–2 (p. 71 Hude) [gonorrhea]

1 Gonorrhea is not a deadly affection, but one that is disagreeable and disgusting even to hear of. For if intemperance [humoural imbalance] and paralysis possess both the liquids and genitals, the semen runs as if through dead parts, nor can it be stopped even in sleep: for whether asleep or awake the discharge is irrestrainable, and there is an unconscious flow of semen. Women also have this disease, but their semen is discharged with titillation of the parts, and with pleasure, and from immodest desires of connection with men. But men are not so greatly itched.

2 The fluid which runs off is thin, cold, colorless, and unfruitful [not semen but pus]. For how could chilled nature evacuate vivifying semen? And even young men, when they suffer this, necessarily become old in constitution, torpid, relaxed, spiritless, timid, stupid, enfeebled, shriveled, inactive, jaundiced [yellowish], whitish, effeminate, anorexic, and frigid; they have heaviness of the members, torpidity of the legs, and are powerless and incapable of all exertion.

[semen makes men hot, strong, hairy, etc., while its lack makes men eunuch-like; gonorrhea is the outcome of saturiasis, the disease of permanent erection: Aretaios 2.12]

4.12.1–11 (pp. 82–84 Hude) [arthritis]

1 Arthritis is a general pain of all the joints; that of the feet we call *podagra*: that of the hip-joint, *schiatika*: that of the hand, *chiragra*. The pain is either sudden, arising from some temporary cause: or the disease lies concealed for a long time, when the pain and the disease are ignited by any slight cause. It is, in short, an affection of all the tendons [*neura*], if the ailment when increased extends to all: the first affected are the tendons which are the ligaments of the joints, and the ones which have their origin and insertion in the bones.

[2–4: normally bones feel no pain, being dense, unless their heat is altered]

[5–9: the onset and spread of arthritis from foot to hand to elbow and knee, to hip, to spine, to all parts of the body]

10 There seems to be a difference between heat and cold, for some cases delight in what is properly loathsome. But I think there's one cause, a chilling of the innate heat, and that there's one disease. But if it speedily swells, and heat appears, there is need of chilling and it delights in such things: this is called the hot species. But if the pain remain internally in the tendons, and the unheated part condenses and doesn't swell, I would call this variety cold, for which one needs hot medicines to recall the heat, of which the very bitter are the best. For heat excites the condensed parts to swelling, and recalls the internal heat, and then there is need of refrigerants. [compare Thessalos, *Remedies* 2 Kronos, in Chapter 4.10]

11 In proof of this, the same things are not always expedient in the same cases, for what is beneficial at one time is harmful at another; in a word, heat is required in the beginning, and cold at the conclusion. Wherefore *podagra* does not often become unremitting; but sometimes it intermits a long time, for it is slight; hence a *podagra* sufferer has won the race in the Olympic games during an intermission.

[men are arthritic more often, but cases in women are more severe; the disease transits to dropsy or asthma]

[4.13: leprosy, called "elephas": see below, Plutarch, Section 16]

(Adams [1856] 261–262, 293–294, 346, 362–365)

11.15 Xenokrates of Aphrodisias (75 ± 20) wrote a variety of pharmacological and dietetic works, especially *Useful Materials from Humans and Animals* (of which extracts survive in Galen, *Powers of Simples* 10) and the work here excerpted. (The numbers are page-references in Thompson [1947].)

Food from Aquatics

1.1–3 [edibility of fishes]

1 Swimming creatures are made as a delightful food for sumptuous pleasures, but there is also a great benefit to health in their consumption.

2 Fishes differ when compared: some are tough-fleshed, some tender. Among the tough-fleshed are the red sea-bream [*Pagrus vulgaris*, 273–274], the toothy sea-bream [*Dentex vulgaris*, 255–256], and the sole [*Pleuronectes solea*, 33–34]; also the flat fishes like the flounder [294–295] and turbot [223]. The tender-fleshed are the wrasse [116–117], the blackbird-wrasse [128], the colored wrasse [276–278], and such like, guaranteed easily digestible. In between are the *oniskos* [181–182], the *bakkhos* [24], and the maigre [*Sciaena aquila*, 241–243].

3 Moreover, some are rock-dwellers, some open-sea dwellers: the sea-dwellers are more nourishing. But fish from the seashore or river mouths have poor humours and taste bad. The deep-sea fishes are excellent: the Adriatic ones are mediocre, but the Turrhenian ones are the sweetest.

[1.4–39: more data on locales and species; 2: the anemone; 3: shellfish; 4–5: salted fish]

*(Ideler [1841/1963] 121)

11.16 Plutarch of Chaironeia (*ca.* 50 to *ca.* 120), writing under the Flavian emperors, Trajan, and in the early years of Hadrian, was a Platonist teacher of philosophy; much of his copious output is in the form of moralizing essays or dialogues.

Table-Talk

8.9.1–5 (731–734) Whether it is possible for new diseases to come into being, and from what causes

1   Philon the physician was maintaining that the disease called elephantiasis [leprosy] had been known for only a short time, since none of the ancient physicians had written a treatise on it, though they expatiated on many others that were minute and petty and obscure to most. I supplied him with an additional witness from philosophy: Athenodoros, who wrote in the first book of his *Epidemics* [ca. 25 ± 75] that both elephantiasis and hydrophobia first made their appearance in the time of Asklepiades [Rufus of Ephesos preserves the account that Straton the secretary of Erasistratos described this disease, calling it *kakochumia*, "evil-humour": Grmek [1989] 168–173]. Those present were surprised that new diseases first came into existence and took shape at that date; but they thought it no less amazing if such striking symptoms had escaped notice for so long. The majority were rather inclined toward the second hypothesis, as more human, for they regarded nature as not at all innovative in such matters – nor likely to foment revolutions in the body as if in some body politic.

[Diogenianus says: no new vice of the soul has appeared, so no new disease of the body is expected; then:]

2   . . . How indeed could the body develop a new malady, or late-born disease, when it does not have, like the soul, its own internal source of motion, but is linked with the rest of nature by common causes and is so tempered in its composition that even its irregularity wanders within limits, like a ship bobbing at anchor? For disease cannot take shape without a cause, into affairs unnaturally introducing genesis from not-being; and to find a new cause for disease would be hard, unless one could show that new air, or strange water, or foods untasted by earlier people, are now for the first time flowing into our world from some other worlds, or from the spaces between them. For it is the things that sustain life which also cause sickness, and there are no special seeds of disease, but it is the disagreement of our food and drink with us, or our mistakes about them, that disturbs our system.

[and he concludes: nature is regular so cannot generate novelty, only change of degree]

[3: Plutarch and Philon assert that a sufficient change of degree *is* a change of kind, so that new diseases are possible]

[4: the number of combinations even of existing entities is very large: see Section 2.5]

[Agatharchides, above, Section 7, is cited]

[Plutarch offers evidence that old diseases no longer occur; then:]

5   As for the introduction of new air or strange water, let us give that up, if Diogenianus doesn't like it: though we do know that the Demokriteans both say and write that when external worlds perish, and foreign atoms flow in from the infinite, then sources of plagues and unusual diseases may fall among us. Let us

also give up the partial destructions that take place on earth, from earthquakes, droughts, and storms – occasions when the winds and earth-born streams must likewise suffer deterioration and change.

But we must not disregard the changes that have occurred in food and cooking and other parts of our diet. Many items that were not eaten or even tasted are now much enjoyed, like wine with honey or sow-womb. They say that the ancients did not even eat brains, which is why Homer said, "I care for him no more than brains" [*Iliad* 9.378, misquoted], so speaking of brains because they found them revolting and so rejected and discarded them; and we know that many older people still cannot eat ripe cucumber, citron, or pepper. Probably the body is affected in a strange way by these, and is altered in its constitution as they quietly produce their own quality or residue. It is also probable that the order and rearrangement of foods makes a great difference; for the what was called "cold course," with oysters, sea-urchins, and raw vegetables, has like a body of light-armed troops been shifted from the rear to the front, and holds first rank instead of last.

The serving of the so-called aperitifs is a great change too. The ancients did not even drink water before dessert, but nowadays people get drunk before eating a thing, and take food after their bodies are soaked and feverish with wine, serving hors-d'oeuvre of light and spicy and sour foods as a stimulant to the appetite and then thus eating heartily the remaining courses. As influential as anything in causing change and new diseases is the multiplication of effects in bathing the body, which, like iron, is made soft and fluid by heat, then plunged into cold water to be tempered.

[Plutarch says that ancient bathing was milder than in his day]

(Minar [1961] 187, 189, 199, 201, 203)

11.17 Rufus of Ephesos (108 ± 7) wrote widely but most works are lost; a few survive in Latin versions (*Joint-Diseases*) or Arabic (*Jaundice*); the *Physician's Queries* describes how to take a case history, while *Saturiasis and Gonorrhea* describes diseases of male genitals.

Kidney and Bladder Diseases

3.11–12 [stones: compare Aretaios 2.9]

In many people rather large stones exist, and the stones create sharp and violent pains, and strangury. For the cavities of the kidneys are not of easy flow, but are extremely small, and the kidney because of its solidity doesn't spread out like the bladder. Now those stones are passed with the urine sooner than those in the bladder, for bladder stones develop rather weak and soft, because they aren't long-lived. Stones in the kidneys are extremely painful both going through the urethra and again when they press against the privates.

3.30–32 [regimen to prevent stones]

It is necessary that water be both sweet and pure, both for the rest of the diet and that in which the drugs will be boiled; avoid river and marsh waters, since they create stones even where there were none. Also the wine must be thin, not too old, and white, for it is more diuretic than red wine (whether dry or sweet). In general, bring the patient to good health by hard work in moderation, massaging both the whole body and the lower back, sometimes rather dry, and sometimes rather well-oiled, and sometimes with medicines, the lees of wine and natron and with a pumice stone.

9.7–12 [surgical treatment]

7   This is how to use the probe. Lay the patient on the back, ordered to bend the legs as much as possible, and spread them as wide apart as seems good, and insert the fingers of the left hand very deep into the anus. Feel the bladder with the fingers and have an attendant press the lower belly until you find the stone.

[8: what to do in special cases]

9   After the stone is grasped, draw it to the urethra, and when it gets there, then hold it the more so it doesn't slip away, and make a transverse incision through the perineum.

10   Now if it's at hand, knock it out with the haft of the scalpel (which is made with a rough haft and a hook at the tip, for best use in this work); but if not, use the instrument invented for such things [tongs].

11   In cutting, don't cut too much: for there's a risk of cutting the bladder itself, which must be entirely avoided. Treat the incision with lint pledgets.

12   This is the safest diagnosis and treatment of bladder stones, and many doctors operating thus achieve success.

*(Sideras [1977] 116, 118, 124, 126, 152)

11.18 Soranos of Ephesos (120 ± 20) wrote also on fevers and drugs, but only his *Gynecology*, and some short works, survive in Greek (a Latin version of his *Acute and Chronic Diseases* is extant).

Gynecology

1.7–13 [uterus]

7   The uterus is situated in the cavity between the hips, between the bladder and the rectum, lying above the rectum, and sometimes completely, sometimes partly, beneath the bladder, because of its variable size. For in children the uterus is smaller than the bladder (and so lies wholly beneath it). But in virgins in their prime, it is equal to the size of the superimposed bladder, whereas in women who

are older and have already been deflowered and even more in those who have already been pregnant, it is so much bigger that in most cases it rests upon the end of the colon. This is even more the case in pregnancy (as can also be perceived by the eye) when the peritoneum and abdomen are greatly distended by the magnitude of the fetus together with its membranes and liquids.

After delivery it contracts, but its size is greater than it was before the pregnancy. Now it's larger than the bladder, but it does not lie evenly beneath it. For anteriorly, the neck of the bladder (lying along the whole vagina and ending in the urethra) is more to the front and proceeds beyond the uterus. But posteriorly, the base of the uterus is higher than the base of the bladder and lies under the navel. Thus the cavity of the bladder rests on the neck of the uterus, whereas its base rests on the hollow of the uterus.

8   By thin membranes the uterus is connected above with the bladder, below with the rectum, laterally and posteriorly with excrescences of the hips and the *os sacrum*. When these membranes are contracted by an inflammation, the uterus is drawn up and inclined, but when they are weakened and relaxed, the uterus prolapses. Now the uterus is not an animal (as some people thought), but it is similar in certain respects, having a sense of touch, so that it is contracted by cooling agents but relaxed by loosening ones. [Aretaios 2.11 preserves the tradition of the "wandering womb"]

[9: uterine nomenclature – see Figure 11.1]

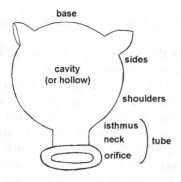

Figure 11.1

10   The orifice [cervix] lies in the middle of the vagina, for the neck of the uterus is enclosed tightly by the inner part of the vagina, while the outer part of the vagina ends in the labia. From the latter the orifice is more or less distant, according to the age (thus in most adult women the distance is five or six *daktuls* [9 to 11 cm]). The orifice becomes more accessible with deliveries since the neck of the uterus elongates. It varies in size too, but in most cases in the natural state the orifice is as large as the external end of the auditory canal. Yet at certain times it dilates, as in the desire of intercourse for the reception of the semen, and in the menses for the excretion of the blood, and in pregnancy in proportion to the

growth of the embryo. In parturition it dilates further, to an extreme degree till it even admits the hand of an adult. In its natural state, the orifice is soft and fleshy, similar to the sponginess of the lung or the softness of the tongue. But in women who have borne children it becomes more callous and, as Herophilos says, "similar to the head of an octopus or to the larynx," being callused by the passage of secretions and children.

[11: the whole uterus is a sinewy structure]

[12: compare above Herophilos, *Anatomy*, book 3 (fr. 61)]

[13: the whole uterus is composed of two layers which are arranged crosswise]

1.36 [best time for conception: Hippokrates, *Nature of the Child* 15 says the same]

Just as for sowing seed outside every season is not propitious for the purpose of bringing forth fruit upon the land, so in humans too not every time is suitable for conception of the seed discharged during intercourse. Now so that the desired end may be attained through the well-timed practice of sex, it's useful to state the proper time. The best time for fruitful intercourse is when menstruation is ending and abating, when urge and appetite for coitus are present, when the body is neither in want nor too congested and heavy from drunkenness and indigestion, and after the body has been rubbed down and a little food been eaten and when a pleasant state exists in every respect.

[prior to menstruation, uterus is congested; at start of menstruation, uterus is expelling]

Consequently, the only suitable time is at the waning of the menses, for the uterus has been lightened, and warmth and moisture moderately imparted. For again, it is not possible for the seed to adhere unless the uterus has first been roughened and plowed as it were in its base. Now just as in sick people food taken during a remission and before the paroxysm is retained, but is ejected by vomiting if taken at the paroxysm, in the same manner the seed too is securely retained if offered when the menses are abating. But if some women have conceived at another time, especially when menstruating a short while, one must not pay attention to the outcome in a few, but must point out the proper time as derived from scientific considerations [*technikê theôría*].

1.39–41 [shaping the fetus]

39 What is one to say concerning the fact that various states of the soul also produce certain changes in the mold of the fetus? For instance, some women, seeing monkeys during intercourse, have borne children resembling monkeys. The tyrant of the Cyprians who was misshapen, compelled his wife to look at beautiful statues during intercourse and became the father of well-shaped children; and horse-breeders, during covering, place noble horses before the

mares. Thus, in order that the offspring may not be rendered misshapen, women must be sober during coitus because in drunkenness the soul becomes the victim of strange fantasies; this furthermore, because the offspring bears some resemblance to the mother as well, not only in body but in soul. Therefore, it is good that the offspring be made to resemble the soul when it is stable and not deranged by drunkenness. Indeed, it is utterly absurd that the farmer takes care not to throw seed upon very moist and flooded land, and that on the other hand people assume nature achieves a good result in generation when seed is deposited in bodies which are very moist and inundated by satiety.

[1.40: post-coital massages aid in conception; 1.41: the waxing moon aids in conception]

1.60–62 [contraception and abortion]

[60: distinction between contraception and abortion]

61 For if it is much more advantageous not to conceive than to destroy the embryo, one must consequently beware of having sex at those times we said were suitable for conception. And during the sexual act, at the critical moment of coitus when the man is about to discharge the seed, the woman must hold her breath and draw herself away a little, so that the seed may not be hurled too deep into the hollow of the uterus. And getting up immediately and squatting down, she should induce sneezing and carefully wipe the vagina all round; she might even have a cold drink.

It also aids in preventing conception to smear the orifice of the uterus all over before with old oil or honey or cedar resin or balsam, alone or together with white lead; or with a moist cerate containing myrtle oil and white lead; or before the act with moist alum, or with galbanum together with wine; or to put a lock of fine wool into the orifice of the uterus; or before coitus to use vaginal suppositories with the power to contract and to condense. For such of these things as are styptic, clogging, and cooling cause the orifice of the uterus to shut before the time of coitus and do not let the seed pass into its base. Such, forever, as are hot and irritating, not only do not allow the seed of the man to remain in the hollow of the uterus, but draw forth as well another liquid from it.

[62: eight more recipes, of which five involve pomegranate rind: compare Riddle [1992]]

(Temkin [1956] 8–11, 34–38, 63–64)

11.19 Galen of Pergamon (129 to 210 ± 5), viewing himself as philosopher and physician, regarded Hippokrates and Plato as the most worthy authorities; he wrote voluminously on medicine and much else, and served as court physician to three emperors: M. Aurelius, Commodus, and Seuerus.

Sects [written 166 ± 2: "schools" of medicine]

1–3, 6 [a "sect" was a system or worldview, such as Platonism or Stoicism;
the five chief medical sects in Galen's day were the Dogmatics or Rationalists,
supposedly founded by Hippokrates; the Erasistrateans (omitted by Galen); the
Empirics (from 250 BCE); the Methodists, apparently founded by Thessalos of
Tralles (see Chapter 4); and the Pneumaticists (omitted by Galen), apparently
founded by Athenaios (d. *ca.* 20 BCE). In *His Own Books*, Galen suggests
this book ought to be among the first to be read.]

1 The aim of the art (*technê*) of medicine is health, but its end is the possession
of health. Doctors must know how to bring about health, when it is absent, and
how to preserve it, when it is present. Those things which bring about health
when it is not there are called medicines and remedies; those things which
preserve it when it is there, healthy regimens [regimen included both dietary and
exercise regulation]. Thus it is also that the ancient account says that medicine is
the science of the healthy and of the diseased, calling healthy those things which
preserve an existing health or restore a ruined health, diseased the opposite of
these. For the doctor needs the knowledge of both, in order to choose the one and
avoid the other.

[the two main sects are Empiricists and Rationalists]

2.1 The empiricists claim that the art comes about in the following way. One
has observed many affections in people. Of these, some occur spontaneously,
both in the sick and the healthy, e.g., nose-bleeding, or sweat, or diarrhea, or
something else of the kind which brings harm or advantage, though what it is
that produced the effect has no perceivable cause. In the case of other affections,
the cause is manifest, but they, too, occur not due to some choice of ours, but by
some chance. Thus it just so happened that somebody fell or was hit or wounded
in some other manner and that then there was a flow of blood, or that somebody
who was ill gratified his appetites and drank cold water or wine or something else
of the kind, each of which had a harmful or beneficial effect.

...

It is this kind of experience which has contributed most to their art. For when
they have imitated, not just twice or three times, but very many times, what was
formerly beneficial, and when they then find out that, for the most part, it has the
same effect in the case of the same diseases, then they call such a memory a
"theorem" and think that it already is trustworthy and forms part of the art. But
when many such theorems had been accumulated by them, the whole
accumulation was the art of medicine, and the person who had accumulated the
theorems was a doctor. Such an accumulation came to be called by them autopsy,
being a certain kind of memory of what is perceived to happen many times in the
same way.

...

2.2 Now, it also sometimes happened that one encountered diseases which had not been seen before or diseases which were known, but which one encountered in areas where there was no ready supply of medicines which had been observed by experience. Hence they turned the "transition to the similar" into a sort of device to find remedies. By means of this device, they often transfer the same remedy from one affection to another and from one place affected to another, and they proceed from a formerly known remedy to one quite similar.

...

3.1 On the other hand, the method which proceeds by means of reason exhorts us to study the nature of the body which one tries to heal and the powers of all the causes which the body encounters daily, as a result of which it becomes healthier or sicker than it was before. Moreover, they say, the doctor also has to be knowledgeable about "airs, waters, places," [referring to the work by Hippokrates] occupations, foods, drinks, and habits, so that he may figure out the causes of all diseases and be able to compare and to calculate the powers of the remedies, i.e., that something which has such and such a power, if applied to this kind of cause, naturally produces that kind of effect. For, they say, it is not possible for him to have an ample choice of remedies to resort to, unless he has been trained in all these things in their many aspects.

...

Thus the disposition itself offers the rationalists the indication of what is beneficial. But this indication in itself is not enough, they say; we also need another indication, derived from the power of the sick person, and another one, from his age, and yet another one, from the particular nature of the patient himself. But in this way, we also obtain a particular indication of what is beneficial from the season of the year, and the nature of the place, and his occupations, and his habits.

6.1 [the new-fangled Methodists] The so-called Methodists, however (for this is how they name themselves, claiming that not even their dogmatic predecessors could practice the art by a method), seem to me not only to disagree with the ancient sects as to the account of the art but, beyond that, also to rearrange the practice of the art in many respects. They claim that neither the part affected has anything useful for an indication of treatment, nor the cause, nor the age, nor the season, nor the place, nor the consideration of the power of the sick person, nor his nature, nor his disposition. They also put aside habits, claiming that the indication of what is beneficial, derived from the affections alone, is enough for them, and not even from these, taken as specific particulars, but assuming them to be common and universal. And hence they also call these affections which pervade all particulars "communities" [common, or universally-shared, features]. And some of them try to show that all diseases which occur because of regimen manifest two communities and a third mixed one, whereas some try to show simply that all diseases are thus. These communities they called "containment" and "flux," and they say that each disease is either "containing" or "fluent" or a combination of both. For, when the natural bodily outflows are checked, they call this "containing"; when they continue rather more they call it "fluent."

(Walzer and Frede [1985] 3–6, 10)

Anatomical Procedures [written about 170]

1.2 [observation and dissection]

. . . Of all living things the ape is most like humans in viscera, muscles, arteries, veins, and nerves, as in the form of the bones. From the nature of these it walks on two legs and uses its fore-limbs as hands, and has the flattest sternum of all quadrupeds, and humanoid clavicles, and a round face with small neck. Given these facts, its muscles could not be otherwise, for they are extended over the bones, reproducing their size and shape. So also arteries, veins, and nerves conform to the bones. [compare Aristotle, *Animal Researches* 2.8–9 (502a16–b27), and Chapter 10.6, Poseidonios]

. . .

Make it your serious endeavor not only to learn precisely from books the form of each bone but also to examine assiduously with your own eyes the human bones themselves. This is quite easy at Alexandria because the physicians there employ ocular demonstration [*autopsia*] in teaching their students. For this reason, if for no other, try to visit Alexandria. But if you cannot, it is still not impossible to see something of human bones. I, at least, have done so very often on the breaking open of a grave or tomb. [two accounts of Galen's use of adventitious human corpses]

If you have not the luck to see anything of this sort, dissect an ape and, having removed the flesh, observe each bone precisely. Choose for this the most humanoid apes, those with short jaws and small canines. You will find other parts also humanoid, for they can walk and run on two feet. (Those, on the other hand, like the dog-faced baboons, with long snouts and large canines, far from walking or running on their hind-legs, can hardly stand upright.) Even the more human sort fall a little short of a precisely erect posture: firstly the head of the femur fits into the socket at the hip-joint rather transversely, and secondly, of the muscles which extend downward to the knee, some go further.

7.15 [vivisection of the heart]

Ascertain precisely in a dead animal the bending places of the ribs and recall them before you start. Arrange the animal on its back as explained [7.12]. The procedure is as follows. Remove the hair where you're going to cut, and make two longitudinal cuts dividing the flexures of the ribs. Next make a transverse incision at right angles across the xiphoid process, where, of course, you will encounter the arteries and veins. (Disregard hemorrhage from them, for you no longer aim at keeping the animal alive.) Now bend back the sternum and make a third cut under it, separating the pericardium from it. If you wound the pericardium without wounding the heart, pay no heed in this case, for your aim is to see if both the ventricles beat, and that together, and not, as some say, only the left. You will see still more clearly now than before and with abundant evidence whether the arteries throughout the whole animal expand and contract alternately or at the same time and rhythm.

All this will be clear to you at once when the heart is exposed. As time passes, the movements of each ventricle become brief, long pauses intervening, and also there becomes apparent the expansion [*diastolē*] of the right ventricle, accomplishing its function according to its own nature, as you will see particularly when those parts approach immobility. For in each ventricle the apex stops moving first and then the part next to it, and so on until only the bases are left still moving. When even these have stopped, an ill-defined and short movement at long intervals is still seen in the "auricles." The cause of this phenomenon we must seek at leisure, for it's not reasonable that its outgrowths should move longer than the heart itself. But here we do not aim to seek causes in this matter, but observed anatomical phenomena only. [compare Aristotle, *Animal Researches* 2.11 (503a15–b28): vivisected chameleon's heart beats]

7.16 [experiment done in about 165, repeating one by Erasistratos; an earlier account in *Blood in the Arteries* 8; Galen disparages various less and more absurd alternate methods; then:]

The procedure is as follows. Of the large arteries near the skin, expose one, such as that by the groin, which is the one that I habitually use for this operation. Ligature it above and compress the artery itself with the fingers of the left hand, choosing as great a length as possible from the ligature devoid of a large branch. Then make in its wall a straight incision long enough for you to insert a tube between the ligature and the fingers. (Have ready a tube a *daktul* long [about 2 cm], such as a writing reed, or bronze pipe made for the purpose.) Obviously there won't be hemorrhage from the incised artery during this step, since the upper part, whence comes the blood, is stopped by the ligature, while the lower part no longer pulsates because of the ligature and because it is compressed by the fingers. Hence you have great leisure to insert the tube into the artery through the incision in its wall, and then tie off artery and reed with fine linen thread. (Take care that no part of the reed go too far beyond the incision of the artery, and that the reed be of a caliber that the arterial coat does not lie slack on it, for we want it to remain in place, neither running up beyond the division in the artery, nor down it.) This done, loosen the noose and, as a precaution, alter the position of the fingers with which you were compressing the artery, to the part round the reed. If the reed be tight and precisely bound, there is no need to control it, and you can observe at ease the uninterrupted part of the artery above the tube still pulsating as before and the lower part quite pulseless.

(Singer [1956/1999] 1–2, 196–197, 199)

Preserving Health [written about 175]

[1.1: health is the balance of the elements]

1.6.1–12 To them, therefore, let us leave the custody of the health they dream up, and let us come to those forms which are obvious; and, assuming that their nature is twofold, as we have just now said, let us assign the specific scope of each; to the perfect state precise care, so far as perceptible; to the imperfect state, imprecise care. For we must try to correct the distempers of health, making moister those conditions which are too dry, and making drier those which are too moist; and similarly purging the excess of those which are too warm, and restraining the excess of those which are too cool. And by what hygienic regimes one would do this, the following account will show.

First I must consider how one would preserve the health of the best constitution; but before this, let us consider what is the best constitution. It is, according to those who describe its nature, that which is best proportioned, which has the conformation of its parts precisely suited to their functions, and which in addition exhibits every number and size and mutual relation of all parts advantageous to their actions.

According to our standards, the precisely correct body weight is midway between thin and corpulent (and there is no difference between "corpulent" and "obese"). And likewise of the other extremes, the precise mean is such that one could not call it either hirsute or bald, soft or hard, white or black, small-veined or large-veined, irascible or apathetic, drowsy or insomniac, sluggish or alert, over-sexed or frigid. And if the exact mean of all the extremes were in all parts of the body, this would be the best to observe as being the symmetry most suitable for all labors. And it will also possess all the other standards of good constitution of each part which we have mentioned in *Constitutions*, book 2. For many bodies are well-constituted in the head, for example, and poorly in the thorax, the abdomen, and the genitalia. And in some the distemper is in the limbs, and in many in some one of the viscera, or in some other one part, or in several, so that in some persons the distemper is in several viscera.

(Green [1951] 20–21)

Advice for an Epileptic Boy [written around 190]

2, 4 [diet to prevent seizures]

2 I shall try to go through as clearly as possible the regimen by which the boy may benefit not a little and may suffer least harm from unexpected daily occurrences. These one must avoid as far as possible. Sometimes, however, he will necessarily encounter frost and extreme heat, strong winds and strenuous baths, repulsive food and whirling wheels, lightning and thunder, sleeplessness and indigestion, distress and anger and weariness and similar things of which the chief characteristic is that they stir up and trouble the body extremely, remind it of the disease, and produce a paroxysm. It is necessary to avoid these carefully, and if ever they occur and a paroxysm follows, then the boy must abstain completely from any motion, he must stay at home, and be put on a very light diet until the body throws off the fatigue resulting from the attack.

...

4   [Galen advises "eat a variety of fruits and vegetables"; then:] Speaking generally, I recommend abstinence from daily or immoderate use of such food as engenders unhealthy humours, or as causes constipation or flatulence and is hard to digest. Such food, if taken constantly or more than is advisable at a time, usually causes harm not in this disease only but in all other diseases too.

Thus far the dietetic remarks would be equally valid for many other diseases; it is, however, peculiar and special to this disease that one must chiefly beware of phlegmatic food. Therefore, it is not good to partake habitually of things which, although harmless otherwise, have a viscous or cold or thick humour, such as orach [Atriplex rosea], blite [Amaranthus blitum] and mallow, and although I do not exclude them, I don't wish them to be eaten always. Gourds belong to the same category and cucumber, apples, and pears even more so, and finally the so-called mushrooms, the worst of all foods with a phlegmatic, thick, and viscous humour. From these I advise complete abstinence, just as from turnips and all other edible roots. For they have a thick humour and are on the whole hard to digest, except if they contain something sharp and warm, like parsnips and radishes. The boy may taste radish every few days, but he should try to abstain from parsnips and especially from turnips. He may have plenty of such food as contains something sharp and pungent, and which does not obviously engender bad humours nor has a smell affecting the head. Those things, however, which by their heat make the head full such as wine, mustard, parsley, parsnips, onions and *smurnion* [Smurnium perfoliatum], belong to this class of exceptions. They overheat and engender bad humours. Mustard, although very apt to separate the humours, must be avoided since it affects the head.

[5: abstain from fatty meat and oysters]

(Temkin [1934] 181, 184–185)

Anatomy of Nerves [written after 195]

1–10

1   All the physicians agree that none of the parts of the animal has either motion (which we call purposeful) or sensation without a nerve [neuron] and that if the nerve be cut, the part immediately becomes motionless and insensitive. It is

not known to all, however, that the origin of the nerves is the brain, and the spinal marrow, and that some grow from the brain itself, and others from the spinal cord, although this is the way they are seen in dissections.

[2: the optic nerve; 3–9: others]

10 Now the muscles common to other parts of the larynx do not always receive nerves from the 6th pair, just as neither do those muscles joining the bone called lambdoid [like "Λ"] or hyoid [like "Y"] to the sternum, about which it is accurately stated in *Anatomical Procedures* where there is also a statement concerning the distribution of the three nerves said to emerge from foramen which is at the end of the lambdoid structure. Although almost everyone thinks they are a single nerve, if we take up in a noose the one running along the arteries (*n. vagus*), the animal immediately becomes speechless. Indeed, the muscles of the larynx also receive offshoots from it. But of the other two, one reaches each of the two muscles of the pharynx and the root of the tongue (*n. glossopharyngeus*), but the other goes to the muscle of the scapula, the flat muscle, and some others there (*n. accessorius*). But it has escaped notice that those nerves, which are not small and run along the arteries, course through the neck and thorax before entering the body of the stomach, to which the principal part of these nerves is attached and distributed. But it is remarkable that they say that some parts of them are distributed to the diaphragm although actually the diaphragm does not receive the least bit from this pair; and again they do not even mention that some parts of these nerves return from the midst of the thorax to certain muscles of the larynx (*n. recurrens*), nor say what power they have. And yet these cause aphonia in animals if injured, as well as the great pair of nerves with the arteries, because these same are parts of the cause of the loss of voice if they are injured. [the facts about speechlessness were discovered by Galen around 160]

(Goss [1966] 328, 330–331)

Venesection [written after 200]

5 [benefits of blood-letting]

Not only do the parts of the animal derive their nourishment from the blood, but the innate heat also owes its continuance to it, just as the fire on the hearth does to the burning of suitable logs, by which we see whole houses made warm. And just as this fire is sometimes harmed if logs are piled on it indiscriminately, and sometimes if, although not too abundant, they are very damp, or if none are put on it at all, or very few – so also the heat in the heart sometimes becomes less than normal because of the excess of blood, or a great shortage of it, or a cold quality; and sometimes more, either because of a warm quality of the blood, or a moderate excess of it. And whatever the heart may suffer as a result of cold or heat, the other parts of the body immediately share in.

[other parts of the body may become hot or cold, and thereby affect adjacent parts, perhaps even the heart]

And once the heart is heated, the whole substance of it readily becomes hot, just as the house round the hearth is heated when the hearth has a large fire on it. The Greeks call such a condition of the body fever. Sometimes the *plēthos* of blood, before it has begun to putrefy, arrives in force at some part, either mortifying it completely, so as to destroy its function, or doing it notable damage. The apoplexies originate in this way, by a concerted rush of a quantity of blood to the governing center of the animal. Similarly, when it descends on some other part, it causes an abnormal swelling in it. Inflammation also comes from this sort of process. When the blood that has descended on the part is too thick and melancholic, the swelling that results is scirrhous, just as it is flabby when the flux is more phlegmatic. When the flux is bilious it leads to *erusipelas*. All these things are precisely classified for you in the works I have recently mentioned.

Now, as I said, applying the things that have been demonstrated to the question before us, I shall show that the argument concerning venesection follows from them. It seems best to start from the fact that *plēthos* is of two kinds. The variety known as dynamic *plēthos* readily goes on to putrefy, and of course also sometimes descends on a part, causing abnormal swellings in parts so affected. The other sort, which is known as *plēthos* by filling, also frequently rushes down into parts leading to swellings, but it is a cause of apoplexies and rupture of veins as well; it is therefore essential to try to evacuate *plēthos* quickly, before it has had a chance to do the patient some grave harm.

(Brain [1986] 72–74)

11.20 Philoumenos of Alexandria (around 180) wrote on gynecology (lost), bowel diseases (partly preserved in Latin), and this work on poisonous animals, often indebted to Nikandros. He offers 10 paragraphs of general advice, then discusses 27 different toxic beasts (from wasps, scorpions, and spiders to snakes); many of his snakes cannot be securely identified. (The source translates the passages out of order because of the particular use being made of them.)

Venomous Animals

17 [echidna or viper]

With respect to those that have been bitten by the male viper *echidna*, we find punctures two or four in number, but more widespread than from the asp. From them blood oozes at first, then a bloody oily fluid like bile along with swollen, reddish, blister-like inflammation, with discoloration from extravasation of fluid having spread. The mouth is parched; there are burning, faintness, and a shivering sensation, then also vomiting of bile, colic, heaviness of the head and backache, vertigo, pallor, retching, fever, rapid breathing, leaden-colored skin, and cold sweat. Death occurs within seven days, most often on the third, especially in those wounded by the *echidna*. [remedies for the bite]

20 [*dipsas*: "thirst-inducer"]

The *dipsas* is called by some theriacists the burner serpent. Its length is one cubit, drawn out from thickness to thinness. Black and orange spots spread along the entire body. The head is very narrow. In those bitten by it there follow at once swelling and corresponding inflammation. Now these things are common in other cases: the sufferer becomes very thirsty and fevered, and though taking much drink expels nothing through urine, sweat, or vomit. They perish from two causes: either overwhelmed by great thirst if they drink nothing, or from much over-filling should they take drink, as in the case of dropsical persons, bursting asunder with accompanying spilling at the groin downwards or toward the lower abdomen. [remedies for the bite]

22–23 [*ammodutes* "sand-burrower" and *sêps* "putrefier"]

22   The *ammodutes* is found of one cubit and no longer. It is sandy in color with black spots distributed over the body. The tail is very hard and is cleft from above. By some it is called the *kechrias* [only here]. Its jaws are rather wide. In most cases of those struck by it destruction follows immediately. The strike of this coiled animal is terrible, thus its name. In those it strikes there occur vomiting, secretion from the wound, then swelling and in a little while it runs with serous matter; drowsiness and swooning follow closely. Erasistratos says that the liver, the bladder, and the colon are affected, that on dissection these parts are found to be destroyed. Death occurs in three days, with some surviving until the seventh day. Death is quicker when one is struck by the female. [remedies for the bite]

23   The so-called *sêps* is found two cubits in length and goes from thickness to thinness. It is straightforward- and slow-moving, its head is flat, its mouth tapering, and over all its body it is sprinkled with white spots. In those struck by this one there happens to flow from the visible perforations blood followed by a little stinking fluid, also swelling and pain. The parts affected mortify and turn white, the skin of all the body leprous, all bodily hair falls out. The end comes in as long as three days. [remedies for the bite]

(Knoefel and Covi [1991] 100, 113, 128–129, 110)

12

"PSYCHOLOGY"

Our own awareness and analysis of the world becomes an object of our wonder and study. Every literate culture records its speculations about the relations connecting mental activities to our surrounding body and world. "Soul," the life and mind in us, is sometimes dormant, and eventually absent (departed or dissipated). Uniquely among animals we speak and grasp *words*. We observe patterns of behavior linked perhaps to family or tribe, and invoke unseen forces to explain anomalies and distortions of character.

Mesopotamians and Egyptians felt that the soul resided in the heart; but in the earliest Greek literature, it was often convenient to describe conflicting feelings and desires as if emanating from multiple semi-autonomous organs: liver, diaphragm (*phrên*), or heart (Pelliccia [1995]). Sêmonides of Amorgos (around –650) interpreted the manifold characters of women in terms of various animals: e.g., the wretched "ferret-woman," who is sex-crazed and thievish. The earliest Greek speculations about the operations of nature assumed we were an integral part thereof. Anaximenes of Miletos (–545 ± 20) supposed that a life-giving *pneuma* (breath or spirit) pervaded the world and us. Xenophanes of Kolophon (–510 ± 30) emphasized the contingency and relativity of human knowledge and sensation (frr. B34, 38 DK). Herakleitos of Ephesos (–500 ± 20) darkly hinted that soul is fire, expended in anger, dampened by wine, eclipsed in sleep, and extinguished by the moist death of a diseased body. Only war-slain souls die a better, because fiery, death.

Pythagoras of Samos (perhaps –530 ± 30) pursued a dualistic path and introduced the notion that souls migrate at death to a new body (human or animal). His later follower Philolaus of Kroton (–420 ± 10) argued that the incarnate soul perceives through internal attunement or harmony (like a resonating string). An earlier Krotoniate, Alkmaion of Kroton (perhaps –500 ± 20), following or accepting Pythagoras, suggested that human character, emotion, and thought might be explained by the mixture of several internal juices, such as blood, phlegm, bile, and *melancholê* (compare Chapter 11), and claimed that the soul was immortal because in perpetual motion.

In the aftermath of the failed Persian invasion, Parmenides of Elea (–470 ± 20) supposed that knowledge and sensation depended on the balance of hot and cold within us, and indeed within everything, since beings perceive by likenesses (heat by heat, cold by cold, etc.), so that all existing objects have some share of perception. Empedokles of Akragas (–455 ± 20) poetically depicted sensation as originating in the interactions of similars, especially in our blood (the best mixture of elements), centered around our heart. He worked out details of elemental shapes and corresponding holes in sense organs (fire in the eyes for sight, a bell in the ear to hear, and *pneuma* in the nose allowing smell). In the cosmology of Anaxagoras of Klazomenai (also –455 ± 20), "Mind" ruled all, and humans had some share thereof enabling thought and perception. Moreover, it is by the interaction of opposites that we perceive, in proportion to our lack of each kind: salt seems saltier the less salty we are. Antiphôn of Athens (–430 ± 20) wrote on dream-interpretation (frr.B78–81 DK). The atomistic system of Demokritos of Abdera (writing –410 ± 30) sought to explain all secondary qualities in terms of mechanical effects of atoms (arrangement, motion, or shape). Sensation is caused by atoms of certain shapes impinging upon the relevant sense organ (e.g., bitter taste is caused by small, smooth, rounded atoms of "sinuous" circumference). Theophrastos records many details (see below). Diogenes of Apollonia (writing around –430 ± 10) attributed all thought and sensation to air and its mixture within us. Clarity of perception and thought is determined by the fineness and dryness of our air. A similar theory is propounded in one of the books in the library attributed to Hippokrates (around –400 ± 30), *Sacred Disease* 16.

In the Hippokratic library, diseases of the mind or "soul" are considered and treated just as bodily ills. The *Epidemics* often lists delirium as a more or less significant symptom of illness: the feverish Erasinos (book 1, case 8), Philistes of Thasos afflicted to death with headache (book 3.1, case 4), the daughter of Euruanax (case 6), or the grief-stricken wife of Delearkes of Thasos (book 3.17, case 15). The *Prognosis* records delirious hand-waving as an inauspicious sign (section 4), and in cases of fever indicates that headache, visual disorders, and convulsions are significant (section 24). Wine is to be avoided in disorders of the head or *phrên* (*Regimen in Acute Diseases* 63; *Diseases* 2.72). One work, *Sacred Disease*, attempts to explain a particular psychic disorder, epilepsy – on the basis of excess phlegm afflicting the brain, which the author is certain is the seat of feelings, thought, sensation, and judgment. Another psychic ill is *melancholê* (depression with other symptoms), described in *Epidemics* 6.8.31 and *Diseases* 1.30 (compare also *Epidemics* 7.86–91). The character of patients can sometimes be deduced from body type (*Epidemics* 2.5.1, 23; 2.6.1, 14; 6.4.19; and *Regimen* 1.36), while the anti-slavery philosopher Antisthenes of Athens (–400 ± 10) is credited with a book on physiognomy.

Although Plato of Athens (writing –365 ± 20) was primarily concerned with ethics, he offers his best guess as to sensation and psychic architecture in the *Timaios*. The eyes are fiery enabling sight by mingling with external fire or inducing sleep when that is unfeasible (45b–46c); colors are produced by the different sizes of flame particles (67c–68d). Hearing exists for the sake of appreciating harmony and rhythm of sound (47de), in which rapid motion is high pitched, and uniform motion is smooth sounding (67bc). Taste happens through various contractions or dilations in the tongue (65c–66c), while smell lacks symmetry and can only be good or bad (66d–67a). Pleasure and pain are respectively natural and violent alterations within us (64a–65b), while mental disorders are imbalances in the psychic architecture (86b–88c). Earlier, in the *Republic* (especially book 4), he had elaborated his belief that the structure of the soul and of the state are necessarily parallel and tripartite: the rational part (brain or philosopher-king) rules the spirited part (heart or "guardians"), and the appetitive part (liver or workers) is constrained to feed the whole. The location of these three souls in our bodies he explains in the *Timaios* (69b–72d): head and brain safely separated from body by neck, heart walled off from lowly liver by diaphragm. In men the penis and in women the womb are almost separate self-willed animals which move about and cause trouble (91ac). He offers a different model in *Republic* 4 (441ab) and *Phaidros* 246, of the human part of the soul as a charioteer driving a pair of winged horses, our animal parts.

Herakleides of Pontos (–340 ± 20) in his *Music* (book 3) described the psychic effects of the various modes of Greek music (fragment preserved in Athenaios, *Deipnosophists* 14 [624c–625b]) – the Dorian mode like that conservative people is stern, intense, dignified, and masculine; the Aeolian mode like its eponymous folk is lofty and bold; while the Ionian mode is severe, hard, and suited to tragedy.

Aristotle of Stageira (writing –335 ± 10) considered the problem of the "physiological" basis of the mind in several of his many works; he advocated the theory that the *hegemonikon* ("center of command") was located in our heart (*Parts of Animals* 2.10 [656a14–b8], 3.3 [665a10–26]), while the soul was not separable from the body (*On the Soul* 1.3 [406a31–b5]) and was tripartite, like Plato's (*On the Soul* 2.3 [414a29–b19], 3.9 [432a15–b8]). In *On the Soul* 3.2 (425b12–426a2) and 3.7 (431a1–431b9) he asserts that the imaginative faculty mediates between perception and understanding. In *Parts of Animals* 2.4 (650b20–27) blood affects thought – thin blood is best for thinking. For Aristotle the soul could not act apart from the body (*On the Soul* 1.1 [403a3–25]), and animals have sensation in common with us (*Animal Researches* 4.8 [532b34–533a17]). He discusses the different senses and explains each as a species of contact actualizing a potential (*On the Soul* 2.7–11 [418a26–424a15]). *On Sensation* discusses sight, hearing, taste, and smell, and Aristotle relates each to an element (water: sight; air: sound; solid earth: touch; and smoky fire: smell), but then asserts that taste and smell are in fact the same (taste being produced by scent particles dissolved in water). *Sleep* explains that digestion, disease, or drugs evaporate intestinal juices which rise to the head and there condense (the brain being a refrigerator), thence dripping down and shutting off the sensorium at the heart, until digestion completes. *Dreams* claims that persistent memories lightly stimulate consciousness but are perceptible only in sleep, and that vision rays (compare Chapter 7) actually affect objects (e.g., menstruating women redden mirrors). Finally, *Prophecy in Sleep* allows that dreams may have medical meaning or may modify subsequent behavior, and events in dreams may be explicable by likeness to waking events (some people being very susceptible to external stimuli).

Members of Aristotle's school composed the *Problems* and the *Physiognomy*. The latter explains (section 1):

> Mental character is not independent of and unaffected by bodily processes, but is conditioned by the state of the body; this is well exemplified by drunkenness and sickness, where altered bodily conditions produce obvious mental modifications. And contrariwise the body is evidently influenced by the affections of the soul – by the emotions of love and fear, and by states of pleasure and pain. But still better instances of the fundamental connection of body and soul and their very extensive interaction may be found in the normal products of nature. There never was an animal with the form of one kind and the mental character of another: the soul and body appropriate to the same kind always go together, and this shows that a specific body involves a specific mental character. Moreover, experts on the animals are always able to judge of character by bodily form: it is thus that a horseman chooses his horse or a sportsman his dogs. (Barnes [1984] 1237)

Three methods are mentioned (comparisons with animals, as in Sêmonides, racial patterns, and facial cues), but the author prefers using similarities of body parts (section 2): movements, color, facial expressions, hair, skin, voice, etc. For example, "soft hair indicates cowardice, coarse hair bravery"; or "rapid motions indicate fervid temper"; or deep voices imply courage (arguing from animal noises). Typical persons are briefly described, e.g.: "the sly man is fat around the face, with wrinkles round his eyes, and he wears a drowsy expression." In the *Problems*, chapter 3 is devoted to the effects of alcohol, parts of chapter 4 to the origin of sexual desire, chapter 7 to sympathic actions (such as responsive yawning), chapter 19 to the effects of music, chapter 27 concerns signs of fear (trembling and urination), and chapter 30 treats *melancholê* (Simon [1978] 229–232); palmistry is discussed in 10.29. Perhaps a little later is the physician and physiognomist Loxos (who survives in late Latin extracts: Misener [1923]).

12.1 Theophrastos of Eresos (lived –370 to –286) succeeded Aristotle as the head of his Lukeion "school," and wrote widely, amassing data questioning Aristotle's system (Keyser [1997c]). The works here excerpted were part of a set of works on natural philosophy (see also Chapters 8–11). Theophrastos' careful and precise distinctions are noteworthy.

On the Senses

1 [basic theory of perception]

The various opinions concerning sense perception, when regarded broadly, fall into two groups. By some investigators it is ascribed to similarity, while by others it is ascribed to contrast: Parmenides, Empedokles, and Plato attribute it to similarity; Anaxagoras and Herakleitos attribute it to contrast.

The one party is persuaded by the thought that other things are, for the most part, best interpreted by similarity; that it is innate to all creatures to know their kin; and furthermore, that sense perception takes place by means of an effluence, and like is borne toward like.

49–58 [Demokritos on vision, hearing and thought: compare below Section 2 (Epikouros)]

49 Demokritos in his account of sense perception does not distinguish whether it is due to contrast or to similarity. For in so far as he ascribes the action of the senses to an alteration, it would seem to depend on contrast; for the like is never altered by the like. On the other hand, sense perception would seem to depend on similarity in so far as he ascribes perception and, in a word, alteration to the fact that something is affected. For things that are not the same cannot be acted upon, he says, but even when things that are different do act, their action is not due to their difference but to there being something the same. Upon such matters he may thus be understood either way. He now undertakes to discuss the senses each in turn.

50 Vision he explains by the reflection in the eye, of which he gives a unique account. For the reflection does not arise immediately in the pupil. But the air between the eye and the object of sight is compressed by the object and the visual organ, and thus becomes imprinted; since there is always some effluence arising from everything. Thereupon this imprinted air, because it is solid and is of a hue contrasting with the pupil, is reflected in the eyes, which are moist. A dense substance does not receive this reflection, but it penetrates what is moist. So moist eyes see better than hard eyes; provided their outer tunic be most fine and close-knit, and the inner tissues be most spongy and free from dense and strong flesh, and free too from thick oily moisture; and provided the ducts connected with the eyes be straight and dry, so they may "perfectly conform" to the entering imprints. For each knows best its kindred.

51 Now in the first place this imprint upon the air is an absurdity. For the substance imprinted must have density and not be "fragile"; just as Demokritos himself, in illustrating the character of the "impression," says that "it is as if one were to take a mold in wax." In the second place, an object could better imprint water than air, since water is denser: while it ought to be more visible, it is less so. In general, why should Demokritos assume this imprint, when in his discussion

of forms he has supposed an effluence that conveys the object's form? For these images due to effluence would be reflected.

52   But if this occurs and the air is molded like wax that is squeezed and pressed, how does the reflection occur, and what is its character? For the imprint here as in other cases will evidently face the object seen. And so it's impossible for a reflection facing us to arise unless this imprint is turned around. The cause and manner of this reversal ought to be shown; for in no other way could vision happen. Moreover when multiple objects are seen in the same place, how can multiple imprints be made in the same air? And again, how could we possibly see each other? For the imprints would inevitably clash, since each of them would be facing whence it came. So this must be examined.

53   Furthermore, why does not each person see himself? For the imprints from ourselves would be reflected in our own eyes just as they are in the eyes of our companions, especially if these imprints directly face us and if the effect here is the same as with an echo (Demokritos says that in the case of the echo the sound is reflected back to the speaker). Indeed imprinting on air is absurd. For we should be forced to believe, from what he says, that all bodies are producing imprints in the air, and that great numbers of them are sending their impressions across one another's path, which is odd and unreasonable. If the impression moreover endures, we ought to see bodies that are out of sight and remote, if not by night, at least by day. (Although these imprints would likely persist no less at night, since then the air is so much cooler.)

. . .

55   [hearing] His explanation of hearing is very much like others'. For the air, he holds, bursts into a cavity and causes motion. While it gains entrance to the body in this same manner at every point, still it enters especially through the ears because there it traverses the largest empty space, where least it "tarries." In consequence no part of the body perceives sounds but this alone. But when it's inside, it's "sent broadcast" by reason of its velocity; for sound arises when air is condensed and forcibly enters the body. So he explains sensation within the body, just as he explains perception external to it, by touch.

56   Hearing is keenest, he maintains, when the outer tunic is tough and the ducts are empty and unusually moisture-free and are well-bored in the rest of the body as well as in the head and ears; when, too, the bones are dense and the brain is well-tempered and its envelope is very dry. For the sound thus enters compact, since it traverses a large, dry, and well-bored cavity, and swiftly the sound is "sent broadcast" evenly through the body and does not again escape.

. . .

58   [thought] Concerning thought, Demokritos says merely that "it arises when the soul's composition is duly proportioned." But if one becomes overly hot or cold, he says thinking alters; and thus also the ancients rightly believed that the mind became "deranged." So it is clear that he explains thought by the

composition of the body: a view perhaps not unreasonable in one who regards the soul itself as corporeal.

(Stratton [1917] odd pp. 67, 109–117)

Plant Etiology

6.1.1–3 [flavors]

1 Touching flavors and odors, since these too belong to plants, we must endeavor to set forth, just as in the preceding discussions, what happens with each type and for what reasons. The nature of each of the two things has been distinguished elsewhere, to this effect: both are mixed (with certain specifications) in a ratio, flavor being the intermixture in what is fluid of the dry and earthy (or the straining through the dry of the fluid by heat; it makes perhaps no difference), odor the intermixture in the transparent of the flavored dry ("transparent" applying in common to air and water); and what has happened in flavor and odor is (one may say) the same, but has not happened in the two cases in the same things. Let these points, then, be laid down in conformity with the foregoing distinction. [dryness and odors already in Aristotle, *Sensation* 5 (444a22–b6); further in Theophrastos, *Plant Etiology* 6.11.1–2, 6.14.8–12, and 6.16.5–8]

2 It is easy to give the species of flavors so far as the number goes: sweet, oily, dry-wine, astringent, pungent, salty, bitter and acid; it is harder to differentiate them by their essence. For at the very beginning there is a point that involves some investigation: the question whether we are to account for the differences by the different effects when the flavors are tasted, or (as Demokritos does) by the several shapes out of which they are composed (unless the shapes are bound up in a certain way with the production of effects and are introduced to account for them), or else by yet some other way there may be of accounting for the difference.

3 By explaining the differences by the different effect when we taste them I mean such account as this:

1. Sweet: the savor with the capacity to expand the native fluid of the tongue, or the savor with the capacity to make smooth or that has fine particles or is smooth;
2. Astringent: the one with the capacity to desiccate or to solidify this fluid gently;
3. Pungent: the one with the capacity to cut, or to separate out, the heat in the native fluid into the region above, or simply the savor with the capacity to burn or heat;
4. Salty: the one with the capacity to irritate and desiccate;
5. Bitter: the one with the capacity to corrupt the fluid, or to melt or irritate, or simply the savor that is rough or roughest;
6. Dry-wine: the one with the capacity to scour the sense organ (or the fluid in

337

them or the fluid on the surface), or to irritate or solidify or desiccate, or simply a gentle and soft kind of astringency.

(Einarson and Link [1990b] odd pp. 201–209)

On Odors

64–68 [sensing smells]

64 What can be the reason why Demokritos, though he assigns various flavors to the sense of taste, yet does not in like manner assign various smells and colors to the senses to which they belong? According to his system he should have done so. Perhaps the same criticism should apply to all who have dealt with the subject? For they all either give the various qualities and distinguish the experiences of this sense alone or at least comparatively neglect the others: thus with colors they distinguish white and black, and with flavors sweet and bitter, yet they make no corresponding classification of smells, but merely class them as "pleasant" or "unpleasant." So too they fail to distinguish different experiences of the sense of touch, whereas several belong immediately to this sense, as hardness, softness, roughness, smoothness.

65 In sounds still more are there differences, as that between shrill and deep. Again some sense-experiences are simple, some compound. Flavors are simple first in the sense that they cannot be resolved into two components: instances are water, oil, phlegm, blood, and in general anything which floats or which causes separation, like vinegar or milk [cream?]. (Where mixture can be produced by pressure or crushing, it is quite a different matter.) Secondly there are flavors which are not miscible for our use, or which even spoil one another if they are mixed, as sea-water, or soapy [nitrôdê] water or bitter waters: these spoil wines or other things that are good to drink, unless they are taken at once.

66 Now the odors which are thus immiscible are numerous, and, speaking generally, it is the unpleasant odors which do not combine with the unpleasant ones. It would indeed be difficult, if not impossible, to find a case in which mixture is an improvement to the odor: in fact one might say that not even every combination of one fragrant thing with another will produce such a quality. But sometimes it's worse and sometimes it's better, as in the case of perfumes: for some admixtures remove excessive strength or harshness, but others make scents feeble and as if watery. With solids, however, all combinations are possible.

67 In fact powders are the better the more ingredients they have. Also the admixture of wine makes some perfumes and things used for incense more fragrant, for instance myrrh. It appears also that perfume sweetens wines, so that some add it in the manufacture, some put it in at the time of drinking. Nor is it unreasonable that between these senses, since they are akin and active in the same substances, there should be some reciprocity: for, to speak generally, no taste lacks smell and no smell lacks taste, since what has no taste produces no smell.

68 It also happens that smells actually change along with tastes, for instance in wine and certain fruits. And in some cases, as with grapes, the change takes place earlier, during the flowering period: while in perfumes it occurs only when they have reached their peak and almost past. Almost all perfumes undergo alteration at certain seasons of the year, and especially the weakest kinds: floral perfumes alter when their flowers are in bloom. [parallels of flavors and odors: *Plant Etiology* 6.9.1–3, 6.14.1–2]

(Hort [1926] 383, 385, 387, 389)

12.2 Epikouros of Samos (–340 to –269, and a citizen of Athens) adapted the theories of Demokritos, although his own goal was not explanation but peace of mind. Thus, his method was to allow for as many models as possible, so long as all were natural, and he rarely offers novel explanations, merely rendering existing ideas into atomist terms. An opinion could be held, he believed, so long as nothing was known to contradict it.

Letter to Herodotos

49–53 [theories of sense perception: for immediately preceding section, see Chapter 7.2]

49–50 And we must indeed suppose that it is on the impingement of something from outside that we see and think of shapes. For external objects would not imprint their own nature, of both color and shape, by means of the air between us and them, or by means of rays or of any effluences passing from us to them, as effectively as they can through certain delineations penetrating us from objects, sharing their color and shape, of a size to fit into our vision or thought, (*50*) and travelling at high speed, with the result that their unity and continuity then results in the impression, and preserves their co-affection all the way from the object because of their uniform bombardment from it, resulting from the vibration of the atoms deep in the solid body. And whatever impression we get by focusing our thought or senses, whether of shape or of properties, that is the shape of the solid body, produced through the image's concentrated succession or after-effect.

But falsehood and error are always located in the opinion which we add.

51 For the portrait-like resemblance of the impressions which we gain either in sleep or through certain other focusings of thought or of the other discriminatory faculties, to the things we call existent and true, would not exist if the things with which we come into contact were not themselves something. And error would not exist if we did not also get a certain other process within ourselves, one which, although causally connected, possesses differentiation. It is through this that, if it is unattested or contested, falsehood arises, and if attested or uncontested, truth.

52 This doctrine too, then, is a very necessary one to grasp, so that the criteria based on self-evident impressions should not be done away with, and so that falsehood should not be treated as equally established and confound everything.

Hearing too results from a sort of wind travelling from the object which speaks, rings, bangs, or produces an auditory sensation in whatever way it may be. This current is dispersed into similarly-constituted particles [sound is a divisible material]. These at the same time preserve a certain co-affection in relation to each other, and a distinctive unity which extends right to the source, and which usually causes the sensory recognition appropriate to that source, or, failing that, just reveals what is external to us.

53 For without a certain co-affection brought back from the source to us such sensory recognition could not occur. We should not, then, hold that the air is shaped by the projected voice, or likewise by the other things classed with voice. For the air will be much less adequate if this is an effect imposed on it by the voice. Rather we should hold that the impact which occurs inside us when we emit our voice immediately squeezes out certain particles constitutive of a wind current in a way which produces the auditory feeling in us. [pseudo-Aristotle, *Audibles* (802a) says that sounds are clear when unimpeded by soft objects]

We must suppose that smell too, just like hearing, would never cause any feeling if there were not certain particles travelling away from the object and with the right dimensions to stimulate this sense, some kinds being disharmonious and unwelcome, others harmonious and welcome.

(Long and Sedley [1987] 73–74)

12.3 Straton of Lampsakos was the third head of Aristotle's school (from –286 to –269), and wrote on a wide variety of scientific topics (so that he was called "the naturalist"): see also Chapter 5.3 and 6.3.

(unknown work)

(fr. 111W paraphrased in Plutarch, Desire and Grief *4 [697b])*
[origin of sensation in the *hegemonikon*]

Some philosophers have ascribed all affections indiscriminately to the soul, like the scientist [*phusikos*] Straton, who declared that not only our desires but also our griefs, not only our fears and envies and malicious pleasures, but also our physical hurts and pleasures and pains and in general all sensations come about in the soul. According to him, everything of this sort is in the soul; we do not have a pain in the foot when we stub our toe, nor in the head when we crack it, nor in the finger when we gash it. Nothing has any sensation except the soul's *hegemonikon*; any blow is quickly relayed thereto, and its sensation is what we call pain. One may compare the way we think that a noise which in fact sounds in our ears is outside us; we add to the sensation an estimate of the distance between the origin of the noise and the *hegemonikon*. Similarly we think that the pain resulting from a wound is, not where it is sensed, but where it originated, as the soul is drawn towards the source that has affected it. Hence, when we bump into something, we often instantly contract our eyebrows [compare below Section 4, Herophilos], and sometimes catch our breaths while the *hegemonikon* rapidly refers the sensation to the part which received the knock. Again, if our limbs are secured by bonds there is no feeling in our extremities, and if we are wounded, we press hard with our hands, resisting the transmission of the injury and squeezing the blow to keep it in the parts that have no feeling, so that it does not become a pain by making contact with the part of us that has understanding.

(Sandbach [1969] 43, 45, 47)

12.4 Herophilos of Chalkedon (–280 ± 25) was the first Greek physician to perform human dissections and even vivisections, on which he founded his anatomy, in which the brain was the seat of thought.

(fr. 137, paraphrased) [hegemonikon]

Concerning the hegemonikon: *Plato and Demokritos locate it in the entire head (or brain); Straton in the space between the eyebrows; Erasistratos in the area of the meninx of the brain, which he calls "on the skull"* (epikranis); *Herophilos in the ventricle of the brain which is also its "base"* (basis); *Parmenides in the entire thorax, and so too Epikouros; Empedokles, Aristotle, and all the Stoics, in the entire heart or in the* pneuma *around the heart.*

(fr. 226, paraphrased) [dreams]

Herophilos says that some dreams are inspired by a god and arise by necessity, while others are natural ones and arise when the soul forms for itself an image (*eidôlon*) of what is to its own advantage and of what will happen next; and still others are mixed and arise spontaneously [or: "accidentally"] according to the impact of the images, whenever we see what we wish, as happens in the case of those who in their sleep make love to the women they love. [compare below Section 11 (Artemidoros), and M. Aurelius, *Meditations* 1.8.17, 1.9.27]

(von Staden [1989] 314–315, 386)

12.5 Chrusippos of Soloi (*ca.* –280 to –206) wrote so comprehensively and authoritatively on the Stoic teachings that his interpretation became standard (however, all his writings are lost). He studied and wrote in Athens for about a quarter of a century, and then was head of the Stoic school from –231. This extract is quoted by Galen.

On the Soul [arguing that the soul resides in the heart; the gaps are of unknown length]

Since anger arises here, it is reasonable that the other desires are here too, and indeed the remaining affections, and deliberations, and all that resembles these things. Most people, cajoled by common usage, and holding close to the tendency mentioned above, truthfully apply such terms to many of these things. For first, to begin from this point, all people conform to this when they say that inflamed anger "rises" in some persons, and when they demand that some should "swallow" their wrath; and again, when we say that certain lacerations are swallowed or not swallowed by them, we are speaking in conformity with a tendency of this kind. In the same way people say that none of these goes down for them, and "He swallowed the remark and went away"; and in reply to the accusation, "You carry all inquiries to your mouth," Zenon answered, "But not all are swallowed." In no other circumstances would it be more appropriate to speak of swallowing remarks, and of their going down, if our governing part, to which all these things are carried, were not in the region of the chest. Thus if our *hegemonikon* were in the head, it would be ridiculous and inappropriate to say that words go down; I fancy that they would more properly be said to go up, not down in the manner described earlier. For since auditory sense-perception is carried down to the region of the mind, if the mind is in the chest, then "descent" will be the proper term; but if it is in the head, that expression will be rather inappropriate.

...

Women indicate this to a somewhat greater degree than the cases just mentioned. For if a remark does not go down with them, they often move their finger to the region of the heart and say that it does not go down there.

...

In keeping with this we say that some people "vomit up" their impressions, and further we call a person "deep," many expressions of this sort being used in a way that accords with our previous discussion. Thus when people have swallowed the statement (let us say) that it is day, and have stored this up in their minds, and then make that other assertion, that it is not day – the circumstances remaining the same – it is not absurd or inappropriate to say that they vomit up.

...

And the tendency that leads us to say that remarks, whether threats or insults, do not go down far enough to reach and touch the persons so that their minds are set in motion, also leads us to say that some persons are "deep," because nothing of that kind reaches them in its descent.

...

[Chrusippos argues from Greek cardiac etymology and metaphors]

...

For the palpitation of the heart in fear is manifest, and the concourse of the whole soul to that place, these effects following not as a more after-effect, as when one part has a natural sympathy with the others, inasmuch as people shrink into themselves, drawing themselves together toward this place as being the *hegemonikon*, and at the heart as being in them the protector of the *hegemonikon*. And the affections of distress arise somewhere there naturally, no other place experiencing sympathy or sharing the affection. For when certain pains associated with the affections of distress become intense, no other place exhibits these physical effects, but the region of the heart does so to a great degree.

[Chrusippos also distinguished between valid "impressions" by sensation on the soul (which truly alter the soul and convey knowledge) and the unreliable and purely internal "phantasms" seen in dreams.]

(deLacy [1984] 201, 203, 205, 209)

12.6 Melampous (date very uncertain, probably –250 ± 50) wrote also on the significance of bodily tremors; in both works his goal is to deduce character from physical characteristics.

Divination from Birthmarks [entire]

If there is a mole on the forehead of a man, he will be the master of many good men. But if there be a mole on a woman's forehead, she will have royal power, or she will be very great. If a mole is above a man's eyebrow, he will obtain a good and beautiful wife. But if a mole is above a woman's eyebrow, and its color be tawny, she will obtain a rich and handsome man. If, however, on a man's eyebrows, he should not marry because he would become the husband of five wives. Likewise for a woman.

If there be a mole on a man's nose, and its color be tawny, he will be insatiable in intercourse, even if he has the mole concealed. But if on a woman's nose, or eye, the same will happen to her as to a man, even if she has the mole concealed. If to the side of a man's nose, he will travel from region to region. But on a woman, she will be weak in her feet, even if she has the mole concealed.

If on a man's cheek, he will avoid riches. But if on a woman, if on the lower part of her jaw, she will be emotional, even if she has her mole toward her paunch. If on a man's tongue, he will obtain a wealthy and beautiful wife. If there be a mole on a man's lips, he will be gluttonous. Likewise for women. But if a man has a mole on his chin, he will be rich in gold and silver. Likewise on a woman, even if she has her mole towards her spleen. If a man has a mole on his ears, he will be wealthy and of good reputation. Likewise on a woman, even if she has her mole towards her femur.

If a man has a mole on his neck, he will be quite wealthy. The same for a woman. If a man has a mole behind the nape of his neck, he will be beheaded.

If on his loins, he will be a beggar and an unlucky burden for his family. Likewise, a woman. If on his upper arm, he will be captured and oppressed. If on his armpit, he will find a wealthy and beautiful spouse. Likewise, a woman. If on his hands, he will be the parent of many children. Likewise, a woman.

If on his chest, he will be a pauper. Likewise, a woman. If above his heart, he will be most wicked. The same thing happens to a woman when she has a mole on her breast. If on the belly, both will be gluttonous. If on the spleen, they will be weak and sickly. If on the paunch, they will be diseased. If concealed, they will be unrestrained in love-making. If on the genitals, a man certainly will be the father of men. A woman, however, the opposite.

If on the femur, they will be rich. If on the knees, a man will acquire a wealthy wife. A woman, however, if the mole is on her right knee, will be virtuous; but if on her left knee, she will be the mother of much offspring. If there is a mole on his heel, a man will have a wife without beauty apart from her dress. But a woman will have a husband with no power. But if they have a mole on their feet, they will be the parents of many children.

Therefore observe in regard to men and women. If there be a mole on the right parts, they will be rich and altogether virtuous. If on the left side, they will be sickly and poor.

*(Franz [1780] 501–508)

12.7 Ptolemaïs of Kurênê (probably –100 ± 60; Barker [1989] 230 prefers first century BCE) is known only from quotations in Porphurios (third century CE); she wrote on music.

Introduction

(fr. in Porphurios, Commentary on Ptolemy's "Harmonics" *23.24–24.6)* [the scale]

Pythagoras and his successors wish to accept perception as a guide for reason at the outset, to provide reason with a spark, as it were; but they treat reason, when it has set out from these beginnings, as working on its own in separation from perception. Hence if the *sustêma* [arrangement of notes] discovered by reason in its investigation no longer accords with perception, they do not retrace their steps, but level accusations, saying that perception is going astray, while reason by itself has discovered what is correct, and refutes perception.

An opposite position to this is held by some of the musical theorists who follow Aristoxenos, those who applied themselves to a theoretical science based in thought, while nevertheless setting out from expertise on instruments. For they treated perception as authoritative, and reason as attending on it, for use only when needed. According to these people, to be sure, it is only to be expected that the rational postulates of the *kanon* [the scale] are not always concordant with the perceptions.

(fr. in Porphurios, Commentary on Ptolemy's "Harmonics" 25.3–26.5)

What is the difference between those who are distinguished in the field of music? Some preferred reason by itself, some perception, some both together. Reason was preferred by those of the Pythagoreans who were especially keen on disputing with the musical theorists, arguing that perception should be thrown out completely, and that reason should be brought in as an autonomous criterion in itself. These people are wholly refuted by their practice of accepting something perceptible at the beginning, and then forgetting that they have done so. The instrumentalists on the other hand preferred perception: they gave no thought at all, or only feeble thought, to theory.

What is the distinction between those who prefer the combination of both? Some accepted both perception and reason in the same way, as being of equal power, while others accepted the one as the leader, the other as the follower. Aristoxenos of Taras accepted both in the same way. For what is perceived cannot be constituted by itself apart from reason, and neither is reason strong enough to establish anything without taking its starting points from perception, and delivering the conclusion of its theorizing (*theorema*) in agreement with perception once again. In what way does he want perception to be in the lead of reason? In order not in power (*dunamis*). For when the perceptible thing, whatever it may be, has been reviewed by perception, then, he says, we must put reason in the lead, for the theoretical study of this percept. Who are those who treat both together alike? Pythagoras and his successors. For they wish to accept perception as a guide for reason at the outset, to provide reason with a spark, as it were; but they treat reason, when it has set out from these beginnings, as working on its own in separation from perception. Hence if the *sustêma* discovered by reason in its investigation no longer accords with perception, they do not retrace their steps, but level accusations, saying that perception is going astray, while reason by itself has discovered what is correct, and refutes perception. Who are in opposition to these? Some of the musical theorists who follow Aristoxenos, those who applied themselves to a theoretical science based in thought, while nevertheless setting out from expertise on instruments. For they treated perception as authoritative, and reason as attending on it, for use only when needed.

(Barker [1989] 240–242)

12.8 Thrasullos perhaps of Mendê (died 36, active 10 ± 25), who wrote on Plato, Demokritos, the harmony of the spheres and the nature of the stars, was the personal astrologer of the emperor Tiberius (reigned 14 to 37; compare the astrologer Balbillos). This extract is paraphrased from his lost work on music.

Music

(fr.) [note is a pitch of an attuned sound]

Thrasullos, then, when discussing the *harmonia* perceptible in an instrument, says that a note is a pitch of an attuned sound. It is called "attuned" if there can be found a note higher than the high one and lower than the low one: the same note is also "intermediate." Thus if we conceived a note that exceeded all height of pitch, it would not be "attuned," for we shall not call the noise of an immense thunder-clap "attuned" – one that is often deadly because of its excessive greatness: "The bloodless wound of a thunder-clap has slain many," as someone says [attributed to Euripides]. And again, if there were some note so low that there was none lower, it would not even be a note, since it would not possess attunement.

(Barker [1989] 212)

12.9 Aretaios of Kappadokia (either about 70 ± 20 or perhaps as late as 170 ± 20: Oberhelman [1994]) was a Pneumaticist who wrote on *Fevers*, on *Female Disorders*, and other works besides the one here excerpted.

Acute and Chronic Diseases

3.4.1–3 (pp. 38–39 Hude) [epilepsy]

1 Epilepsy is a manifold and horrible illness: beastly in the paroxysms, acute and deadly; for sometimes one paroxysm has killed. Or if the person is used to bearing it, he lives enduring shame, ignominy, and pain: and the disease does not readily depart, but inhabits the better periods and the prime of life, and dwells with boys and young men. Sometimes it is luckily expelled in another later period of life, when it departs along with the beauty of youth; and then, having rendered them deformed, it destroys youths from envy of their beauty, either by dis-empowerment of a hand, or by distortion of the face, or by the deprivation of some sensation.

2 But if the illness lurk there until rooted, it will yield neither to physician nor changes of age, but lives with the patient until death. And sometimes the disease is painful with convulsions and distortions of limbs and of face; and sometimes it turns the mind to madness. The sight of a paroxysm is unpleasant, and its departure disgusting with spontaneous evacuations of urine and bowels. But its general character is disreputable, for it seems to afflict sinners against the Moon: whence some have called it the sacred (*hieros*) disease. But also for other reasons, either from the greatness of the evil, for *hieros* means great, or because its cure is not human, but divine, or from the opinion that it's due to the entrance of a *daimôn* [spirit intermediate between gods and mortals] into the person, or from all these causes together, it has been called sacred.

346

3 Such symptoms as occur in the acute form of the disease I have explained. But if it becomes chronic, the patients are not unharmed even in the intervals, but are languid, spiritless, downcast, inhuman, unsociable, and unfriendly at any age; sleepless, with many nightmares, anorexic, with bad digestion, pale, lead-colored; slow to learn from torpidity of wit and senses; dull of hearing, have noises and buzzing in the head; speech indistinct and bewildered, either from the nature of the disease, or from wounds during attacks: the tongue is wildly rolled about in the mouth. The disease also sometimes disturbs the understanding, so that the patient becomes altogether fatuous. But the cause of these affections is coldness with wetness [same core notion in Hippokrates, *Sacred Disease*].

3.5.4–7 (pp. 40–41 Hude) [melancholê]

4 But how and from what parts of the body most of these complaints originate, I will now explain. If the cause remain in the *hupochondrion* [lower belly], it collects about the diaphragm, and the bile passes upwards, or downwards in *melancholê*. But if it also affects the head from *sumpatheia*, and the abnormal irritability of temper change to laughter and joy for most of their life, these become mad from the increase of the disease rather than from pain of the affection. Dryness is the cause of both.

5 Adult males, therefore, or even younger, are subject to mania and *melancholê*. Women are more maddened than men. As to age, towards and actually in the prime of life; the seasons of summer and autumn birth it, and spring brings it to *krisis*. The evidence is not obscure, for the patients are quiet or stern, downcast or unreasonably torpid, for no reason. And they also become peevish, dispirited, sleepless, and start up from a disturbed sleep.

6 Absurd fear also seizes them, if the disease is increasing, when their dreams are true, terrifying, and clear: for whatever evil they fear when awake rushes upon their visions in sleep. They are prone to change their mind readily, to become base, petty, stingy, and soon simple, profligate, munificent, not from any virtue of the soul, but from the mutability of the disease. But if the illness presses harder there is hatred, agoraphobia, vain lamentations; they complain of life, and desire death. In many, the understanding turns to insensibility and folly, so that they become ignorant of everything, or forgetful of themselves, and live beastly lives.

7 The habit of the body also becomes perverted; color darkish-green, if bile doesn't pass downward, but is diffused everywhere with the blood. They are voracious yet emaciated; for their sleep does not strengthen their limbs with food or drink, but insomnia disperses the nourishment. Therefore the bowels are dry, and discharge nothing; or if they do, the feces are dried, round, in a dark and bilious fluid; urine scanty, acrid, bile-tinged. They are flatulent about the *hupochondrion*; the eructations smelly and stinky, like brine from salt; and sometimes an acrid liquid refluxes with bile. Pulse for the most part small, torpid, feeble, rapid, like that from cold.

3.6.6 *(p.42 Hude)* [a mad carpenter]

This story also is told: A carpenter was a skilful artisan while in the house, would measure, chop, plane, nail, and mortise wood, and finish the work of the house correctly, would associate with the workmen, agree with them, and reward their work with suitable pay. While at the work-site, he thus possessed his understanding. But if at any time he went away to the market, the bath, or any other necessity, having laid down his tools, he would first groan, then shrug his shoulders as he went out. When he had got out of sight of the servants, or of the work and the site, he became completely mad; yet if he returned he quickly recovered his reason. Such a bond was there between the site and his understanding.

(Adams [1856] 296–297, 299–300, 302–303)

12.10 Antonius Polemon of Laodikaia on the Lukos (born 88 ± 3, died 144) wrote a (lost) history, some speeches, and this work (around 135 ± 2) which survives only in Arabic and a late paraphrase; he was close to the emperors Trajan and Hadrian (Gleason [1995] 21–67).

Physiognomics

26 *The nose* [pseudo-Aristotle, *Physiognomy* 6 compares animal noses]

Thinness of the tip of the nose indicates intensity of anger. A depression and thickness of the nose designates mockery of its possessor. A nose that is fat, long, round, and robust, shows vigor and power and magnanimity, such as are according to the habits of dogs and monkeys [or "lions"?]. But if a nose is long and thin, as accords with the characteristics of birds, ascribe the rest of their customs to him. When you see a nose long and correctly and properly joined with the forehead, praise its possessor and ascribe to him strength and the best mode of thinking and knowledge. To the opposite nose, which is similar to the noses of women, ascribe very little cogitation and knowledge. [gap in the text] Don't praise a short, small nose, for it designates thievery and base souls. A curved nose denotes great thought. A nose like a vulture [gap in the text] or a monkey shows fornication and a love of sex [or "marriage"]. Ascribe stupidity to a round nose with wide nostrils. To have a strong nose indicates vigor and thoughtfulness.

27 *The forehead and brow* [pseudo-Aristotle, *Physiognomy* 6 compares animal brows]

Narrowness of the forehead indicates lack of intelligence. Nor do I praise a wide forehead, since its possessors are senseless and slow. Length in a forehead designates much knowledge and mental agility. But I do not also praise shortness in a forehead, since it is a sign of feebleness; nor prominence and roundness, since that shows little shame and incapacity for learning. A hard rough forehead, on which are projections and which is uneven, is a sign of fraud, deceit, and impiety; occasionally there is even stupidity and foolishness in its possessor. But nonetheless I praise a symmetrical forehead, suitable to its owner's face, which compared with it is neither too big nor too small, having four angles [perhaps "orthogonal"?]. Indeed I praise such a forehead: for you will find its owner wise and perspicacious, excellent in thought, strong and great in endeavor and magnanimous. If you see one whose space between the eyebrows is not level, ascribe to him little contemplation and prudence. He whose space between the eyebrows is stretched, attribute leisure to him. A lot of flesh on the forehead with its slight prominences indicates thoughtfulness, planning, and those who care about things.

36 *The color of the whole body* [pseudo-Aristotle, *Physiognomy* 6 is similar]

The color black indicates cowardice, long-lived anxiety, and gloominess; of this kind are southerners like Ethiopians, the blacks [Arabic "zanj"], the Egyptians, and those close to them. A fair white color turning towards ruddiness designates boldness and violent irascibility. Complete whiteness denotes feebleness. Ruddiness of the face and body indicates deceit and much cogitation. The color in which there is some yellow mixed in indicates evil, enthusiasm, fright, and cowardice, unless the yellow arises from disease. If you see yellow veering towards black without disease, it designates timidity, voracity, a lot of talk, irritability, and a long tongue. The color in which there is a little ruddiness indicates much intelligence.

*(Foerster [1893] 228–231, 244–245)

12.11 Artemidoros of Ephesos (160 ± 30) designated himself "of Daldis" after his maternal hometown (in Ludia); his interpretations assume *sumpatheia* between images and events (compare Price [1986] 9–31). His dream-book is one of only two extant (the other being the Hippokratic *Dreams*); see also Alexander of Mundos, Chapter 10.9.

Dreams

1.1–2 [oneiros vs. enhupnion]

1 *Oneiros* differs from *enhupnion* in that the first indicates the future, while the other indicates the present. To put it more plainly, it is the nature of certain experiences to run their course in proximity to the mind and to subordinate themselves to its dictates, and so to cause nocturnal emissions. For example, it is natural for a lover to seem to be with his beloved in a dream and for the

frightened to see what he fears, or for the hungry to eat and the thirsty to drink and, again, for one who has stuffed himself with food either to vomit or to choke. It is possible, therefore, to view those experiences as containing not a prediction of the future but rather a reminder of the present. This being so, you can see that some things pertain only to the body, some things pertain only to the mind [or "soul"] (and some are common to both body and mind). For example, when a lover dreams that he is in the company of his beloved, or a patient that he is being treated and is seeing doctors, these are common to both body and mind. On the other hand, vomiting and sleeping and, again, drinking and eating must be regarded as pertaining to the body, just as feeling either joy or grief pertains to the mind. It is clear from these examples that, of those dreams which pertain to the body, some are seen because of a lack while others are seen because of an excess, and that of those dreams which pertain to the mind, some are due to one's fears, while others are due to one's hopes. [compare Herodotos 7.16.2]

2   ... *Oneiros* is a movement or condition of mind, manifold and signifying good or bad in the future. Since this is so, the mind predicts everything that will happen in the future, later or sooner, by means of natural proprietary images called elements. It does this because it assumes that, in the interim, we can be taught to learn the future through reasoning. But whenever the actual occur- rences admit of no delay whatsoever, because whoever it is that guides us causes them to happen without delay, the mind, thinking prediction useless to us unless we grasp the truth before learning through experience, shows these things directly, without waiting for anything external to show us the meaning. In a way it cries out to each of us, "Look at this and be attentive, for you must learn from me as best you can." Everyone agrees and no one will ever deny that, directly after the vision itself or without even a short interval of time, such things happen. And, indeed, some of them come true at the very moment of perception, so to speak, while we are still under the sway of the dream-vision. Hence, it bears its name not without reason, since it is simultaneously seen and comes true.

...

On these same grounds, it is also possible to detect that some "common" dreams (involving both self and others) have results pertaining to self rather than to self and others. For the first part of this branch of the subject, the early writers mostly hold good; the second part on occasion, though rarely, causes confusion to experts by happening in the way I have described. We must draw the following distinction. All dreams pertaining to self which do not involve others, which arise in and for the dreamers alone, and which do not operate with regard to and by means of others, have a result for the dreamers alone: for example, speaking, singing, dancing, and also boxing, competing for a prize, hanging oneself, dying, impalement, diving, finding a treasure, making love, vomiting, defecating, sleeping, laughing, crying, speaking with the gods, and so on. But those which involve the body or a body-part, or external objects such as beds, boxes, or baskets, as well as other furnishings, apparel, and so on, although they are personal, often have the tendency to affect others too, depending on the closeness of the relationship. For example, the head indicates father; the foot indicates a slave; the right hand indicates father, son, friend, or brother; the left hand indicates wife, mother, mistress, daughter, or sister; the penis indicates parents, wife, or children; the shin indicates wife or mistress. And (not to belabor the point) each of the other objects must be regarded likewise. All "common" and "alien" dreams which arise for and through us must be regarded as pertaining only to ourselves, but those which do not arise in this way will come true for others. And yet, if these others are our friends and something good is indicated, the joy and pleasure will come to them and partly to us. But if the dream signifies something bad, misfortunes will be theirs and grief ours, not altogether from their misfortunes but, indeed, a personal grief. But if they are our enemies, one must draw the opposite conclusion.

1.64 [bathing dreams]

Very early writers thought that dreams about washing were not bad, since they knew nothing of public baths but washed in so-called private bathing-tubs [Hippokrates, *Regimen in Acute Diseases* 65–68]. Later writers, since the public baths had by then already come into existence, thought that dreams in which a person washed himself or saw a bath, even if not bathing, were bad. They thought that the bath signified turmoil because of the noise made in them, harm because of the excreted sweat, anguish and mental ["of the soul"] anxiety because the color and surface of the body change in the bath.

Some modern writers subscribe to this old view, use the same criteria, and thus err in judgment and don't correspond to experience. Long ago it was reasonable for the baths to be considered unlucky, since people did not wash regularly and did not have so many baths. Rather, they washed themselves when they returned from war or when they ceased some strenuous activity. (Thus, the bath and the act of bathing were, to them, a reminder of toil or war.)

But nowadays some people do not even eat unless they have first bathed. Others, moreover, also bathe afterwards. Then they wash when they are about to take supper. Therefore now the bath is nothing but a road to luxury. And thus washing in baths that are beautiful, bright, and of moderate temperature, is auspicious: it signifies wealth and success in business for the healthy and health for the sick. For healthy people wash themselves even when unnecessary.

If a dreamer were to wash in an unusual manner, it would not be good for him. For example, if a man enters the hot baths with his clothes on, it signifies sickness and great anguish for him. The sick enter the baths clothed and, furthermore, people who are anxious about important affairs sweat in their clothes.

[further cases of abnormal and hence inauspicious bathing; empty baths signify failure; baths in pure water are auspicious; swimming is dangerous]

1.79 [dreams about one's mother]

That the case of one's mother is both complex and manifold and admits of many different interpretations not all dream interpreters have realized. The fact is that sex by itself is not enough to show what is portended. Rather, the manner of the embraces and the various positions of the bodies indicate different outcomes.

First, then, we will discuss face-to-face intercourse between a dreamer and his living mother, since a living mother does not mean the same as a dead mother. Therefore, if anyone possesses his mother face-to-face (which some also call the "natural" method), if she is still alive and his father is in good health, it means that he and his father will become enemies because of the jealousy that generally arises between rivals. But if his father is sick, he will die, since the dreamer will take care of his mother both as a son and as a husband.

But it is lucky for every craftsman and laborer. For we ordinarily call a person's trade his "mother." And what else would having intercourse with her mean if not to be occupied with and profit from one's art?

It is also lucky for every demagogue and politician. For a mother signifies one's homeland. And just as a man who follows the precepts of Aphrodite when he makes love completely governs the body of his obedient and willing partner, the dreamer will control all the affairs of the city.

And if the dreamer is estranged from his mother, they will become friends again because of the sex. For sex is also called "friendship." But frequently this dream has indicated that people who live apart will be brought together into the same place and will live together. It also signifies, therefore, that a son will return from a foreign country to his homeland, if his mother lives there [compare Herodotos 6.107]. If she does not, the dream signifies that he will journey to wherever she lives. If the dreamer is a poor man who is lacking the necessities of life but one whose mother is rich, he will receive from her whatever he wishes, or she will die shortly afterwards and he will inherit and, in this way, he will take pleasure in his mother. On the other hand, many sons have taken care of and provided for their mothers and so the mothers have taken pleasure in their sons. The dream indicates that the sick will recover and return to their natural state, since nature is the common mother of all things and we say that the healthy, not the sick, are in their natural condition. (Apollodoros of Telmêssos [a city famed for its seers], a well-known authority [otherwise unknown], also mentions this point.)

[dreams of sex with mother: by sick man who's mother is dead means he will himself die; by man in lawsuit over land means he will win; by man abroad means he will return home]

[sex with mother from behind or kneeling or standing is bad; from beneath is ambiguous; dreaming of using multiple positions is very bad]

2.68 [flying dreams]

If someone dreams that he is flying a little above the earth and upright, it means good luck for the dreamer. The greater the distance above the earth, the higher his position will be in regard to those who walk beneath him. For we always call the more prosperous "higher ones." It is good if this does not happen to someone in his own country, since it signifies emigration because the person does not set his foot upon the ground. For the dream is saying to some extent that the dreamer's native land is inaccessible to him.

Flying with wings is auspicious for everyone alike. The dream signifies freedom for slaves, since all birds that fly are without a master and have no ruler. It means that the poor will acquire much money. For just as money raises people up, wings raise birds up. It signifies offices for the rich and very powerful. For just as flyers are above those crawling upon the earth, rulers are above private citizens. But to dream that one is flying without wings and very far above the earth signifies danger and fear for the dreamer.

Flying around tiled roofs, houses, and blocks of houses signifies confusion and disturbance of the soul. For slaves, dreaming that one is flying up into the heavens always signifies that they will pass into more distinguished homes and often even that they will pass into the court of a king. I have often observed that free persons, even against their will, have journeyed to Italy. For just as the sky is the home of the gods, Italy is the home of kings. But the dream indicates that those who wish to hide and conceal themselves will be discovered. For everything in the sky is clear and easily visible to everyone.

[flying with birds portends travel; flying while pursued is dangerous; inability to fly or flying head-downwards is very inauspicious]

(White [1975/1990] 14–15, 17, 48–49, 61–62, 132)

12.12 Aelius Aristides (160 ± 20), a literary hypochondriac who spent most of the last half-century of his life in the Asklepieion at Pergamon, recorded his illnesses and dreams about and rituals to Asklepios; he also wrote speeches, especially a panegyric of Rome (see Chapter 1, p. 15).

Sacred Tales

3.21–22 [dream about healing; August 148, in Pergamon]

21 Immediately after Asklepios cured my difficulty in breathing, he healed in the following way the trouble about my neck and the tension in my ears and the *opisthotonos* ["back-bending" of spine], which was now fully developed.

> *He said that there was a royal ointment. It was necessary to get it from his wife. And somehow after this, a servant of the palace, clad in white and girdled, appeared at Telesphoros' Temple and statue, and escorted by a herald, went out by the doors where the statue of Artemis is, and bore the remainder of the ointment to the Emperor.*

This more or less was the dream, to recollect it unclearly.

22 When I entered the Temple and was walking about in the direction of the statue of Telesphoros, the temple warden Asklepiakos came up to me. And while he happened to stand by the statue, I told him the vision which I had, and I asked him, "What might the ointment be, or who should use it?" But when he had listened and marveled, as he was accustomed to do, he said, "The search is not far nor need there be much traveling, but I shall bring it to you right from here. For it lies by the feet of Hugieia ["health," the wife of Asklepios], since Tuchê herself just now put it there, as soon as the Temple was opened." [Iulia] Tuchê was a noble lady. And going to the Temple of Hugieia, he brought the ointment. And I anointed myself, where I happened to stand. The ointment also had a wonderful smell, and its power was immediately manifest. For faster than I have said it, the tension relaxed. [Aelius obtains the recipe and keeps using the ointment.]

(Behr [1986] 312)

12.13 Ptolemy (Claudius Ptolemaeus) of Alexandria (*ca.* 100 to *ca.* 175) wrote on a wide variety of "mathematical" topics (see Chapters 3–8).

Harmonics

1.1 [reason aids the senses in making distinctions]

Harmonic knowledge is the power that grasps the distinctions related to high and low pitch in sounds: sound is a modification (*pathos*) of air that has been struck (this is the first and most fundamental of things heard): and the criteria of *harmonia* are hearing and reason, not however in the same way. Rather, hearing is concerned with the matter and the modification, reason with the form and the cause, since it is in general characteristic of the senses to discover what is approximate and to adopt from elsewhere what is accurate, and of reason to adopt from elsewhere what is approximate, and to discover what is accurate.

. . .

Thus just as a circle constructed by eye alone often appears to be accurate, until the circle formed by means of reason brings the eye to a recognition of the one that is really accurate, so if some specified difference between sounds is constructed by hearing alone, it will commonly seem at first to be neither less nor more than what is proper: but when there is tuned against it the one that is constructed according to its proper ratio, it will often be proved not to be so, when the hearing, through the comparison, recognizes the more accurate as legitimate, as it were, beside the bastardy of the other; for judging is in general easier than doing the same thing (as it is easier to judge wrestling than to wrestle, to judge dancing than to dance, to judge flute-playing than to play the flute, and to judge singing than to sing). This sort of deficiency in perceptions does not miss

354

the truth by much when it is simply a question of recognizing whether there is or is not a difference between them, nor does it in detecting the amounts by which differing things exceed one another, so long as the amounts in question consist in larger parts of the things to which they belong. But in the case of comparisons concerned with lesser parts the deficiency accumulates and becomes greater, and in these comparisons it is plainly evident, the more so as the things compared have finer parts. The reason is that the deviation from truth, being very small when taken just once, cannot yet make the accumulation of this small amount perceptible when only a few comparisons have been made, but when more have been made it is obvious and altogether easy to detect.

3.3 [the power of harmonia]

The power [dunamis] of harmonia employs as its instruments and servants the highest and most marvelous of the senses, sight and hearing, which, of all the senses, are most closely tied to the hegemonikon, and which are the only senses that assess their objects not only by the standard of pleasure but also, much more importantly, by that of beauty. For on the basis of each of the senses one can discover, in perceptibles, distinctions proper to that sense alone, such as white and black, for example, in the field of things seen, high and low pitch in that of things heard, good and bad odor in that of things smelled, sweet and bitter in that of things tasted, and soft, for instance, and hard in that of things touched; and one can also discover, of course, the agreeable or disagreeable nature of each of these different qualities. But no one would classify the beautiful or the ugly as belonging to things touched or tasted or smelled, but only to things seen and things heard, such as shape and melody, or the movements of the heavenly bodies, or human actions; and hence these, alone among the senses, give assistance with one another's impressions in many ways through the agency of the rational part of the soul, just as if they were really sisters. It is only hearing that displays things seen, by means of explanations, and only sight that reports things heard, by means of writings, and the result is often clearer than if either of the two had announced the same things by itself – as when things transmitted by speech are made easier for us to learn and remember when accompanied by diagrams or letters, and things recognizable by sight are displayed more vividly through a poetic representation, the appearance of waves, for instance, or scenery or battles or the external circumstances of pathos, so that our souls are brought into sympathy with the forms of the things reported, as if we were seeing them. It is therefore not just by each one's grasping what is proper to it, but also by their working together in some way to learn and understand the things that are completed according to the logos [right ratio], that these senses themselves, and the most rational of the sciences that depend on them, penetrate progressively into what is beautiful and what is useful. Related to sight, and to the movements in place of the things that are only seen (i.e., heavenly bodies) is astronomy: related to hearing and to the movements in place, once again, of the things that

355

are only heard (i.e., sounds) is harmonics. They employ both arithmetic and geometry, as instruments of indisputable authority, to discover the quantity and quality of the primary movements; and they are as it were cousins, born of the sisters, sight and hearing, and brought up by arithmetic and geometry as children most closely related in their stock.

3.5 [three primary parts of the soul]

There are three primary parts of the soul, the intellectual the perceptive and the animating, and there are also three primary forms of homophone and concord, the homophone of the octave, and the concords of the fifth and the fourth. Hence the octave is attuned to the intellectual part, since in each of these there is the greatest degree of simplicity, equality, and stability; the fifth to the perceptive part; and the fourth to the animating part. For the fifth is closer to the octave than is the fourth, since it is more concordant, due to the fact that the difference between its notes is closer to equality's and the perceptive part is closer to the intellectual than is the animating part, because it too partakes in a kind of apprehension. Now things that have animation do not always have perception, and neither do things that have perception always have intellect: things that have perception, conversely, always do have animation, and things that have intellect always have both animation and perception. In just the same way, where there is a fourth there is not always a fifth, and neither is there always an octave where there is a fifth: where there is a fifth, conversely, there is always a fourth too, and where there is an octave, there are always both a fifth and a fourth. The reason is that the former are made up of the less perfect melodic intervals and combinations, the latter of the more perfect.

One can say that the animating part of the soul has three forms, equal in number to the forms [or "species"] of the fourth, related respectively to growth, maturity, and decline – for these are its primary powers. The perceptive part has four, equal in number to those of the concord of the fifth, related respectively to sight, hearing, smell, and taste (if we treat the sense of touch as being common to them all, since it is by touching the perceptibles in one way or another that they produce our perceptions of them). The intellectual part, finally, has at most seven different species, equal in number to the species of the octave: these are imagination [phantasia, "imaging"] (concerned with the reception of communications from perceptibles), intellect (concerned with the first stamping-in of an impression), reflection [ennoia] (concerned with the retention and memory of the stamped impressions), thought (concerned with recollection and enquiry), opinion (concerned with superficial conjecture), reason [logos] (concerned with correct judgement), and knowledge [epistêmê, "scientific knowledge"] (concerned with truth and understanding).

[further analogies follow, derived in part from Plato, Republic 4]

(Barker [1989] 276–277, 372–373, 375–376)

12.14 Galen of Pergamon (129 to 210 ± 5), viewing himself as philosopher and physician, considered Hippokrates and Plato as the most worthy authorities; he wrote voluminously on medicine and much else, and served as court physician to three emperors: M. Aurelius, Commodus, and Seuerus.

Soul's Dependence on Body

3, 5, 8, 11 [effects on soul from body]

3 . . . The bodily mixture does not just alter the soul's activities, as I have said, but can also cause its separation from the body. What other conclusion is possible when one sees drugs with the powers of cooking or overheating causing the immediate demise of their takers? And the poisons of wild beasts belong in this category too. The bite of the asp is observed to cause instant death, similar to that caused by hemlock, since the effect of this too is a cooling one.

So one is bound to admit, even if one wishes to posit a separate substance for the soul, at least that it is slave to the mixtures of the body: these have the power to separate it, to make it lose its wits, to destroy its memory and understanding, to make it more timid, lacking in confidence and energy, as happens in cases of *melancholê* – or the opposite of these qualities, as in the case of the moderate drinker of wine.

5 . . . Now, loss of memory, of intelligence, of motion, or of feeling resulting from the stated causes can be regarded as impairments of the soul's ability to employ its natural functions. But cases of people seeing things which are not there or hearing things that no one said, or making obscene or blasphemous or indeed completely meaningless utterances, seem to indicate not just a loss of faculties which the soul innately possessed, but the presence of some opposite faculty. Such a consideration may in itself cast doubt on the non-bodily nature of the soul as a whole. For if the soul were not some quality, form, affection, or faculty of the body, how could it actually acquire a nature opposite to its own, just by communion with the body? I pass over this point, to avoid a digression which might be longer than our main subject of discussion. At any rate, the overwhelming effect on the soul of the ills of the body is clearly demonstrated by the case of people suffering from *melancholê, phrênitis,* or mania.

8 [Galen is explaining Hippokrates, *Epidemics* 2.5] . . . Men in whom the artery in the elbow produces a very strong movement are manic. For the ancients referred to arteries too as veins, as I have frequently demonstrated; and they did not refer to every kind of pulse as "pulse," but only to that kind which is clearly noticeable, that is a strong one. Hippokrates was in fact the initiator of the usage which later came to prevail, talking of all arterial motion, of whatever kind, as "pulse." But in the treatise quoted he is still using the old terminology, and what he means is that a powerful motion of the artery is evidence of a manic and sharp spirited man. This is because what causes such pulsation is an abundance of warmth in the heart. Such warmth makes people manic and sharp-spirited, whereas a cold mixture makes people lethargic, heavy, and slow to move.

11 [from the conclusion] . . . But such phenomena in the rational part of the soul as shrewdness, or various degrees of foolishness, are dependent on mixture; and mixture itself is due to birth and to good-humored regimen. And these conditions give rise to vicious or beneficent circles: the sharp-spirited become so because of the hot mixture, but then by their sharpness of spirit inflame their inborn heat. And, similarly, people with a well-proportioned mixture enjoy well-proportioned motions of the soul, and are assisted towards good spirits.

(Singer [1997] 155, 159–160, 167, 175)

Differential Diagnosis of Symptoms

3 [damage to the *hegemonikon*]

I want to discuss now the damage to the activity of the *hegemonikon*, and especially of its perceptive activity. Its damage is called a form of paralysis or it is defined as stupor or as seizure. There is another form, the faulty or disturbed movement ["out of tune"], which is called *paraphrosunê* [mental derangement] (and this is like a weakening or slackening as in *kôma* or *lethargos* [two conditions of deep sleep or inactivity]). When the thinking faculty itself is paralyzed, it is called want of understanding; if its movements are defective we deal with folly and dementia; and when it is out of tune, it is called *paraphrosunê*. Mostly both disturbances exist in mental derangement, when simultaneously both happens: wrong perceptions are formed and the reasoning becomes improper.

Sometimes only one of the functions is affected, as in the illness of the physician Theophilos, since he was otherwise able correctly to discuss and judge about everything present, but assumed that some flute players had seized the corner of his house where he slept, and that they continuously played the flute and made crashing sounds and gazed upon him, some standing there, others sitting, and continuously played the flute without interruption during the night, nor was there the slightest quietness during the day; he shouted continuously, demanding they be expelled from his house. This was a form of *paraphrosunê*. And when he finally recovered and escaped this disease, he described in detail what everybody had said and had done who had entered the house, and he even remembered the phantasm concerning the flute players. [compare Hippokrates, *Epidemics* 7.86]

However, to some people such phantasms do not occur, but they are unable to reason correctly, since the reasoning power of their soul is affected. Take for example, the patient who, when suffering from *phrênitis*, closed all doors from the inside and held all his household utensils out the window, asking by-passers to command him to throw it down. But he identified each of the things by its right name and thus was evidently not deranged in recognizing each of these things nor in recalling its name. Why did he therefore want to throw it all down from upstairs and break it? He was unable himself to understand it, but manifestly he demonstrated his disturbance by his acts.

The same occurs in regard to the symptoms of the soul's recollective faculty in those people who are sick or even already recovered, as we can learn from Thoukudides, who wrote that some of those who were saved from the great plague [of Athens during the Peloponnesian war: "Thucydides" 2.49.8] had forgotten everything until that time, so that they were not only unable to recognize their family but even themselves.

(Siegel [1973] 163–164)

BIBLIOGRAPHY

Sources of translations quoted

Adams, F. (1856) *The Extant Works of Aretaeus the Cappadocian*, London: Sydenham Society.

Barker, A. (1989) *Greek Musical Writings*, vol. 2, Cambridge: Cambridge University Press.

Behr, C. A. (1986) *P. Aelius Aristides: The Complete Works*, 2 vols., Leiden: Brill.

Brain, P. (1986) *Galen on Bloodletting*, Cambridge: Cambridge University Press.

Bruin, F. and A. Vondjidis (1971) *The Books of Autolycus*, Beirut: The American University of Beirut.

Burstein, S. M. (1978) *The Babyloniaca of Berossus*, SANE 1/5, Malibu, CA: Undena.

Burton, H. E. (1945) "Euclid's Optics," *Journal of the Optical Society of America* 35: 357–372.

Caley, E. R. (1926) "The Leyden Papyrus X," *Journal of Chemical Education* 3: 1149–1166 [using the numbering of Halleux].

Caley, E. R. (1927) "The Stockholm Papyrus," *Journal of Chemical Education* 4: 979–1002 [using the numbering of Halleux].

Caley, E. R. and J. F. C. Richards (1956) *Theophrastus on Stones*, Columbus, OH: Ohio State University Press.

Casson, L. (1989) *The Periplus Maris Erythraei*, Princeton, NJ: Princeton University Press.

Cherniss, H. (1976) *Plutarch's Moralia*, vol. 13.1, Cambridge, MA: Harvard University Press.

Cherniss, H. and W. C. Helmbold (1957) *Plutarch's Moralia*, vol. 12, Cambridge, MA: Harvard University Press.

Cohen, M. R. and I. E. Drabkin (1958) *A Source Book in Greek Science*, Cambridge, MA: Harvard University Press.

Condos, T. (1997) *Star Myths of the Greeks and Romans*, Grand Rapids, MI: Phanes.

Coutant, V. (1971) *Theophrastus De Igne*, Assen: Van Gorcum.

Coutant, V. B. and L. V. Eichenlaub (1975) *Theophrastus De Ventis*, Notre Dame, IN: University of Notre Dame Press.

deLacy, P. H. (1984) *Galen: On the Doctrines of Hippocrates and Plato, Corpus Medicorum Graecorum* 5.4.1, pt. 2, Berlin: Akademie Verlag.

Dicks, D. R. (1960) *Geographical Fragments of Hipparchus*, London: Athlone.

D'Ooge, M. L. (1926) *Nicomachus of Gerasa: Introduction to Arithmetic*, New York: The Macmillan Co.

Drachmann, A. G. (1963) *Mechanical Technology of Greek and Roman Antiquity*, Copenhagen: Munskgaard.

360

Einarson, B. and G. K. K. Link (1926,1990a,1990b) *Theophrastus: Causes of Plants*, 3 vols., Cambridge, MA: Harvard University Press.

Furley, D. J. (1955) *Aristotle: On the Cosmos*, Cambridge, MA: Harvard University Press.

Goldstein, B. R. (1967) "The Arabic Version of Ptolemy's Planetary Hypotheses," *Transactions of the American Philosophical Society* 57.4.

Goss, C. M. (1966) "On Anatomy of Nerves by Galen of Pergamon," *American Journal of Anatomy* 118: 327–335.

Gow, A. S. F. and A. F. Scholfield (1953) *Nicander: The Poems and Poetical Fragments*, Cambridge: Cambridge University Press.

Green, R. M. (1951) *A Translation of Galen's Hygiene*, Springfield, IL: Charles C. Thomas.

Gulick, C. B. (1927, 1929, 1930, 1941) *Athenaeus Deipnosophistae*, vols. 2, 3, 4, and 7, Cambridge, MA: Harvard University Press.

Heath, T. L. (1913/1966/1997) *Aristarchus of Samos*, Oxford: Oxford University Press.

Heath, T. L. (1926/1956) *The Thirteen Books of Euclid's Elements*, 2nd edn, Cambridge: Cambridge University Press.

Heath, T. L. (1932) *Greek Astronomy*, London: J. M. Dent.

Hort, A. F. (1916, 1926) *Theophrastus Enquiry into Plants*, 2 vols, Cambridge, MA: Harvard University Press.

Jones, A. (1990) "Ptolemy's First Commentator," *Transactions of the American Philosophical Society* 80.7.

Jones, H. L. (1917, 1923, 1930) *Strabo*, vols. 1, 2, and 7, Cambridge, MA: Harvard University Press.

Kidd, D. (1997) *Aratus' Phaenomena*, Cambridge: Cambridge University Press.

Kidd, I. G. (1999) *Posidonius*, vol. 3, Cambridge: Cambridge University Press.

Knoefel, P. K. and M. C. Covi (1991) *A Hellenistic Treatise on Poisonous Animals*, New York: Edwin Mellen Press.

Lloyd, G. E. R. (1973) *Greek Science After Aristotle*, New York: W.W. Norton.

Long, A. A. and D. N. Sedley (1987) *The Hellenistic Philosophers*, vol. 1, Cambridge: Cambridge University Press.

Longrigg, J. (1998) *Greek Medicine: From the Heroic to the Hellenistic Age: A Source Book*, London: Routledge.

Marsden, E. W. (1971/1999) *Greek and Roman Artillery: Technical Treatises*, Oxford: Oxford University Press.

Minar, E. L. (1961) *Plutarch's Moralia*, vol. 9, Cambridge, MA: Harvard University Press.

Murphy, S. (1995) "Heron of Alexandria's On Automaton-Making," *History of Technology* 17: 1–44.

Neugebauer, O. and H. B. vanHoesen (1959) *Greek Horoscopes*, Philadelphia: The American Philosophical Society.

Paton, W. R. (1922) *Polybius* vol. 1, Cambridge, MA: Harvard University Press.

Pearson, L. and F. H. Sandbach (1965) *Plutarch's Moralia*, vol. 11, Cambridge, MA: Harvard University Press.

Phillips, A. A. and M. M. Willcock (1999) *Xenophon and Arrian, On Hunting*, Warminster: Aris & Phillips.

Pingree, D. (1976) *Dorothei Sidonii Carmina*, Leipzig: Teubner.

Prager, F. D. (1974) *Pneumatica, The First Treatise on Experimental Physics: Western Version and Eastern Version. Facsimile and Transcript of the Latin Manuscript, CLM 534. Translation and Illustrations of the Arabic Manuscript, A. S. 3713, Aya-Sofya, Istanbul*, Wiesbaden: Reichert.

361

Rescher, N. and M. E. Marmura (1965) *The Refutation by Alexander of Aphrodisias of Galen's Treatise on the Theory of Motion*, Islamabad: Islamic Research Institute.

Robbins, F. E. (1940) *Ptolemy Tetrabiblos*, Cambridge, MA: Harvard University Press.

Sandbach, F. H. (1927/1969) *Plutarch's Moralia*, vol. 15, Cambridge, MA: Harvard University Press.

Scarborough, J. and V. Nutton (1982) "The Preface of Dioscourides' *Materia Medica*: Introduction, Translation, and Commentary," *Transactions and Studies of the College of Physicians of Philadelphia*, new series, 4: 187–227.

Schoff, W. H. (1912) *The Periplus of Hanno*, Philadelphia: Commercial Museum.

Scholfield, A. F. (1958a,1958b,1959) *Aelian: On the Characteristics of Animals*, 3 vols., Cambridge, MA: Harvard University Press.

Shapiro, A. E. (1975) "Archimedes's Measurement of the Sun's Apparent Diameter," *Journal of the History of Astronomy*, 6: 75–83.

Siegel, R. E. (1973) *Galen on Psychology, Psychopathology, and Function and Diseases of the Nervous System; an Analysis of his Doctrines, Observations and Experiments*, Basel: Karger.

Singer, C. (1956/1999) *Galen on Anatomical Procedures*, Oxford: Oxford University Press.

Singer, P. N. (1997) *Galen: Selected Works*, Oxford: Oxford University Press.

Smith, A. M. (1996) *Ptolemy's Theory of Visual Perception*, Philadelphia: The American Philosophical Society.

Steckerl, F. (1958) *The Fragments of Praxagoras of Cos and his School*, Leiden: Brill.

Stevenson, E. L. (1932) *Geography of Claudius Ptolemy*, New York: New York Public Library.

Stratton, G. M. (1917) *Theophrastus and the Greek Physiological Psychology before Aristotle*, New York: The Macmillan Co.

Temkin, O. (1934) "Galen's 'Advice for an Epileptic Boy'," *Bulletin of the Institute of the History of Medicine* 2: 179–189.

Temkin, O. (1956) *Soranus' Gynaecology*, Baltimore: Johns Hopkins University Press.

Terian, A. (1981) *Philonis Alexandrini de Animalibus*, Chico, CA: Scholars Press.

Thomas, I. (1939, 1941) *Selections Illustrating the History of Greek Mathematics*, 2 vols., Cambridge, MA: Harvard University Press.

Todd, R. B. (1976) *Alexander of Aphrodisias on Stoic Physics*, Leiden: Brill.

Toomer, G. J. (1976) *Diocles on Burning Mirrors*, New York and Berlin: Springer.

Toomer G. J. (1984b) *Ptolemy's Almagest*, New York and Berlin: Springer.

Vallance, J. T. (1990) *The Lost Theory of Asclepiades of Bithynia*, Oxford: Oxford University Press.

von Staden, H. (1989) *Herophilus: The Art of Medicine in Early Alexandria*, Cambridge: Cambridge University Press.

Walzer, R. and M. Frede (1985) *Three Treatises on the Nature of Science*, Indianapolis, IN: Hackett.

White, R. J. (1975/1990) *The Interpretation of Dreams: Oneirocritica by Artemidorus*, Park Ridge, NJ: Noyes.

Texts newly translated

Aujac, G. (1975) *Géminos: Introduction aux Phénomènes*, Paris: Les Belles Lettres.

Berthelot, M. (1888/1967) *Collection des Anciens Alchimistes Grecs*, vol. 2, Paris: G. Steinheil.

Boudreaux, P. (1922) *Catalogus Codicum Astrologorum Graecorum* 8.4, Brussels: Lamertin.

deFalco, V. and M. Krause (1966) *Hypsikles: Die Aufgangszeiten der Gestirne*, Göttingen: Vandenhoeck & Ruprecht.

Deichgräber, K. (1965) *Die griechische Empirikerschule*, 2nd edn, Berlin: Weidmanns.

Dupuis, J. (1892/1966) *Théon de Smyrne*, Paris: Hachette; Brussels: Lamertin.

Foerster, R. (1893) *Scriptores Physiognomonici Graeci et Latini*, 2 vols., Leipzig: Teubner.

Franz, J. G. F. (1780) *Scriptores Physiognomoniae Veteres ex recensione Camilli Perusci et Frid. Sylburgii Graece et Latine recensuit, animadversiones Sylburgii et Dan. Guil. Trilleri v. ill. in Melampodem emendationes addidit suasque adspersit notas Iohannes Georgius Fridericus Franzius*, Altenburg: Emanuel Richter.

Friedrich, H.-V. (1968) *Thessalos von Tralles: griechisch und lateinisch*, Meisenheim-am-Glan: Hain.

Garyza, A. (1955/1957) "Paraphrasis Dionysii poematis de aucupio," *Byzantion* 25/27: 201–239.

Gottschalk, H. B. (1965) "Strato of Lampsacus: Some texts," *Proceedings of the Leeds Literary Society: Literary-Historical Section* 1.6.

Gruner, C. G. (1814) *Zosimi Panopolitani de Zythorum Confectione Fragmentum*, Sulzbach: Seidelianis.

Heiberg, J. L. (1894) *Commentaria in Aristotelem Graeca* 7, Berlin: G. Reimeri.

Heiberg, J. L. (1901) "Anatolius sur les dix premiers nombres," *Annales Internationales d'Histoire: Congrès de Paris: 5 Section, Histoire des Sciences*, Paris: 27–57.

Heiberg, J. L. (1907) *Claudii Ptolemaei Opera Omnia*, vol. 2, Leipzig: Teubner.

Heiberg, J. L. (1913/1972) *Archimedis Opera Omnia*, vol. 2, Leipzig: Teubner.

Heiberg, J. L. (1927) *Theodosius Tripolites Sphaerica*, Berlin: Weidmanns.

Henry, R. (1974) *Photius Bibliothèque* vol. 7, Paris: Les Belles Lettres.

Ideler, J. L. (1841/1963) *Physici et Medici Graeci Minores*, Berlin: G. Reimeri; Amsterdam: Adolf M. Hakkert.

Keller, O. (1877) *Rerum Naturalium Scriptores Graeci Minores*, Leipzig: Teubner.

Koechly, H. A. T. (1858) *Manethonis Apotelesmaticorum*, Leipzig: Teubner.

Kollesch, J. and F. Kudlien (1965) *Apollonios von Kition Kommentar zu Hippokrates über das Einrenken der Gelenke, Corpus Medicorum Graecorum* 11.1.1, Berlin: Akademie Verlag.

Krause, M. (1936) *Die Sphaerik von Menelaos aus Alexandrien in der Verbesserung von Abu Nasr b. 'ali b. 'Iraq*, Berlin: Weidmanns.

Kroll, W. (1903) *Catalogus Codicum Astrologorum Graecorum* 6, Brussels: Lamertin.

Kühn, K. G. (1821–1833/1964–1965) *Claudii Galeni Opera Omnia*, Hildesheim: Olms.

Lloyd-Jones, H. and P. Parsons (1983) *Supplementum hellenisticum*, Berlin: De Gruyter.

Mair, A. W. (1928) *Oppian*, Cambridge, MA: Harvard University Press.

Manitius, C. (1894) *Hipparchus: In Arati et Eudoxi Phaenomena Commentariorum Libri iii*, Leipzig: Teubner.

Muller, K. (1855) *Geographi Graeci Minores*, Paris: Didot.

Müller, K. (1883) *Claudii Ptolemaei Geographia*, Paris: Didot.

Nix, L. and W. Schmidt (1900/1976) *Heronis Alexandrini Opera Omnia*, vol. 2, Leipzig: Teubner.

Pingree, D. (1976) *Dorothei Sidonii Carmina*, Leipzig: Teubner.

Powell, J. U. (1925) *Collectanea Alexandrina*, Oxford: Clarendon Press.

Riess, E. (1891/1893) "Nechepsonis et Petosiridis Fragmenta magica," *Philologus* suppl. 6: 325–394.

Rome, A. (1922/1923) "Le problème de la distance entre deux villes dans la dioptra de Héron," *Annales de la société scientifique de Bruxelles* 42: 234–258.

Schmidt, W. (1899/1976) *Heronis Alexandrini Opera Omnia*, vol. 1, Leipzig: Teubner.

Schneider, R. (1912) *Griechische Poliorketiker*, vol. 3, Berlin: Weidmanns.

Schoene, H. (1903/1976) *Heronis Alexandrini Opera Omnia*, vol. 3, Leipzig: Teubner.

Sedley, D. (1976) "Epicurus and the mathematicians of Cyzicus," *Cronache Ercolanesi* 6: 23–54.

Sideras, A. (1977) *Rufus von Ephesos über die Nieren- und Blasenleiden, Corpus Medicorum Graecorum* 3.1, Berlin: Akademie Verlag.

Silberman, A. (1995) *Arrian: Périple du Pont-Euxin*, Paris: Les Belles Lettres.

Wellmann, M. (1906, 1907, 1914 [*sic*]; repr. 1958) *Dioscuridis Anazarbei De Materia Medica*, 3 vols., Berlin: Weidmanns.

Wellmann, M. (1921) *Die Georgika des Demokritos* = *Abhandlungen der preussischen Akademie der Wissenschaften, philosophisch-historische Klasse,* Berlin: Akademie der Wissenschaften.

Wurschmidt, J. (1925) "Die Schrift des Menelaus über die Bestimmungen der Zusammensetzung der Legierungen," *Philologus* 80: 377–409.

Works cited

Barnes, J. (1984) *The Complete Works of Aristotle*, Princeton, NJ: Princeton University Press.

Barton, T. (1994) *Ancient Astrology*, London: Routledge.

Beavis, I. (1988) *Insects and Other Invertebrates in Classical Antiquity*, Exeter: University of Exeter.

Berggren, J. L. (1991) "The relations of Greek Spherics to early Greek astronomy," in A. C. Bowen (ed.) (1991) *Science and Philosophy in Classical Greece*, New York: Garland, 227–248.

Berryman, S. (1998) "Euclid and the Sceptic," *Phronesis* 43: 176–196.

Boegehold, A. L. and M. Crosby (1995) "Kleroterion," in *The Athenian Agora, v. 28: The Lawcourts at Athens*, A. L. Boegehold (ed.), Princeton, NJ: Princeton University Press: 230–234.

Bolzan, J. E. (1976) "Chemical combination according to Aristotle," *Ambix* 23: 134–144.

Butler, A. R. and D. Needham (1981) "An experimental comparison of the East Asian, Hellenistic and Indian Stills," *Ambix* 27: 67–76.

Casson, L. (1986) *Ships and Seamanship in the Ancient World*, Princeton, NJ: Princeton University Press.

Collinder, P. (1964) "Dicaearchus and the Lysimachian measurement of the Earth," *Sudhoffs Archiv für Geschichte der Medizin und der Naturwissenschaften* 48: 63–78.

Cramer, F. H. (1954) *Astrology in Roman Law and Politics*, Philadelphia: The American Philosophical Society.

Dicks, D. R. (1970) *Early Greek Astronomy to Aristotle*, Ithaca, N.Y.: Cornell University Press.

Dijksterhuis, E. J. (1956/1987) *Archimedes*, Copenhagen: Munksgaard; Princeton, NJ: Princeton University Press.

Drachmann, A. G. (1948) *Ktesibios, Philon and Heron, a Study in Ancient Pneumatics*, Copenhagen: E. Munksgaard.

Drachmann, A. G. (1950) "Heron and Ptolemaios," *Centaurus* 1: 117–131.

Drachmann, A. G. (1976) "Ktesibios' Water-clock and Heron's Adjustable Siphon," *Centaurus* 20: 1–10.

Evans, J. (1998) *The History and Practice of Ancient Astronomy*, Oxford: Oxford University Press.

Foley, V., W. Soedel, J. Turner, and B. Wilhoit (1982) "The Origin of Gearing," *History of Technology* 7: 101–129.

Fowler, D. H. (1992) "An invitation to read Book X of Euclid's Elements," *Historia Mathematica* 19: 233–264.

Furley, D. J. (1985) "Strato's theory of the void," *Arisoteles: Werk und Wirkung*, J. Wiesner (ed.), Berlin: De Gruyter, 594–609.

Furley, D. J. and J. S. Wilkie (1984) *Galen on Respiration and the Arteries*, Princeton, NJ: Princeton University Press.

Gillispie, C. C. (ed.) (1970–1980) *Dictionary of Scientific Biography*, 16 vols., New York: Scribner.

Gleason, M. W. (1995) *Making Men*, Princeton, NJ: Princeton University Press.

Goldstein, B. R. and A. C. Bowen (1991) "The Introduction of Dated Observations and Precise Measurement in Greek Astronomy," *Archive for History of Exact Sciences* 43: 93–132.

Golomb, S. W. (1985) "The Invincible Primes," *Sciences* 25.2: 50–57.

Grmek, M. D. (1989) *Diseases in the Ancient Greek World*, Baltimore: Johns Hopkins University Press.

Gutas, D. (1998) *Greek Thought, Arabic Culture*, London: Routledge.

Hopkins, A. J. (1938) "A study of the kerotakis process as given by Zosimus," *Isis* 29: 326–354.

Humphrey, J. W., J. P. Oleson, and A. N. Sherwood (1998) *Greek and Roman Technology: A Sourcebook*, London: Routledge.

Irby-Massie, G. L. (1993) "Women in Ancient Science," in *Woman's Power, Man's Game: Essays on Classical Antiquity in Honor of Joy King*, Mary DeForest (ed.) Wauconda, IL: Bolchazy-Carducci.

Johnson, S. J. (1882) "On a Probable Assyrian transit of Venus," *Monthly Notices of the Royal Astronomical Society* 43: 41–42.

Jones, A. (1991) "Hipparchus' computations of solar longitudes," *Journal of the History of Astronomy* 22: 101–126.

Keyser, P. T. (1988) "Suetonius *Nero* 41.2 and the Date of Heron Mechanicus of Alexandria," *Classical Philology* 83: 218–220.

Keyser, P. T. (1990) "Alchemy in the Ancient World: From Science to Magic," *Illinois Classical Studies* 15: 353–378.

Keyser, P. T. (1992a) "A New Look at Heron's Steam Engine," *Archive for History of Exact Sciences* 44: 107–124.

Keyser, P. T. (1992b) "Xenophanes' Sun on Trojan Ida," *Mnemosyne* 45: 299–311.

Keyser, P. T. (1993a) "From Myth to Map: The Blessed Isles in the First Century B.C.," *Ancient World* 24: 149–168.

Keyser, P. T. (1994b) "On Cometary Theory and Typology from Nechepso-Petosiris through Apuleius to Servius," *Mnemosyne* 47: 625–651.

Keyser, P. T. (1995/1996) "Greco-Roman Alchemy and Coins of Imitation Silver," *American Journal of Numismatics* 7/8: 209–234, plates 28–32.

Keyser, P. T. (1997a) "Science and Magic in Galen's Recipes (Sympathy and Efficacy)," in *Galen on Pharmacology: Philosophy, History, and Medicine*, Armelle Debru (ed.), Leiden: Brill: 175–198.

Keyser, P. T. (1997c) "Theophrastos," in *Dictionary of Literary Biography* vol. 176, *Ancient Greek Authors*, Ward W. Briggs (ed.), Detroit: Gale Research: 371–380.

Keyser, P. T. (1998) "Orreries, the Date of [Plato] *Letter* ii, and Eudoros of Alexandria," *Archiv für Geschichte der Philosophie* 80: 241–267.

Keyser, P. T. (2001) "The Geographical Work of Dikaiarchos (frr. 104–115 W.)," in *Rutgers University Studies in Classical Humanities* vol. 10, W. W. Fortenbaugh and D. G. Mirhady (eds), Piscataway, NJ: Rutgers University Press.

Kilinksi, K. (1986) "Boeotian Trick Vases," *American Journal of Archaeology* 90: 153–158, plates 9–11.

Knorr, W. R. (1982) "Technique of Fractions in Ancient Egypt and Greece," *Historia Mathematica* 9: 133–171.

Knorr, W. R. (1985) "Archimedes and the Pseudo-Euclidean *Catoptrics*: Early Stages in the Ancient Geometric Theory of Mirrors," *Archives internationales d'histoire des Sciences* 35: 27–105.

Knorr, W. R. (1986) *The Ancient Tradition of Geometrical Problems*, Boston: Birkhäuser.

Knorr, W. R. (1993) "Arithmetikê Stoicheiôsis: On Diophantus and Hero of Alexandria," *Historia Mathematica* 20: 180–192.

Lebedev, A. (1993) "Alcmaeon on plants: a new fragment in Nicolaus Damascenus," *Parola del Passato* 48: 456–460.

Lewis, M. J. T. (1999) "When Was Biton?", *Mnemosyne* 52: 159–168.

Lloyd, G. E. R. (1979) *Magic, Reason, and Experience*, Cambridge: Cambridge University Press.

Lloyd, G. E. R. (1983) *Science, Folklore and Ideology*, Cambridge: Cambridge University Press.

Lloyd, G. E. R. (1987) *The Revolutions of Wisdom: Studies in the Claims and Practice of Ancient Greek Science*, Berkeley, CA: University of California Press.

Lloyd, G. E. R. (1991) "The Social Background of Early Greek Philosophy and Science," in *Methods and Problems in Greek Science*, Cambridge: Cambridge University Press: 121–140.

Lloyd, G. E. R. (1992) "Democracy, Philosophy, and Science in Ancient Greece," in J. Dunn (ed.) *Democracy: the Unfinished Journey* (Oxford: Oxford University Press) 41–56.

Longrigg, J. (1993) *Greek Rational Medicine*, London: Routledge.

McCarthy, D. (1995/6) "The Lunar and Paschal tables attributed to Anatolius of Laodicea," *Archive for History of Exact Sciences* 49: 285–320.

Meeus, J. (1989) *Transits*, Richmond, VA: Willmann Bell.

Misener, G. (1923) "Loxus: physician and physiognomist," *Classical Philology* 18: 1–22.

Moss, A. A. (1953) "Niello," *Studies in Conservation* 1: 49–62.

Murray, W. M. (1987) "Do modern winds equal ancient winds?" *Mediterranean Historical Review* 2: 139–167.

Murschel, A. (1995) "The Structure and Function of Ptolemy's Physical Hypotheses of Planetary Motion," *Journal of the History of Astronomy* 26: 33–61.

Negbi, M. (1995) "Male and Female in Theophrastus' Botanical Works," *Journal of the History of Biology* 28: 317–322.

Negbi, O. (1992) "Early Phoenician presence in the Mediterranean islands: A reappraisal," *American Journal of Archaeology* 96: 599–615.

Neugebauer, O. (1955) "The Egyptian 'Decans'," *Vistas in Astronomy* 1: 47–51; repr. in Neugebauer (1983) 205–209.

Neugebauer, O. (1975) *History of Ancient Mathematical Astronomy*, Berlin, New York: Springer.

Neugebauer, O. (1980) "Orientation of the Pyramids," *Centaurus* 24: 1–3; repr. in Neugebauer (1983): 211–213.

Neugebauer, O. (1983) *Astronomy and History: Selected Essays*, New York: Springer.

Oberhelman, S. M. (1994) "On the Chronology and Pneumatism of Aretaios of Cappadocia," in H. Temporini (ed.) (1993–1995), *Aufstieg und Niedergang der römischen Welt II*, vol. 37.2, Berlin: De Gruyter, 941–966.

Patai, R. (1982) "Maria the Jewess: Founding Mother of Alchemy," *Ambix* 29: 177–197.

Pelliccia, H. N. (1995) *Mind, Body, and Speech in Homer and Pindar*, Göttingen: Vandenhoeck & Ruprecht.

Phillips, E. D. (1973/1987) *Aspects of Greek Medicine*, Philadelphia: Charles Press.

Pollard, J. (1977) *Birds in Greek Life and Myth*, London: Thames and Hudson.

Price, S. R. F. (1986) "The future of dreams," *Past and Present* (Oxford) 113: 3–37.

Riddle, J. M. (1992) *Contraception and Abortion from the Ancient World to the Renaissance*, Cambridge, MA: Harvard University Press.

Rihll, T. (1999) *Greek Science*, Oxford: Oxford University Press.

Roccatagliata, G. (1986) *A History of Ancient Psychiatry*, New York: Greenwood.

Romm, J. (1992) *The Edges of the Earth in Ancient Thought*, Princeton, NJ: Princeton University Press.

Sabra, A. I. (1987) "Psychology versus Mathematics: Ptolemy and Alhazen on the Moon Illusion," in *Mathematics and its Applications to Science and Natural Philosophy in the Middle Ages*, E. Grant and J. Murdoch (eds.), Cambridge: Cambridge University Press: 217–247.

Sambursky, S. (1962/1987) *The Physical World of Late Antiquity*, Princeton, NJ: Princeton University Press.

Schmandt-Besserat, D. (1992) *Before Writing*, Austin: University of Texas Press.

Simon, B. (1978) *Mind and Madness in Ancient Greece*, Ithaca, NY: Cornell University Press.

Sines, G. and Y. A. Sakellarakis (1987) "Lenses in antiquity," *American Journal of Archaeology* 91: 191–196.

Smith, A. M. (1982) "Ptolemy's Search for a Law of Refraction: A Case-study in the Classical Methodology of Saving the Appearances and its Limitations," *Archive for History of Exact Sciences* 26: 221–240.

Sorabji, R. (1972) "Aristotle, Mathematics, and Color," *Classical Quarterly* 22: 293–308.

Temporini, H. (1993–1995) *Aufstieg und Niedergang der römischen Welt II*, 37, Berlin: De Gruyter.

Thiel, J. H. (1966) *Eudoxus of Cyzicus*, Utrecht: Groningen.

Thompson, D. W. (1936) *Glossary of Greek Birds*, London: Oxford University Press.

Thompson, D. W. (1947) *Glossary of Greek Fishes*, London: Oxford University Press.

Thurston, H. (1994) *Early Astronomy*, New York: Springer.

Toomer, G. J. (1972) "The Mathematician Zenodorus," *Greek, Roman and Byzantine Studies* 13: 177–192.

Toomer, G. J. (1973) "The Chord Table of Hipparchus and the Early History of Greek Trigonometry," *Centaurus* 18: 6–28.

Toomer, G. J. (1975) "Hipparchus on the Distances of the Sun and Moon," *Archive for History of Exact Sciences* 14: 126–142.

Tóth, I. (1969) "Non-Euclidean Geometry before Euclid," *Scientific American* 221.5 (November) 87–92, 94, 98.

367

Vandiver, P. B., O. Soffer, B. Klima, and J. Svoboda (1989) "The Origins of Ceramic Technology at Dolni Vestonice, Czechoslovakia," *Science* 246: 1002–1008.

Wagman, M. (1992) "Hercules the Champion," *Journal for the History of Astronomy* 23: 134–136.

Weinstock, St. (1948) "The Author of Pseudo-Galen's *prognostica de decubitu*," *Classical Quarterly* 42: 41–43.

Wilson, C. A. (1984) *Philosophers, Iôsis, and the Water of Life*, Leeds: Leeds Philosophical and Literary Society.

Wilson, C. A. (1998) "Pythagorean Theory and Dionysian Practice: The Cultic and Practical Background to Chemical Experimentation in Hellenistic Egypt," *Ambix* 45: 14–33.

Select further reading
(* translation)

Barrett, A. A. (1978) "Observations of Comets in Greek and Roman Sources before AD 410," *Journal of the Royal Astronomical Society of Canada* 72: 81–106.

Berggren, J. L. (1984) "History of Greek Mathematics: A Survey of Recent Research," *Historia Mathematica* 11: 394–410.

*Berggren, J. L. and A. Jones (2000) *Ptolemy's Geography: An Annotated Translation of the Theoretical Chapters*, Princeton, NJ: Princeton University Press.

*Berggren, J. L. and R. S. D. Thomas (1991) *Euclid's Phenomena*, New York: Garland.

Bodson, L. (1991) "Alexander the Great and the Scientific Exploration of the Oriental Part of his Empire," *Ancient Society* 22: 127–138.

Bowen, A. C. (ed.) (1991) *Science and Philosophy in Classical Greece*, New York: Garland.

*Burstein, S. M. (1989) *Agatharchides of Cnidus*, London: Hakluyt.

Cole, T. (1967) *Democritus and the Sources of Greek Anthropology*, Chapel Hill, NC: Western Reserve University.

*deLacy, P. H. (1992) "Galen on Semen," *Corpus Medicorum Graecorum* 5.3.1, Berlin: Akademie Verlag.

Dilke, O. A. W. (1985) *Greek and Roman Maps*, Ithaca, NY: Cornell University Press.

Dilke, O. A. W. (1987) *Mathematics and Measurement*, Berkeley: University of California Press; London: British Museum.

*Duckworth, W. L. H. (1962) *Galen on Anatomical Procedures, the Later Books*, Cambridge: Cambridge University Press.

*Falconer, W. and T. Falconer (1805) *Arrian's Voyage round the Euxine Sea*, Oxford: J. Crook.

Fowler, D. H. (1991) "Ratio and Proportion in Early Greek Mathematics," in A. C. Bowen (ed.) (1991) *Science and Philosophy in Classical Greece*, New York: Garland: 98–118.

Fraser, P. M. (1972) *Ptolemaic Alexandria*, Oxford: Clarendon Press.

*Frede, M. and R. Walzer (1944) *Galen on Medical Experience*, Oxford: Wellcome Institute.

Furley, D. J. (1987) *The Greek Cosmologists*, vol. 1, Cambridge: Cambridge University Press.

*Garofalo, I. and B. Fuchs (1997) *Anonymi Medici de Morbis Acutis et Chroniis*, Leiden: Brill.

*Grant, M. (2000) *Galen on Food and Diet*, London: Routledge.

*Greenwood, J. G. and B. Woodcroft (1851/1971) *The Pneumatics of Hero of Alexandria*, London: Macdonald; New York: Elsevier.

*Hankinson, R. J. (1991) *Galen on the Therapeutic Method, books I and II*, Oxford: Oxford University Press.

Harley, J. B. and D. Woodward (eds.) (1987) *The History of Cartography* vol.1, Chicago: University of Chicago Press.

*Heath, T. L. (1897/1953) *Works of Archimedes*, Cambridge: Cambridge University Press.

*Humbach, H. and S. Ziegler (1998) *Ptolemy, Geography, book 6 : Middle East, Central and North Asia, China*, Wiesbaden: Reichert.

Keyser, P. T. (1993b) "The Purpose of the Parthian Galvanic Cells: A First-Century A.D. Electric Battery Used for Analgesia," *Journal of Near Eastern Studies* 52: 81–98.

Keyser, P. T. (1994a) "The Use of Artillery by Philip II and Alexander the Great," *Ancient World* 25: 27–59.

Keyser, P. T. (1997b) "Sallust's *Historiae*, Dioskorides and the Sites of the Korykos Captured by P. Servilius Vatia," *Historia* 46: 64–79.

*Kheirandish, E. (1999) *The Arabic version of Euclid's Optics*, New York: Springer.

Kingsley, P. (1994) "From Pythagoras to the Turba Philosoporum," *Journal of the Warburg and Courtauld Institutes* 57: 1–13.

Kirk, G. S., J. E. Raven, and M. Schofield (1983) *The Presocratic Philosophers*, 2nd edn, Cambridge: Cambridge University Press.

Knorr, W. R. (1994) "Pseudo-Euclidean Reflections in Ancient Optics," *Physis* 31: 1–45.

*Lawlor, R. and D. Lawlor (1979) *Theon: Mathematics Useful for Understanding Plato*, San Diego: Wizards Bookshelf.

Longrigg, J. (1988) "Anatomy in Alexandria in the Third Century BC," *British Journal of the History of Science* 21: 455–488.

*Macierowski, E. M. (1987) *Apollonios of Perge: On Cutting Off a Ratio*, Fairfield, CN: Golden Hind Press.

*May, M. T. (1968) *Galen: On the Usefulness of the Parts of the Body*, 2 vols., Ithaca, NY: Cornell University Press.

Meiggs, R. (1982) *Trees and Timber in the Ancient Mediterranean World*, Oxford: Oxford University Press.

*Nutton, V. (1999) "Galen: On my own Opinions," *Corpus Medicorum Graecorum* 5.3.1, Berlin: Akademie Verlag.

Partington, J. R. (1970) *A History of Chemistry* 1.1, London: Macmillan.

*Ramin, J. (1976) *Le Périple d'Hannon: The Periplus of Hanno*, Oxford: British Archaeological Reports, supplementary series 3.

Raschke, M. G. (1978) "New Studies in Roman Commerce with the East," in H. Temporini (ed.) (1978), *Aufstieg und Niedergang der römischen Welt* 2.9.2: Berlin: De Gruyter: 604–1361.

*Roseman, C. H. (1994) *Pytheas On the Ocean*, Chicago: Ares.

Sallares, R. (1991) *The Ecology of the Ancient Greek World*, London: Duckworth.

Sambursky, S. (1956/1987) *The Physical World of the Greeks*, London; Princeton, NJ: Princeton University Press.

Sambursky, S. (1959) *Physics of the Stoics*, New York: Macmillan.

Sarton, G. (1939) "Lunar Influences on Living Things," *Isis* 30: 495–507.

Scullard, H. H. (1974) *The Elephant in the Greek and Roman World*, Ithaca, NY: Cornell University Press.

*Sesiano, J. (1982) *Books IV to VII of Diophantus' Arithmetica in the Arabic Translation Attributed to Qusta ben Luqa*, New York: Springer.

BIBLIOGRAPHY

Talbert, R. J. A. (2000) *Barrington Atlas of the Greek and Roman World*, Princeton, NJ: Princeton University Press.

Tester, S. J. (1987) *History of Western Astrology*. Woodbridge, Suffolk: Boydell.

Thomson, J. O. (1948) *History of Ancient Geography*. Cambridge: Cambridge University Press.

Thorndike, L. (1955) "The True Place of Astrology in the History of Science." *Isis* 46: 273–278.

Tobin, R. (1990) "Ancient Perspective and Euclid's Optics," *Journal of the Warburg and Courtauld Institutes* 53: 14–41.

Toomer, G. J. (1984a) "Lost Greek Mathematical Works in Arabic Translation," *Mathematical Intelligencer* 6.2: 32–38.

*Toomer, G. J. (1990) *Apollonius, Conics, books V to VII: the Arabic Translation of the lost Greek Original in the Version of Banu Musa*, New York: Springer.

Unguru, S. (1979) "History of Ancient Mathematics: Some Reflections on the State of the Art," *Isis* 70: 555–565.

vanPaassen, C. (1957) *The Classical Tradition of Geography*. Utrecht: Groningen.

Wright, M. R. (1995) *Cosmology in Antiquity*. London: Routledge.

INDEXES

INDEX OF TERMS

371

INDEX OF ALL METALS,
STONES, PLANTS, ANIMALS

INDEX OF PEOPLE (NOT INCLUDING EXTRACTED AUTHORS)

Dates are mainly periods of activity; legendary personages and gods are omitted.

INDEX OF PLACES

The sites marked + are marked on the maps of Greece and the Mediterranean, on pp. xxxvi-xxxvii. Note that all authors' places of origin are also marked on the appropriate one of those two maps. Two kinds of sites have been omitted from this index: (1) all authors' places of origin which are not explicitly mentioned in an extract or our commentary, and (2) all sites in Ptolemy's maps of Sardinia and Corsica, pp. 146-149, which are not explicitly mentioned in another extract or our commentary.

CONCORDANCE OF
PASSAGES CITED
(NOT EXCERPTED)